KU-609-009

Biology of Sleep Substances

Author
Shojiro Inoué, D.Sc.
Professor and Head
Division of Biocybernetics
Institute for Medical and Dental Engineering
Tokyo Medical and Dental University
Tokyo, Japan

CRC Press, Inc.
Boca Raton, Florida

Library of Congress Cataloging-in-Publication Data

Inoue, Shōjirō, 1935-
Biology of sleep substances / author, Shojiro Inoué.
p. cm.
Bibliography: p.
Includes index.
ISBN 0-8493-4822-6
1. Sleep—Physiological aspects. 2. Biochemistry.
3.Neuropeptides—Physiological effect. I. Title.
QP425.I56 1989
612'.821—dc19 88-19049
CIP

Direct all inquiries to CRC Press, Inc., 2000 Corporate Blvd., N.W., Boca Raton, Florida, 33431.

International Standard Book Number 0-8493-4822-6
Library of Congress Card Number 88-19049
Printed in the United States

DEDICATION

To the experimental animals who supported our studies.

PREFACE

Biology of Sleep Substances may offer you an unfinished story of the everlasting search for the charming and challenging ''Spirit of Sleep''. After a long history of failure to unveil the sleep hormone, the first candidate substance, delta-sleep-inducing peptide (DSIP), was isolated and finally identified as a novel nonapeptide 11 years ago. Since then, an increasing number of endogenous substances, researchers, and papers have rushed to appear. For example, only four papers dealing with sleep substances were presented at the 3rd International Congress of Sleep Research held in Tokyo in 1979. Surprisingly, the 5th Congress held in Copenhagen in 1987 offered a symposium on sleep peptides with 6 speakers and accepted 27 free communications focused on sleep substances. Drs. M. V. Graf and A. J. Kastin published a review article on DSIP with 155 references in 1984. Their second article on the same subject appeared 2 years later with 248 references.

The rapid growth of our knowledge has undoubtedly brought about better and fairer understandings of sleep substances and sleep regulatory mechanisms on one hand but some chaotic confusions on the other. It is written:

''For with much wisdom comes much sorrow; the more knowledge, the more grief.''

- The Old Testament -

> ''There are two kinds of worldly passions that defile and cover the purity of Buddha-nature. The first is the passion for analysis and discussion by which people become confused in judgement. The second is the passion for emotional experience by which people's values become confused.''

- The Teaching of Buddha -

Under such situations, it was a hard task for me to completely overview through the clouds of ignorance and objectively integrate the current state of the art. Provided that I had not been involved in studies on the search for Sleep-Promoting Substance (SPS) from the very beginning and still engaged in researches in the forefront, and provided that I had no experience to compare several important sleep substances including DSIP, muramyl peptides, prostaglandins, and others by myself, I should have hesitated to write this book in spite of the kind recommendations by my respected colleagues. Another difficulty I had to confront was writing in English, a non-native language. Thus, I should like to ask you for your generosity and patience in reading monotonous noncritical descriptions.

The humble wish of the author is that this book may become a milestone of scientific studies on sleep substances, just as Dr. Nathaniel Kleitman's ''Sleep and Wakefulness'' (1939; 1963) has accumulated the literature on sleep research and greatly contributed to the development of this field of science.

Shojiro Inoué

ACKNOWLEDGMENTS

For cooperative studies: Osamu Hayaishi, Akifumi Higashi, Kazuki Honda, Hajime Ichikawa, Akira Inokuchi, Masami Iriki, Masayuki Ishikawa, Mayumi Kimura, Toshiyuki Kimura, Yasuo Komoda, Vladimir M. Kovalzon, Hitoshi Matsumura, Ayako Miura, Hiroaki Nagasaki, Kaori Naito, Satoshi Nishida, Yutaka Oomura, Yasuhisa Okano, Hiroyoshi Osama, Tsuyoshi Ozaki, Young Ho Rhee, Koji Uchizono, Ryuji Ueno, Otas E. Upkonmwan, Ikuo Yamamoto, Masako Yanagawa, and Isamu Yanagisawa.

For friendship and discussions: Joelle Adrien, Alexander A. Borbély, Raymond Cespuglio, Jaber Danguir, Mihailo R. Dzoljic, René R. Drucker-Colín, Dieter Gillessen, J. Christian Gillin, Markus V. Graf, Michel M. Jouvet, Ernst G. Herzog, Yutaka Honda, Kenichi Homma, James A. Horne, Masao Ito, André Kahn, Nobumasa Kato, Werner P. Koella, James M. Krueger, Shiyi Liu, Wallace B. Mendelson, Takeshi Miyata, Jaime M. Monti, Takao Mori, Yutaka Motohashi, Lev M. Mukhametov, Hachiro Nakagawa, Ferenc Obál, Yoshiro Okano, Nobuyuki Okudaira, Teruo Okuma, Tarja Pokka-Heiskanen, Oscar Prospéro-García, Miodrag Radulovacki, Russel J. Reiter, Kazuya Sakai, Richard Scherschlicht, Dietrich Schneider-Helmert, Guido A. Schoenenberger, Hartmut Schulz, Dag Stenberg, Kurt Stephan, Shigeo Suzuki, Hiroshi Takagi, Yasuro Takahashi, Irene Tobler, Shizuo Torii, Reidun Ursin, Herman Van Belle, André M. A. Van Dijk, Albert Wauquier, and Shunichi Yamagishi.

For the permission of using figures: Asahi Simbunsha, Mihail Coculescu, Hinrich Cramer, Jaber Danguir, Elsevier Scientific Publishers, Markus V. Graf, Japan Scientific Societies Press, S. Karger AG, Abba J. Kastin, James M. Krueger, Takeshi Miyata, Marcel Monnier, Stylianos Nocolaidis, John R. Pappenheimer, Pergamon Press, P. Polc, Raven Press, Guido A. Schoenenberger, Schwabe & Co. Verlag, Springer-Verlag, and the Zoological Society of Japan.

For sound sleep: my wife Eiko, my son Zen, and my daughter Kei.

THE AUTHOR

Shojiro Inoué, D.Sc., is Professor and Head of the Division of Biocybernetics, Institute for Medical and Dental Engineering, Tokyo Medical and Dental University, and concurrently Professor of Physiology at the Graduate School of Medicine of the same University.

Prof. Inoué graduated in 1960 from University of Tokyo, Japan, with a B.Sc. degree in biology and obtained his D.Sc. degree in 1965 from the same University.

He is a member of the Association for Study of Biological Rhythms, European Sleep Research Society, Japan Society of Medical Electronics and Biological Engineering, Japan Neuroscience Society, Japanese Sleep Research Society, Japanese Society of Biometeorology, International Society of Biometeorology, International Society for Neuroethology, Physiological Society of Japan, Sleep Research Society, Zoological Society of Japan, etc., and a honorary member of the Societus Physico-Medica Erlangensis.

Prof. Inoué was a recipient of the NIH fellowship in 1967 and the Alexander von Humboldt fellowship in 1967—1968. He was Visiting Professor at University of Erlangen-Nueremberg, West Germany (1973 and 1980), University of Ife, Nigeria (1975—1976), and University of Witten-Herdecke, West Germany (1986), and part-time lecturer at several domestic universities. He has received the Sakkokai Award for research in biological engineering, and the Suzuken Award and the Naito Award for research in sleep substances.

Prof. Inoué has presented 12 invited lectures at international meetings, over 30 invited lectures at national meetings, and approximately 50 guest lectures at foreign and domestic universities and institutes. He has published more than 100 research papers, more than 60 review articles, and 6 books. His current major research interests include the regulatory mechanisms of sleep and the function of endogenous sleep substances.

TABLE OF CONTENTS

Chapter 1
Candidate Substances, Research Background, and Biological Roles

Chapter 2
Techniques for Isolation and Evaluation

Chapter 3

Delta-Sleep-Inducing Peptide and Its Derivatives

Chapter 4

Sleep-Promoting Substance (SPS) and Nucleosides

Chapter 1

CANDIDATE SUBSTANCES, RESEARCH BACKGROUND, AND BIOLOGICAL ROLES

I. WHAT IS AN ENDOGENOUS SLEEP SUBSTANCE?

Sleep, one of the most sophisticated integrative functions in higher animals, appears to be regulated by a variety of endogenous humoral factors. These factors are called ''sleep substances''. However, it is still difficult to unanimously define an endogenous sleep substance since we cannot yet fully understand the mechanism involved in the regulation of sleep. Until several years ago, even those investigators who presumed the existence of a humorally transmitted somnogenic factor regarded that the number of the genuine factors must be solely one or at most two, i.e., one for the regulation of rapid-eye-movement (REM) sleep and the other for that of non-REM sleep. However, our recent increasing knowledge indicates that multiple factors are actually participating in the induction, maintenance, and termination of physiological sleep.[1,2] Thus, it seems unlikely that a single sleep hormone does regulate every aspect of sleep. As a consequence, the definition of a sleep substance and the concept of the humoral theory on sleep have been gradually revised.

Borbély and Tobler[3,4] first proposed criteria essential for a *specific* sleep substance. According to them, an endogenous sleep substance should satisfy the following six requirements: (1) the substance should reliably and distinctly induce and/or maintain physiological sleep, (2) a dose-effect relationship should prevail within a certain dosage range, (3) the action of the substance should be similar with respect to the time course and the effect on sleep states in different species, (4) the substance should be present in the organism, (5) spontaneous or induced changes in the vigilance level should be associated with alternations in the endogenous concentration and/or turnover of the substance, (6) the substance should be chemically identified.

Jouvet[5-7] differentiated ''sleep-inducing or sleep-promoting factors'' from ''sleep-facilitating factors''. According to him, the distinction is as follows:

Sleep-inducing (sleep-promoting) factors: (1) they should be endogenous, in the brain, cerebrospinal fluid (CSF) or blood; (2) they should increase during instrumental deprivation of sleep; (3) they act directly upon executive mechanisms of sleep; (4) they are probably different for slow-wave sleep (SWS) and paradoxical sleep (PS), but they may have the same ancestor molecule; (5) their administration at the receptor level should trigger and increase SWS and/or PS, provided that permissive systems are blocked; (6) inactivation of these factors (at the production or receptor level) should result in either total insomnia or selective suppression of SWS or PS; such suppression could be obtained for a long time and should not be followed by any secondary rebound; (7) insomnia provoked by suppression of sleep-inducing factors should not be reversed by sleep-facilitating factors; (8) they are probably identical in all mammals.

Sleep-facilitating factors: (1) they may be exogenous or endogenous; (2) they do not obligatorily increase during sleep deprivation; (3) they do not act directly upon executive mechanisms of sleep, but control the permissive mechanisms that impair sleep onset; (4) their administration should increase either SWS or PS or both; (5) in case of inactivation of sleep-inducing factors, the administration of sleep-facilitating factors may induce sedation or drowsiness, but not physiological sleep; (6) their facilitating effect upon sleep may be dose-dependent, but this is not obligatory; (7) inactivating these factors may delay sleep onset temporarily, but a secondary rebound of sleep should occur later; (8) they are probably different in different mammals according to their sleep ecology.

Ursin[8,9] defined an endogenous sleep factor as a substance that is produced by the organism and that affects sleep. Her definition allows a wide latitude of possibilities as to how such a factor may work. In addition to its primary or specific effect on sleep-waking mechanisms, its actions may include unusual or unspecific sleep modulations, which neither occur in natural sleep nor result from some secondary effect of the substance.

This author[10] provisionally defined an endogenous sleep factor as an active substance which, under a high physiological demand for sleep in the organism, is produced in the brain and transferred to the whole brain via the body fluid (especially CSF) to induce or maintain sleep. Inoué[11] and Inoué et al.[12] also proposed that, if the physiological demand for sleep is achieved, an exogenously supplied sleep substance should not cause unnatural extra sleep.

Mendelson et al.,[13] and more recently Gillin,[14] modified the proposed criteria of Borbély and Tobler[3,4] as follows: (1) the sleep substance should be purified and identified; (2) the substance normally should be present in the organism: its synthetic system, location, storage form, and concentrations should be clarified; (3) it should be demonstrated convincingly that endogenous administration of the substance promotes sleep or a specific component such as delta electroencephalogram (EEG) activity or REM-sleep compared to adequate controls; (4) the sleep promoted by the substance should be physiologically and psychologically "normal" within a broadly defined range, i.e., autonomic function, body temperature, CSF pressure, EEG wave forms, cyclic succession of non-REM and REM sleep, behavioral arousal, dreaming, and so on; (5) the effects should occur at physiological concentrations at the critical receptor areas, and these concentrations should vary appropriately with the behavioral state, i.e., low in wakefulness and high in sleep; (6) tolerance should not develop with repeated, daily administration.

A large number of investigators with broad disciplines have now become interested in studies on humoral aspects of sleep regulation. Within the past decade, a great variety of substances have been nominated as endogenous sleep factors, although they can hardly satisfy the required criteria. Since neither the term "sleep" nor "sleep substance" could be unambiguously defined to date,[15] it appears infructuous to define too strictly a subject that we do not know quite well. Hence, a rather loose and flexible definition like that of Ursin[8,9] might be preferable and practical at present.

There are several peptides or proteins in chemical nature that are often called "sleep peptides".[16-18] However, all sleep substances are not necessarily peptidergic. The terms "sleep-enhancing (-inducing, -modulating, or -promoting) substance", "sleep inducer (modulator or promoter)", "sleep hormone", and "somnogenic or hypnogenic substance (somnogen or hypnogen)" have been used in the literature. They have a wide range of meanings from an endogenous sleep substance in the above strict sense to a merely sleep-related biochemical factor, sometimes causing confusion in terminology. On the other hand, it is rather clear to discriminate between sleep-related neurotransmitters in a classical sense and sleep-related neuromodulators.[20] In this book, sleep-related neuromodulators (but not sleep-related neurotransmitters) will be dealt with in depth. The term "sleep substances" will be broadly applied to various categories of the putative endogenous sleep factors which have been proposed to date.

It is absolutely necessary to integrate knowledge and to clarify the interacting mechanisms involved in the regulation of sleep. The present chaotic situations in our understanding of sleep substances, however, seem to indicate that multiple sleep factors do really exist for sleep regulation and that further ingenious research can unveil their complicated actions.

II. ORIGIN OF HUMORAL THEORIES OF SLEEP REGULATION

A. Ancient Theories

In the mythical world, sleep may be induced by a mysterious power coming from outside of the organism. An exogenously supplied material like the extract of poppy flowers and other herbs can transfer the hypnogenic power into the body.[21] Buddha[22] regarded sleep as the fifth troops of the evil demon, which may invade the human mind to cover the soul. The evil power was believed to be counteracted by some materials, such as tea. Thus, the state of vigilance seemed to be influenced by the materials existing inside the body. According to the Yin-Yang theory, the interacting positive and negative principles in ancient China, the cause of sleep was ascribed to the congestion of the circulation of spirit and humor.[23,24] The oriental medicine at that time regarded the heart as the most important organ for the regulation of sleep.

The ancient theory of humors contained the idea that chemical substances are transported in the body, and that their presence, abundance, deficiencies, and balances influence the nature of body function, the quality of man's disposition, and his total reaction pattern.[25] Hippocrates regarded that sleep after a meal warms and moistens the sleeper as the food spreads nourishment throughout the body.[26,27] Aristotle[28] focused his attention on the circulation of heat produced by nutritional process. He clearly stated in his essay on sleep and wakefulness (*De somno et vigilia*) in his *Parva Naturalia* as follows:[29] "Sleep arises from the evaporation attendant upon the process of nutrition. The matter evaporated must be driven onwards to a certain point, then turn back, and change its current to and fro, like a tide-race in a narrow strait. Now, in every animal the hot naturally tends to move upwards... This explains why fits of drowsiness are especially apt to come on after meals; for the matter, both the liquid and the corporeal, which is borne upwards in a mass, is then of considerable quantity... It also follows certain forms of fatigue; for fatigue operates as a solvent, and the dissolved matter acts, if not cold, like food prior to digestion... Extreme youth also has this effect; infants, for example, sleep a great deal, because of the food being borne upwards — a mark whereof appears in the disproportionally large size of the upper parts compared to the lower during infancy..."

European philosophers in the late Renaissance gradually formed the idea that the withdrawal of the animal spirit from the brain is the cause of sleep, which occurs as a result of the disturbance of blood circulation in the "cold and moist" brain.[27] It should be noted here that in 1566 the Italian physician Johannes Argenterius attributed the cause of sleep to the existence of the innate warm substance (*calida substantia*) dynamically circulating in the body.[26]

Ekken Kaibara,[30] a Japanese physician, integrated the ideas in the oriental medicine up to his era, and published *Yojokun* (Advice for Healthy Life) in 1713 at the age of 84 years. He explained that the disturbance of circulation of humor or spirit is not the cause, but the result, of sleep. However, he did not mention the cause of sleep.

The sleep-related vital force was replaced by a physico-chemical term in modern science. Alexander von Humboldt,[31] for example, suggested in 1797 that sleep may be due to the reduction of the cerebral oxygen level. A hypothesis that the period of wakefulness causes the accumulation of sleep-inducing metabolic products in the organism, and that these substances are removed during sleep to recover the waking state, seems to have gradually grown mature towards the beginning of the 20th century. Physiologists in the 19th century preferentially referred to fatigue substances such as lactic acid or carbon dioxide as the principal cause of sleep (cf. Kleitman[32]).

The presence of a blood-borne messenger is evidenced by the discovery of secretin[33] in 1902. The term "hormone" and the new concept "humoral control of body function" stimulated enormously endocrinological approaches to various physiological functions. Search for a humorally transmitted hypnogenic substance must have been one of the most attractive new fields of sleep research.

B. Early Experimental Attempts

An experimental approach to the humoral mechanism of sleep was first done independently in Japan and in France soon after the discovery of secretin. Interestingly, the investigators

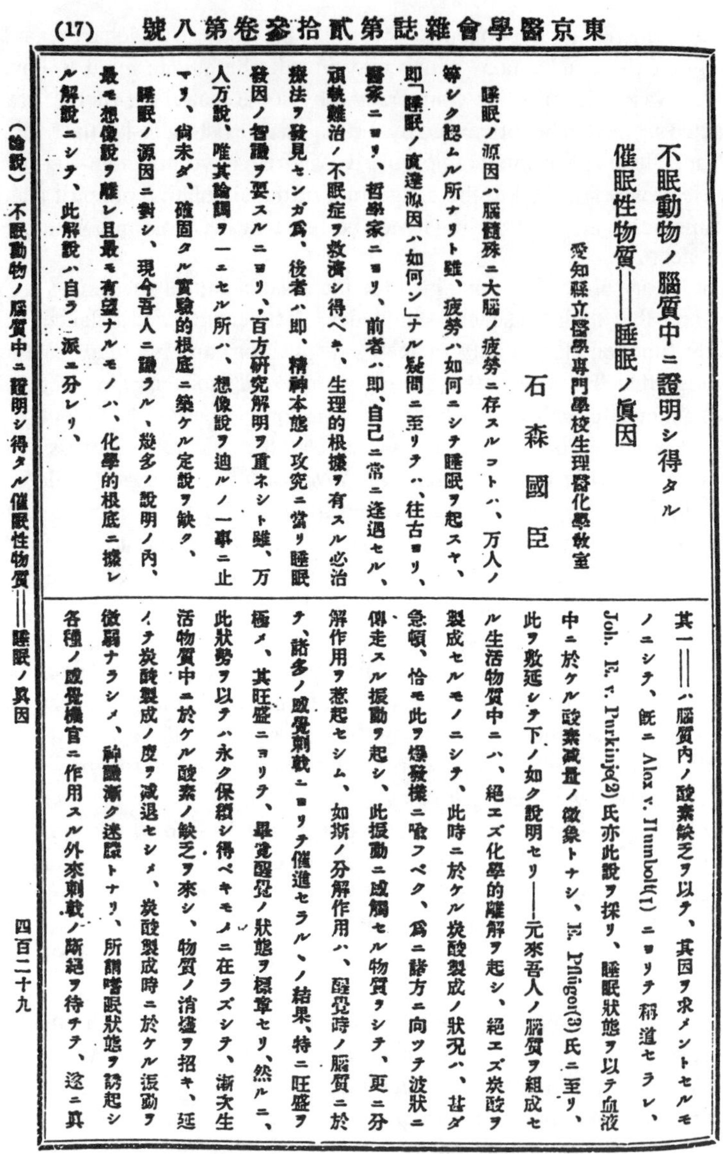

東京醫學會雜誌第貳拾參卷第八號 (17)

不眠動物ノ腦質中ニ證明シ得タル
催眠性物質――睡眠ノ眞因

愛知縣立醫學專門學校生理醫化學教室
石 森 國 臣

睡眠ノ源因ハ腦髓殊ニ大腦ノ疲勞ニ存スルコトハ、万人ノ等シク認ムル所ナリト雖、疲勞ハ如何ニシテ睡眠ヲ起スヤ、即「睡眠ノ直達源因ハ如何ン」ナル疑問ニ至リテハ、往古ヨリ、醫家ニヨリ、哲學家ニヨリ、前者ハ即、自己ニ常ニ逢遇セル、頑執難治ノ不眠症ヲ救濟シ得ベキ、生理的根據ヲ有スル必治療法ヲ發見センガ爲、後者ハ即、精神本態ノ攻究ニ當リ睡眠發因ノ智識ヲ要スルニヨリ、百方研究解明ヲ重ネシト雖、万人万說、唯其論調ヲ一ニセル所ハ、想像說ヲ迪ルノ一事ニ止マリ、尙未ダ、確固タル實驗的根底ニ築ケル定說ヲ缺ク、

睡眠ノ源因ニ對シ、現今吾人ニ識ラル、幾多ノ說明ノ內、最モ想像說ヲ離レ且最モ有望ナルモノハ、化學的根底ニ據レル解說ニシテ、此解說ハ自ラ二派ニ分レリ、

其一――ハ腦質內ノ酸素缺乏ヲ以テ、其因ヲ求メントセルモノニシテ、既ニ Alex v. Humbolt(1) ニヨリテ稱道セラレ、Joh. E. v. Purkinje(2) 氏亦此說ヲ採リ、睡眠狀態ヲ以テ血液中ニ於ケル酸素減量ノ徵象トナシ、E. Pflüger(3) 氏ニ至リ、此ヲ敷延シテ下ノ如ク說明セリ――元來吾人ノ腦質ヲ組成セル生活物質中ニハ、絕エズ化學的離解ヲ起シ、絕エズ炭酸ヲ製成セルモノニシテ、此時ニ於ケル炭酸製成ノ狀況ハ、甚ダ急頓、恰モ此ヲ爆發機ニ喩フベク、爲ニ諸方ニ向ツテ波狀ニ傳走スル振動ヲ起シ、此振動ニ感觸セル物質ヲシテ、更ニ分解作用ヲ惹起セシム、如斯ノ分解作用ハ、醒覺時ノ腦質ニ於テ、諸多ノ感覺刺戟ニヨリテ催進セラルヽノ結果、特ニ旺盛ヲ極メ、其旺盛ニヨリテ、畢竟醒覺ノ狀態ヲ標章セリ、然ルニ、此狀勢ヲ以テハ永ク保續シ得ベキモノニ在ラズシテ、漸次生活物質中ニ於ケル酸素ノ缺乏ヲ來シ、物質ノ消盡ヲ招キ、延イテ炭酸製成ノ度ヲ減退セシメ、炭酸製成時ニ於ケル振動ヲ微弱ナラシメ、神識漸ク迷蒙トナリ、所謂嗜眠狀態ヲ誘起シ各種ノ感覺機官ニ作用スル外來刺戟ノ斷絕ヲ待チテ、遂ニ眞

(論說) 不眠動物ノ腦質中ニ證明シ得タル催眠性物質――睡眠ノ眞因 四百二十九

FIGURE 1. The first page of Ishimori's paper. (From Ishimori, K., *Tokyo Igakkai Zasshi,* 23, 429, 1909. With permission.)

in both countries deprived dogs of sleep and tried to extract a hypnogenic factor from the tissues of various organs and the body fluid, with an assumption that continued wakefulness or fatigue may bring about the accumulation of such a factor inside the body. In 1909, Ishimori[34] published a 29-page full paper in Japanese (see Figure 1), while in 1913 Legendre and Piéron[35] published a 28-page, 4-plate full paper in French (see Figure 2).

Ishimori,[34] with the assistance of 27 students, deprived immature dogs of total sleep for 24-113 h. Their litter mates, which were allowed to behave freely, served as controls. After the treatment, the whole brain was extirpated, homogenized, and dialyzed several times in 80-85% alcohol and ether. A brown-colored extract was obtained after evaporating the media at 45°C. The dry weight of the extract from the treated animals was heavier by 30 to 250 mg/brain than that of the controls. Subcutaneous (s.c.) injection of the extract (0.15 g

A

B

C

FIGURE 2. Sleep-deprived donor dogs in Legendre and Piéron's study. (A) Ainette, (B) Bengant, (C) Tunis. (From Legendre, R. and Piéron, H., *Z. Allg. Physiol.*, 14, 235, 1913. With permission.)

dissolved in warm water) from the sleep-deprived dogs dramatically induced behavioral sleep in active immature dogs. Ishimori describes the scene: "From about 30 min after the injection on, the recipient dog frequently shed tears and saliva, sometimes shivered the whole body, became markedly sluggish without willing to walk, and crouched on the same place, never responding to the calls by his name. He had no interest in food we gave occasionally. When the dog was forced to walk, he tottered and preferred to go into a corner of the room or beneath a desk. If we gave him no more stimulation, gradually he closed the eyes and lied down on the floor to fall asleep." The effects lasted 2 h or longer. However, no abnormal behavior was observed the next morning. Lacrimation and salivation, but no sleep, were also induced by the extract from the control dogs. In conclusion, it is discussed: "The fact that, as evidenced by the above-mentioned series of experiments, the cerebral matter of sleep-deprived animals contains a potent hypnogenic substance which can not be detected in the brain of normally sleeping animals, leads to clarify the cause of induction of our normal sleep. The state that we call wakefulness implies uninterrupted chemical activities in the cerebral tissue which are mainly achieved by various kinds of stimuli coming from the outer environment. As a result of such activities, a special substance, presumably a kind of leucomaine, which can be dialyzed in warmed alcohol, is gradually produced to elicit a certain toxic action on the neurons in the cerebrum. This substance eventually reduces the neural function and induces so-called state of drowsiness, and finally causes natural sleep if the reduction or cessation of the external stimuli simultaneously takes place. After sleeping, the hypnogenic substance accumulated in the cerebral matter is gradually exterminated by means of blood circulation to recover the state of wakefulness, which is triggered by the external stimuli.''

Legendre and Piéron[35-40] deprived adult dogs of total sleep for 150 to 293 h. They histologically observed that morphological changes, i.e., chromatolysis, vacuolization, etc., occurred in neurons located in the frontal and occipital cortex in proportion to the duration of sleep deprivation. They also found slight differences in the viscosity and density of blood before and after the treatment. Their most dramatic finding that has given an impact on the subsequent sleep research was the transmission of the demand for sleep occurring in the sleep-deprived animals to the normal animals.

Normal dogs, which were injected intravenously (i.v.) or intracerebroventricularly (i.c.v.) into the fourth ventricle with either the CSF (4 to 6 ml i.c.v.) or the blood (200 ml i.v.) or the serum (60 ml i.v.; 3 to 8 ml i.c.v.) or the emulsion of cerebral tissue (50 ml i.v.) taken from the sleep-deprived dogs, exhibited signs of sleep for a 2-h observation period prior to sacrifice. Histological examinations of the cerebral tissues revealed that the similar morphological changes to those occurring after sleep deprivation existed in some cases. On the other hand, the pretreatment of the serum and CSF by heating at 55 or 65°C, dialyzing with ultrafiltration, and exposing to oxygen or alcohol, diminished the hypnogenic activity. Normal dogs similarly injected with the blood, serum, or CSF of normally behaving dogs showed no behavioral change.

From the results, Legendre and Piéron reached the conclusion that a hypnotoxic substance, i.e., ''hypnotoxin'', occurring in the brain was responsible for the elevated demand for sleep and the morphological changes in neurons. In spite of their genuine experimentation, the earliest humoral theory could not be correctly evaluated by the contemporary authorities. Unfortunately, the ''hypnotoxic'' state looked like narcosis, being far from physiological sleep. Economo[41] comments as follows: ''Since Piéron's experiments it can be accepted as a certainty that through activity during the waking state not only carbon dioxide and lactic acid but also special fatigue substances are produced which exert a narcotic action upon the brain and may inhibit its function. The accumulation of these residues could very well explain the periodicity of sleep. During sleep these residues are again excreted by the organism. As compared with the previously mentioned vasomotor theory and theory of interruption this chemical theory has at least the great advantage of being partially proved. But they, however, cannot be the exclusive cause of sleep.''

In the meantime, the neural theories on sleep steadily developed to rule out the humoral theories. Economo[41] studied lethargic encephalitis and reached the conclusion that sleep and wakefulness were regulated by the brain center differentially localized in the anterior lateral wall and the posterior wall, respectively, of the third ventricle. Pavlov[42] attributed the cause of sleep to the internal inhibition. The discovery of EEG by Berger[43] contributed greatly to electrophysiological approaches to sleep. Subsequent electrophysiological and neuroanatomical investigations by Hess,[44] Bremen,[45] Nauta,[46] Moruzzi and Magoun,[47] and many others firmly established the neural basis of sleep regulation. On the contrary, the humoral theories only produced a number of short-lived new candidates, one after another.

However, it should be noted that Ivy and Schnedorf[48,49] repeated experiments almost

Table 1
PROPOSED SLEEP-RELATED SUBSTANCES UP TO THE 1950s

Name	Author(s), Year	Ref.
Acetylcholine	Dikshit, 1934	58
Alcohol	Kalter and Katzenstein, 1932	59
Anterior hypothalamic extracts	Garcia, 1950	57
Antidiuretic factor	Schütz, 1944	60
	Faltz et al., 1946	61
Bromine hormone (tetrabrom-des-iodo-thyroxin)	Zondek and Bier, 1932	52—54
Calcium ion	Cloetta, 1936	cf. 32
	Demole, 1927	62
Carbon dioxide	Dubois, 1901	63
Cerebrospinal fluid factors	Stern, 1937	cf. 32
Cholesterol	Brissmoret and Joanin, 1912	64
Excito-anabolic autonomo-tonic hormone complex	Mingazini and Barbara	cf. 41
Fatigue substance	Economo, 1928	65
Fatigue toxin	Cabitto, 1923	cf. 32
Hypophysial hypnotic hormone	Salmon, 1923	66
Hypnogenic substance	Ishimori, 1909	34
Hypnotoxin	Legendre and Piéron, 1913	35
	Schnedorf and Ivy, 1937	49
Lactic acid	Preyer	cf. 32
Leucomaines	Errera, 1891	67
Liver extracts	Forsgren, 1930	50
	Holmgren, 1931	51
Sleep substance	Kroll, 1933	55,56
Stimulating substance	Friede, 1951	68
Ureides	Kalter and Katzenstein, 1932	59
Urotoxin	Bouchard, 1911	69
X, Y (adrenalin)	Bancroft and Rutzler, 1932	70
Wakefulness-producing hormone	Gélyi, 1922	cf. 32

similar to those of Legendre and Piéron,[35] and largely confirmed their results. They also observed a marked elevation of body temperature and intracranial pressure. They believed that "the change or changes in 'fatigue'cerebrospinal fluid, if they are found, will only be a reflection of the underlying chemical changes in the brain conducive to fatigue or sleep."

Thus, in spite of the criticism against the toxin theories, a number of investigators were still challenged to search for an endogenous sleep substance in a trial-and-error manner. They might have been greatly encouraged by the successful development of endocrinology and the discovery of various peripheral hormones during the 1930s. However, only Forsgren,[50] Holmgren,[51] Zondek and Bier,[52-54] Kroll,[55,56] and Garcia[57] tried in vain to isolate an active principle directly from body fluids or bodily tissues of different animals or humans under various physiological conditions. Zondek and Bier[54] even synthesized a bromine compound, tetrabrom-des-iodo-thyroxin, which might be a plausible candidate to their hypophyseal sleep hormone. The other researchers examined known substances in relation to sleep or they suggested the possible involvement of known and unknown humoral factors in the regulation of sleep and wakefulness. The list of substances dealt with up to the 1950s are summarized in Table 1.

The failure in the above-mentioned trials may be attributed to several technical reasons. First, it was difficult to prepare enough materials for the extraction and purification, since the quantity of an active principle in the organism may be extremely small and dynamically changing. Second, no adequate technique was yet available to isolate and identify peptidergic compounds. Third, sleep is too labile to manipulate artificially. Accordingly, there was no reliable assay to quantify the sleep-modulatory effect. Manipulation, such as through i.c.v. administration of sleep-unrelated substances per se, may nonspecifically bring about a change in vigilance level. Behavioral indices were too subjective to discriminate sleep from sleep-like states, such as narcosis or coma.

In addition, there were arguments apparently against the toxin theories as mentioned by Kleitman,[32] First, the demand for sleep does not lineally increase during sleep deprivation. Second, two humorally connected individuals (nondisjointed twins) and cross-circulation in two animals or parabiotic pairs do not show simultaneous occurrence of sleep. For example, Alekseyeva[71] observed that two pairs of infant nondisjointed twins exhibited mutually independent sleep-waking behavior. She concluded that hypnotoxin may appear in the blood only after a prolonged insomnia or fatigue, and that the simultaneous onset of sleep in nondisjointed twins is due mainly to the formation of the Pavlovian conditioning.

It may also be pointed out that the concept of "sleep hormone", which was obviously based on the classic concept of "peripheral hormone" as proposed by Bayliss and Starling[33] or on that of the "neurosecretion" as proposed by Scharrer and Scharrer,[72] could hardly match the actual sleep substances in our current sense. It seems almost improbable to discover a hormone-like sleep factor, since putative sleep substances we now know are less differentiated, less specific, and less broadly distributed than those the former investigators pursued.

III. PIONEERING APPROACHES WITH MODERN TECHNIQUES

A. Sleep Substances from Sleeping Animals

The modern study by Kornmüller et al.[73] opened a new epoch of research for sleep substances. By that time, the objective technique of EEG monitoring was available. Two different states of sleep, i.e., non-REM sleep or SWS and REM sleep or PS, were clearly defined by electrophysiological and other correlates (EEG sleep).

Kornmüller and colleagues performed a blood transfusion between two cats by crossing the carotid arteries. Sleep was artificially induced by the electric stimulation (15-ms, 1-V impulses at 4-Hz frequency for 30-s duration) of the thalamus (Nucleus centralis lateralis thalami, N. ventralis medialis thalami, and/or N. caudatus) of one cat. As rapidly as 20 to 30 s after the initiation of the thalamic stimulation, EEG sleep was induced to occur in the other animal. On the basis of this observation, the German investigators suggested the existence of a blood-borne "sleep hormone". Interestingly, the cross-circulation technique they adopted had been used to deny the humoral theories by several investigators in the former decades.

Monnier (see Figure 3) and associates[74] confirmed the observations of Kornmüller et al.[73] by similar experiments in rabbits. Electrical stimulation of the mediocentral intralaminary thalamus in the donor induced a statistically significant rise in EEG delta activity in both the donor and the recipient (see Figure 4). In contrast, electrical stimulation of the midbrain reticular system induced an EEG arousal in the donor as well as in the recipient.

Since then, the Swiss researchers established an extracorporal dialyzing technique[74-78] (see Chapter 2) to obtain hemodialyzates from the venous blood of sleeping rabbits electrically stimulated in the thalamic "sleep center". They extensively performed a series of experiments to isolate and characterize a sleep factor ("sleep-inducing factor delta").[79-85] In 1977, they were finally successful in identifying a novel nonapeptide, "delta-sleep-inducing peptide" (DSIP), as the active substance in the hemodialyzates.[86,87] DSIP was the first candidate for a "sleep peptide" originally extracted and identified from the organism. The physiological properties of DSIP will be described in detail in Chapter 3.

In 1976, Sachs et al.[88] suggested the existence of humoral factors in the CSF of rats, which may serve to regulate the level of motor activity and vigilance. CSF taken from normal rats during the light period, i.e., their resting phase, and infused into the lateral

FIGURE 3. Portrait of Marcel Monnier.

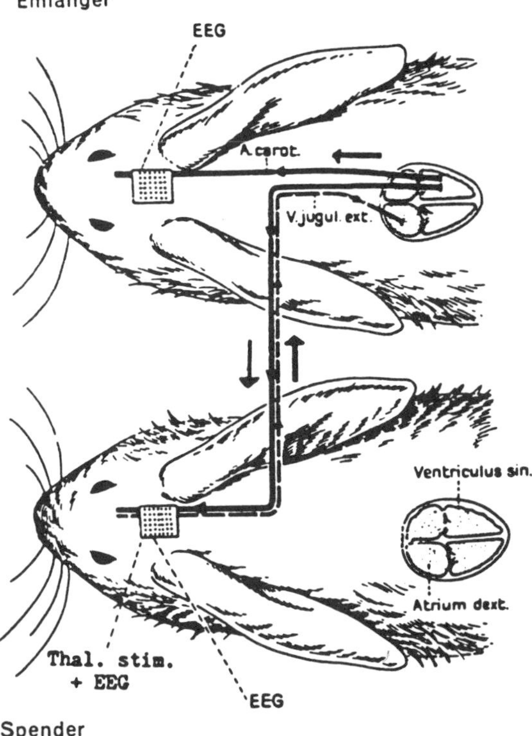

FIGURE 4. Crossed circulation experiment in rabbits. (From Monnier, M. and Schoenenberger, G. A., *Schweiz. Med. Wochenschr.*, 103, 1733, 1973. With permission.)

FIGURE 5. Portrait of John Pappenheimer.

cerebral ventricle of recipient rats during the dark period, i.e., their active phase, reduced the motor activity. Conversely, CSF taken from darkness-exposed donor rats and i.c.v. infused in recipients during their light period induced an increase in the motor activity. No further analysis was reported as to the chemical constituents of the CSF.

B. Sleep Substances from Sleep-Deprived Animals

1. Total Sleep Deprivation

In the middle 1960s in the U.S., Pappenheimer (see Figure 5) reinvestigated the earlier works by Legendre and Piéron[35] and Schnedorf and Ivy.[49] He recollects the days as follows:[89] "In our first experiments we judged the results by simple observation, just as Piéron had 55 years earlier. We did not really expect positive results. Yet all of us who participated in these pilot experiments had the strong impression that cats infused with fluid from goats that had been deprived of sleep for 48 hours were abnormally sleepy in comparison with the same cats infused with fluid taken from the same goats when the goats had not been deprived of sleep. Indeed, the results were so striking and of such potential physiological significance that we put other plans aside in order to devote full time to the systemic exploration of the Piéron phenomenon." Pappenheimer and co-workers[90,91] further found that infusion of 0.1 ml CSF withdrawn from sleep-deprived goats into the lateral ventricle of rats reduced locomotor activity for 6 to 12 h. By means of ultrafiltration, they separated 2 fractions (Factor S with a molecular weight of less than 500, and Factor E with a molecular weight range of 500 to 10,000) from CSF of sleep-deprived goats. The sleep-promoting Factor S was found only in the CSF of sleep-deprived goats, whereas the hyperactivity-inducing Factor E was present in CSF from both normal and sleep-deprived goats.

Both Factors S and E appeared to be peptides.[91,92] However, since it was not feasible to collect CSF in the amounts needed for systematic studies of the chemistry and physiology of Factor S, brain tissue from sleep-deprived goats, sheep, and rabbits, and that of slaugh-

FIGURE 6. Portrait of Koji Uchizono.

terhouse cattle, were adopted for the subsequent investigations.[93,94] Nevertheless, these attempts also proved inadequate to obtain sufficient materials for further analyses.

Shortly after the first reports on humoral factors were published from Monnier's and Pappenheimer's laboratories, Ringle and Herndon[95,96] tried to do similar experiments and detected no sleep-enhancing property in the blood and CSF of sleep-deprived rabbits. They deprived rabbits of total sleep for 72 h in revolving cages. Heparinized plasma samples obtained by cardiac puncture were dialyzed against water. The dialyzates, which were lyophilized, dissolved in water, and i.v. injected, did not affect EEG activity in rabbits and rats or ambulatory activity in mice. The CSF (0.1 ml) withdrawn from the cisterna magna of sleep-deprived rabbits was infused into the left lateral ventricle of conscious unrestrained rats for more than 28.6 min. Neither the cortical EEG nor the behaviors indicated sleep promotion in the recipient rats. Thus, these American investigators reached the negative conclusions that sleep deprivation does not promote the appearance of dialyzable sleep-promoting material in the blood of rabbits, and that the appearance of a sleep-promoting material in the CSF following sleep deprivation is limited to certain species, not occurring in all mammals.

Moruzzi,[97] a prominent sleep researcher in the modern age and the discoverer of the midbrain reticular activating system, comments as follows: ''It is puzzling that sleep-promoting substances may appear in the blood or in the cerebrospinal fluid both as a consequence of prolonged wakefulness and of induced sleep. Moreover, the physiological significance of these experiments appears still uncertain...''

In the early 1970s in Japan, Uchizono (see Figure 6) and associates,[98] including this author, extracted a sleep-promoting factor from brainstems of sleep-deprived rats. The crude extract prepared by dialyzing the homogenized brainstems of 24-h sleep-deprived rats against water (see Chapter 2 for details) reduced locomotor activity, increased EEG delta activity, and prolonged total sleep time in recipient rats (6 brainstem equivalent units i.p.). No effect was found in the extract from nontreated rats. Subsequently, we established techniques for chronic infusion of test materials into the third cerebral ventricle of freely moving rats under concurrent monitorings of EEG, electromyogram (EMG), brain temperature, and behaviors.[99-101]

Purified fractions from the dialyzate of the brainstems of sleep-deprived rats caused reductions of neuronal firing activity in the crayfish abdominal ganglion and stretch receptor (0.0002 to 0.02 brainstem units *in vitro*),[102-105] and increases in daily EEG sleep in mice (0.5 to 3 units i.c.v.;[104] 0.05 to 1 units i.p.[106]), and reductions in nocturnal locomotor activity and increases in nocturnal EEG sleep in rats (2 to 5 units i.c.v.),[100-102,104,105,107,108] The plural number of active fractions has been called sleep-promoting substance (SPS) in a general term.[107,108] In 1983, one of the SPS components was identified as uridine.[109] The other components are under the process of further purification. The details on SPS will be dealt with in Chapters 2 and 4.

Mendelson et al.[13,110] reinvestigated sleep-inducing properties of some candidate substances including DSIP, Factor S, and SPS by their own methods and concluded that none of the currently available factors have been convincingly shown to be physiologic mediators of sleep.

2. Selective Sleep Deprivation

In the late 1960s in Mexico, Drucker-Colín and collaborators[111] undertook experiments to detect a sleep factor from push-pull perfusates of the midbrain reticular formation in cats deprived of REM sleep by the water-tank method. The perfusates induced sleep if transfused into the lateral ventricle of waking cats. Similarly prepared perfusates taken from sleeping donors increased the duration and decreased the latency of SWS in recipient cats.[112] Further studies revealed that the active factor occurring in the perfusates of REM sleep-deprived cats appeared to be proteins in chemical nature.[113-115] Spanis et al.[116] reported that the REM sleep-related perfusates contained two different proteins with molecular weights of 45,000 and 73,000. On the basis of these observations, the Mexican researchers suggested that "REM sleep proteins" might be responsible for the regulation of REM sleep. Although no attempt has been made to isolate the proteins, Prospéro-Garcia et al.[117] suggested that the CSF of REM sleep-deprived donor cats contains a vasoactive intestinal polypeptide (VIP)-like factor. This factor will be dealt with further in Chapter 7.

C. Sleep Substances from Human CSF and Urine

1. Bromine Compound (γ-Br)

In the course of investigations on bromine intoxication in human patients, Yanagisawa and Yoshikawa[118] found that, in addition to exogenous bromine, a bromine-containing organic substance occurred physiologically in both CSF and blood. This substance was isolated from a large amount of pooled human CSF and finally identified as 1-methyl-heptyl-γ-bromoacetoacetate (γ-Br or MHBAA).[119] Subsequent studies revealed that γ-Br (0.1 to 5 mg i.v.) selectively promoted PS in encéphale isolé cats.[120] This will be described in detail in Chapters 2 and 7.

2. Urinary Factor S

Since Factor S of Pappenheimer et al.[93] appears to be stable, both *in vivo* and *in vitro,* Factor S or its derivatives might be absorbed in the blood with the normal resorption of CSF, and eventually be concentrated and excreted by the kidney. Under such an assumption, Krueger et al.[121] collected human urine of first morning micturition from healthy male adults. Partially purified extracts designated as urinary Factor S (see Chapter 2) revealed chemical similarities to Factor S. I.c.v. infusion of urinary Factor S induced excess SWS in both rats and rabbits.[121-123]

Interestingly, the urinary factor was proven to be a small glycopeptide with muramic acid, resembling bacterial peptideglycans.[123] In 1984, Martin et al.[124] and Krueger et al.[125] reported that urinary Factor S was finally identified as muramyl peptides (MP), i.e., *N*-acetylglucosamyl-1,6-anhydro-*N*-acetylmuramyl-Ala-γ-Glu-diaminopimelyl-Ala and that compound lacking the terminal alanine. The former substance induced prolonged excess sleep in rabbits (1 pmol i.c.v.). The fact that MPs are indeed pyrogenic bacterial cell-wall peptideglycans has eventually triggered unexpected development of a new field of sleep research, i.e., toward the immune theory of sleep (see Chapter 5).

D. Hibernation Inducers

As early as 1933 and 1952, Kroll[55,56] extracted from the brain tissues of hibernating hamsters, hedgehogs, and bats a "sleep hormone" that could induce "experimental winter sleep". Since then, experimental approaches to a hibernation inducer by several investigators evidenced the presence of "hibernation triggers" in several species (for review, see Reference 126). Recently, Oeltgen et al.[127] and Spurrier et al.[128] reported that an endogenous opioid-like peptide was detected in the blood of hibernating woodchucks, and that i.c.v. infusion of this unidentified factor (designated as HT) caused hypothermia, behavioral depression, bradycardia, and aphasia in conscious rhesus monkeys. They also reported that summer hibernation in 13-lined ground squirrels was induced by i.v. injection of whole plasma or its albumin fraction from winter-torpid black bears.[129] The active factor has been called hibernation-induction trigger (HIT). The bear HIT seems to be not an opioid itself but a precursor of endogenous opioids or a potent releaser of them.[130,131] Stanton et al.[132] demonstrated that continuous i.c.v. infusion of melatonin at doses of 200 and 400 μg/0.5 μl/h using an osmotic minipump resulted in a dose-related prolongation of hibernation bout duration in golden-mantled ground squirrels.

Mammalian hibernation is closely connected with sleep.[133] For example, hibernation of golden-mantled squirrels is characterized by a continuation of SWS at moderate temperatures.[134] Daily shallow torpor with a lowering of body temperature and metabolism can be observable even in nonhibernating animals like fasting doves[135] and small-sized marsupials.[136,137] However, since the biological function of sleep may not be identical to that of sleep, no further discussion will be focused on the hibernation inducers.

E. Sleep-Suppressive Substances

Antagonistic to the actions of sleep substances, some endogenous factors may reduce sleep and increase the vigilance level. Gélyi in 1922 suggested that the exhaustion of a wakefulness-producing hormone causes sleep (see Reference 32). The electric stimulation of the midbrain or bulbar reticular activating system resulted in an elevation of vigilance level in animals humorally connected by cross circulation.[73,138,139] This fact suggests that a blood-borne, arousal-inducing factor may be involved in the phenomenon. Both Monnier et al.[74,75,139] and Fencl et al.[91] found a wakefulness-promoting fraction in their preparations from the hemodialyzates of sleeping rabbits and the CSF of sleep-deprived goats, respectively. Toh[140,141] extracted "nerveside" from dog and ox brains. This chemically unidentified phosphopeptide (150 and 300 μg i.c.v.) decreased EEG sleep and increased wakefulness in rats.[142] During the process of purifying SPS preparations, a few fractions exerted sleep-suppressing effects.[142a]

Riou et al.[143] reported several endogenous peptides, such as angiotensin II (100 ng i.c.v.), arginine vasotocin (AVT) (10 and 50 μg/kg i.p.), renin (0.01 unit i.c.v.), and substance P (100 ng i.c.v.) that selectively decreased PS in rats. The sleep-suppressive effects of other substances, including adrenocorticotropic hormone (ACTH), benzodiazepine receptor antagonists, prostaglandin E_2 (PGE_2), and thyrotropin-releasing factor, will be dealt with in later chapters.

F. Sleep-Modulatory Properties of Known Substances

Apart from the attempt to extract a sleep-inducing factor from the body fluids or tissues, several workers examined the sleep-modulatory properties of known compounds. Accordingly, a variety of substances of various chemical nature have been mentioned as a putative sleep substance. It is well known that growth hormone (GH) is secreted during delta sleep at the first few periods of sleep cycles in humans.[144] In 1975, Stern et al.[145] proposed that GH may play a role in the subsequent appearance of REM sleep, and showed that bovine GH (50 to 1000 μg i.p.) increased REM sleep in cats and reduced REM-sleep latency. These results were confirmed by later studies.[13,14,146]

In 1976, Pavel et al.[147] reported that i.c.v.-injected AVT, a nonapeptide synthesized or purified from the pineal gland, induced SWS and suppressed PS in cats at a dose level as low as 10^{-18} g corresponding to 10^{-21} mol or 600 molecules. The Romanian researchers[148-150] also reported that, in contrast to the effects in cats, i.v. or s.c. injection of AVT decreased the total time of non-REM sleep and increased that of REM sleep in humans. Later studies did not confirm their observations[4,13,110,151] except for the report that a transient increase of SWS occurred within 1 h after i.c.v. administration of 25 pmol AVT in two rabbits.[121] Another pineal hormone, melatonin, is also repeatedly referred to as a putative sleep-inducing factor. This and other substances, including several nutrients, piperidine, and steroid hormones, will be dealt with in Chapter 8.

In 1982, Ueno et al.[152] reported that microinjection of prostaglandin D_2 (PGD_2) into the preoptic area induced SWS in rats under restrained conditions. Subsequent extensive studies have revealed that PGD_2 is one of the most promising candidates for an endogenous sleep substance (see Chapter 6).

IV. OVERVIEW OF THE STATUS QUO

A. Revival of Humoral Theory

Since the isolation and identification of DSIP by Schoenenberger et al.,[86,87] the concept of a humoral control of sleep has been vigorously revived and its validity rapidly established. In spite of critical comments[153] or reports on nonconfirmative data,[4,13,95,96,110] which might be based largely on the unawareness of an appropriate methodology required for determining optimal dosage, a route of administration, recipient animals, the time of day for treatments, the duration of recording time period, indices for evaluation, and so forth, both an SPS component and Factor S were chemically identified. In addition, an increasing number of naturally occurring substances have been newly added to the list of candidates (see Table 2). They can also exhibit more or less pharmacological pathological aspects of sleep-modulatory properties and concomitantly extra-sleep physiological activities.

As mentioned in the first part of this chapter, a working hypothesis that a single hormone-like substance regulates all aspects of sleep has become highly improbable. Instead, multiple factors play either specifically or interrelatedly a limited role in the regulation of some aspects of sleep. No one knows the full story. As Akimoto,[154] a pioneer of modern sleep research in Japan, states in his essay on the history of sleep research, the *alchemists* who pursued *the* sleep substance should now challenge as *scientists* the old simple but unsolved question as to why we sleep.

B. The Roles of Sleep Substances in Sleep Regulation

On the basis of recent advances in physiology, biochemistry, pharmacology, and ethology of sleep substances, novel concepts and genuine theories have gradually been developed by several investigators who are playing a leading role in this field of sleep research. Although the hypotheses are still in their infancy, their outlines can be summarized as follows.

Table 2
LIST OF PUTATIVE SLEEP SUBSTANCES UP TO 1988

Name	Description in this book
Adenosine	Chapter 4
Arginine vasotocin (AVT)	Chapters 1,7
Benzodiazepine (BDZ) receptor agonists	Chapter 8
Cholecystokinin (CCK)	Chapter 8
Corticotropin-like intermediate lobe peptide (CLIP)	Chapter 8
Corticotropin-releasing factor (CRF)	Chapter 8
Delta-sleep-inducing peptide (DSIP)	Chapters 1—3,9,10
Deoxycytidine	Chapters 4,7
Desacetyl-α-melanocyte-stimulating hormone (des-α-MSH)	Chapter 8
Endotoxin	Chapter 5
Growth hormone (GH)	Chapter 7
Growth hormone-releasing factor (GRF)	Chapter 8
Insulin	Chapter 8
Interferon 2 (IFN)	Chapter 5
Interleukin-1 (IL1)	Chapter 5
Lipid A	Chapter 5
Melatonin	Chapter 8
1-Methylheptyl-γ-bromoacetoacetate (γ-Br, MHBAA)	Chapters 1,2,7,
Muramyl peptides (MP)	Chapters 1,2,5,9
Neuropeptide Y (NPY)	Chapter 8
Piperidine	Chapter 8
Prolactin (PRL)	Chapter 7
Prostaglandin D_2 (PGD_2)	Chapter 6
PS factors	Chapter 7
Sleep-promoting substance (SPS)	Chapters 1,2,4
Somatostatin (SRIF)	Chapters 7,8
Steroid hormones	Chapter 8
Tumor necrosis factor (TNF)	Chapter 5
Uridine	Chapters 2,4,9,10
Vasoactive intestinal polypeptide (VIP)	Chapter 8

1. Programming Theory

It has been uncovered that DSIP not only affects sleep, but also modulate a variety of physiological activities in humans and animals.[155-158] Especially, taking the normalizing effects of DSIP on behavioral and psychophysiological functions, the Basel researchers have developed a concept of a ''programming'' substance. This will be dealt with in detail in Chapter 3.

2. Analyses by Modeling Sleep-Regulatory Mechanisms

Borbély and associates[159-161] formulated a two-process model of sleep regulation on the basis of experimental data in humans and animals. The model rests on the assumption that sleep is regulated by two separate processes, i.e., a sleep-dependent process (Process S) and a circadian, pacemaker-dependent process (Process C). The level of Process S, which increases during wakefulness and declines during sleep, is assumed to be regulated by a sleep substance. Kobayashi[162] has recently formulated a model of human SWS regulation and emphasized the possible involvement of a sleep substance. According to him, the amount of SWS (= stage 3 + 4 sleep) distributed in successive sleep cycles depends on the amount of the sleep substance in the preceding cycle.

Koella[163-165] schematized the input-output relations in the sleep-waking regulatory system. He endowed sleep factors with a role of the feedback information carrier from the tissue. Sleep factors may carry information about the present state and the recent history of the central nervous system and also chronological signals delivered by the biological clock.

3. Series- or Parallel-Process Theories

Jouvet[5-7] proposed that sleep-inducing (hypnogenic) factors should take part in biochemical sequential chain reactions from A to B, from B to C, from C to ..., leading to the final cause of sleep, whereas sleep-facilitating factors control the effector mechanisms of SWS and PS. On such a theoretical consideration he revised his classical monoamine theory[166,167] to formulate the hydraulic theory of PS. In a similar fashion, Danguir[168] suggested that digestive processes serially occurring after food intake finally result in sleep induction (ischymetric sleep theory; see Chapter 8). Cespuglio and colleagues[169] and Chastrette et al.[170] have focused their attention on the proopiomelanocortin family of peptides. A specific situation of waking state like that of stress induces the release of ACTH, which in turn produces desacetyl-α-melanocyte-stimulating hormone (α-MSH) and corticotropin-like intermediate lobe peptide (CLIP). ACTH, α-MSH, and CLIP can sequentially promote wakefulness, SWS, and PS, respectively (see also Chapter 8). In contrast, Drucker-Colín et al.[171] postulated that many sleep factors can independently in parallel affect or impinge upon a mechanism which is the veritable element responsible for producing sleep.

4. Immune Theory for SWS

Krueger and co-workers[172] have found close relationships between immune reactions and SWS occurrence. The considerations to their mutual causality have eventually led them toward the immune theory of sleep (see Chapter 5). The impact of this theory has gradually forced sleep researchers to revisit the classical restitution theories[173-175] on the function of sleep.

5. Multiple-Factors Theory or Interactions Theory

Recently, it has been broadly accepted that a great variety of putative sleep substances (see Table 2) are involved in regulatory processes of either SWS or PS or both. One of the most important tasks we confront is related to the analysis and synthesis of interrelationships among the numerous ''state-regulating'' substances.[176-179] The interactions of multiple sleep factors will be described in detail in Chapter 9.

6. Reciprocal Two-Substance Theory

Ehlers and Kupfer[180] have proposed that growth hormone-releasing factor and corticotropin-releasing factor reciprocally interact in the regulation of circadian sleep-waking rhythm (see Chapter 8). Ernst and Schoenenberger[181] have pointed out a crucial reciprocal relationship between DSIP and its analogue or metabolite, phosphorylated DSIP, in the regulation of sleep-wakefulness (see Chapter 3). Two different prostaglandins (PG), PGD_2 and PGE_2, can modulate sleep and wakefulness, and their actions are mutually reciprocal. Hence we speculate that a PG-dependent sleep regulatory system may exist and play an important role in the regulation of sleep and wakefulness[182] (see Chapter 6).

7. Other Related Theories and New Prospects

Horne[183] has postulated that sleep substances are mainly concerned with the regulation not of ''core sleep'' but of ''optional sleep'', an unobligatory portion of sleep, by elevating or maintaining sleep pressure. Turek[184] referred to the existence of endogenous substances that can influence the mammalian circadian system. These substances may affect a central circadian pacemaker to regulate the entrainment, generation, and expression of circadian rhythms, thus modifying the timing of sleep (see Chapter 10).

Rhyner et al.[185] took a novel approach for the selective isolation of rat forebrain mRNA transcripts by generating a tenfold enriched recombinant library after the subtraction of cerebellar sequence. It is suggested that sleep-related proteins may be identified through the search for possible transcripts specifically induced by sleep deprivation.

Table 3
LIST OF REVIEWS ON ENDOGENOUS SLEEP SUBSTANCES PUBLISHED IN THE 1980s

Author(s)/editor(s)	Remarks
Adrien[189]	Proceedings of the symposium "Endogenous sleep factors: a critical approach" held at 8th European Congress of Sleep Research, Szeged, September 4, 1986; 7 minipapers
Borbély[190]	Reports on the workshop "Endogenous sleep factors" held at 6th European Congress of Sleep Research, Zürich, March 23, 1982
Borbély[191]	Up to 1985; 101 references
Borbély and Tobler[3]	Up to 1980; 12 references
Drucker-Colín[17]	Up to 1980; 85 references
Drucker-Colín and Valverde-R.[192]	Up to 1980; 241 references
Graf and Kastin[155]	Up to 1984; 155 references
Graf and Kastin[156]	Up to 1986; 248 references
Inoué[11]	Up to 1983; 111 references
Inoué[10]	Up to 1983; 20 references
Inoué[176]	Up to 1985; 46 references
Inoué[193]	Monograph on endogenous sleep substances as a whole
Inoué[18]	Up to 1986; 20 references
Inoué[177]	Up to 1986; 5 references
Inoué[194]	Up to 1988; 5 references
Inoué and Borbély[195]	Proceedings of 8th Taniguchi International Symposium on Brain Sciences, "Humoral control of sleep and its evolution", Kyoto/Otsu, October 21—25, 1984; 21 reviews
Inoué et al.[196]	Up to 1982; 54 references
Inoué and Schneider-Helmert[19]	Monograph based on the presentations of the symposium "Sleep regulation by sleep peptides" held at 5th International Congress of Sleep Research, Copenhagen, July 3, 1987; 7 reviews
Komoda[197]	Up to 1984; 27 references
Kovalzon[198]	Up to 1983; 14 references
Krueger et al.[172]	Up to 1984; 151 references
Liu[24]	Up to 1982; 49 references
Monnier and Schoenenberger[157]	Up to 1982; 80 references
Pappenheimer[199]	Up to 1980; 6 references
Takahashi and Kato[200]	Up to 1983; 66 references
Schoenenberger[158]	Up to 1983; 49 references
Uchizono[201]	Up to 1983; 72 references
Ursin[9]	Up to 1983; 35 references
Ursin and Borbély[202]	Proceedings of the workshop "Endogenous sleep factors" held at 6th European Congress of Sleep Research, Zürich, March 23, 1982; 8 minipapers
Wauquier et al.[20]	Proceedings of the international workshop "Sleep: neurotransmitters and neuromodulators", Antwerp, August 30—September 1, 1984; 6 out of 26 papers deal with sleep substances

We have mathematically analyzed the effect of sleep substances on the dynamic properties of the time series of state transfers.[186,187] The joint probability of inter- and intrastate transfers and the Markovian dependency[188] of the second- and third-order values are significantly modified after the administration of uridine and other sleep substances.

Finally, recently published review articles (summarized in Table 3) may reflect interesting and important developmental changes in the evaluation of humoral theories of sleep.

REFERENCES

1. **Inoué, S.,** Sleep substances: their roles and evolution, in *Endogenous Sleep Substances and Sleep Regulation,* Inoué, S. and Borbély, A. A., Eds., Japan Scientific Societies Press, Tokyo, 1985, 3.
2. **Inoué, S.,** Multifactorial humoral regulation of sleep, *Clin. Neuropharmacol.,* 9 (Suppl. 4), 470, 1986.
3. **Borbély, A. A. and Tobler, I.,** The search for an endogenous 'sleep-substance', *Trends Pharmacol. Sci.,* 1, 356, 1980.
4. **Tobler, I. and Borbély, A. A.,** Effect of delta sleep inducing peptide (DSIP) and arginine vasotocin (AVT) on sleep and locomotor activity in the rat, *Waking Sleeping,* 4, 139, 1980.
5. **Jouvet, M.,** Hypnogenic indolamine-dependent factors and paradoxical sleep rebound, in *Sleep 1982,* Koella, W. P., Ed., S. Karger, Basel, 1983, 2.
6. **Jouvet, M.,** Indolamines and sleep-inducing factors, *Exp. Brain Res.,* Suppl. 8, 81, 1984.
7. **Jouvet, M.,** Hypnogenic indolamine-dependent factors and paradoxical sleep rebound, in *Functions of the Nervous System,* Vol. 4, Monnier, M. and Meulders, M., Eds., Elsevier, Amsterdam, 1983, 147.
8. **Ursin, R.,** Endogenous sleep factors: introduction, in *Sleep 1982,* Koella, W. P., Ed., S. Karger, Basel, 1983, 106.
9. **Ursin, R.,** Endogenous sleep factors, *Exp. Brain Res.,* Suppl. 8, 118, 1984.
10. **Inoué, S.,** Sleep substances, *Iden,* 37, 125, 1983.
11. **Inoué, S.,** Sleep substances: their roles and evolution, *Seikagaku,* 55, 445, 1983.
12. **Inoué, S., Honda, K., Komoda, Y., Uchizono, K., Ueno, R., and Hayaishi, O.,** Little sleep-promoting effect of three sleep substances diurnally infused in unrestrained rats, *Neurosci. Lett.,* 49, 207, 1984.
13. **Mendelson, W. B., Wyatt, R. J., and Gillin, J. C.,** Whither the sleep factors?, in *Sleep Disorders: Basic and Clinical Research,* Chase, M. H. and Weitzman, E. D., Eds., Spectrum, New York, 1983, 281.
14. **Gillin, J. C.,** Endogenous sleep promoting substances: an introduction and overview, *Clin. Neuropharmacol.,* 9 (Suppl 4), 456, 1986.
15. **Inoué, S. and Borbély, A. A.,** Preface, in *Endogenous Sleep Substances and Sleep Regulation,* Inoué, S. and Borbély, A. A., Eds., Japan Scientific Societies Press, Tokyo/VNU Science Press BV, Utrecht, 1985, v.
16. **Uchizono, K.,** Sleep peptides, *J. Clin. Sci.,* 13, 1188, 1977.
17. **Drucker-Colín, R.,** Endogenous sleep peptides, in *Psychobiology of Sleep,* Wheatley, D., Ed., Raven Press, New York, 1981, 53.
18. **Inoué, S.,** Sleep peptides, *J. Clin. Sci.,* 22, 1203, 1986.
19. **Inoué, S. and Schneider-Helmert, D., Eds.,** *Sleep Peptides: Basic and Clinical Approaches,* Japan Scientific Societies Press, Tokyo/Springer-Verlag, Berlin, 1988.
20. **Wauquier, A., Gaillard, J. M., Monti, J. M., and Radulovacki, M., Eds.,** *Sleep: Neurotransmitters and Neuromodulators,* Raven Press, New York, 1985.
21. **Bulfinch, T.,** *The Age of Fable* (Japanese transl.), Iwanami Shoten, Tokyo, 1978, 109.
22. *Sutta-nipata* (Japanese transl.), Iwanami Shoten, Tokyo, 1984, 89.
23. **Morita, D.,** *Research on Ancient Chinese Medical Thought,* Yuzankaku Shuppan, Tokyo, 1985.
24. **Liu, S.,** Studies on the brain mechanisms of sleep, *Acta Psychol. Sinica,* 1, 19, 1982.
25. **Brooks, C. H. and Levey, H. A.,** Humorally-transported integrators of body function and the development of endocrinology, in *The Historical Development of Physiological Thought,* Brooks, C. H. and Granefield, P. F., Eds., Hafner Publishing, New York, 1959, 183.
26. **Wittern, R.,** Der Schlaf als medizinisches Problem am Beginn der Neuzeit, *Habilitationsschrift,* Ludwig-Maximilians-Universität München, Munich, West Germany, 1978.
27. **Dannenfeldt, K. H.,** Sleep: theory and practice in late renaissance, *J. Hist. Med. Allied Sci.,* 41, 415, 1986.
28. **Aristotle,** Parva naturalia (Japanese transl.), in *Complete Works of Aristotle, Vol. 6,* Iwanami Shoten, Tokyo, 1968, 240.
29. **Aristotle,** *Parva naturalia* (Beare, J. I., Transl.), cited from **Kleitman, N.,** *Sleep and Wakefulness,* The University of Chicago Press, Chicago, 1963, 342.
30. **Kaibara, E.,** *Yojokun,* Iwanami Shoten, Tokyo, 1961.
31. **Humboldt, A. v.,** *Versuche über die gereizte Muskel- und Nervenfaser nebst Vermuthungen über chemischen Proceß des Lebens in der Thier- und Pflanzenwelt,* Posen, 1797.
32. **Kleitman, N.,** *Sleep and Wakefulness,* The University of Chicago Press, Chicago, 1963.
33. **Bayliss, W. M. and Starling, E. H.,** The mechanism of pancreatic secretion, *J. Physiol. (London),* 28, 325, 1902.
34. **Ishimori, K.,** True cause of sleep — a hypnogenic substance as evidenced in the brain of sleep-deprived animals, *Tokyo Igakkai Zasshi,* 23, 429, 1909.
35. **Legendre, R. and Piéron, H.,** Recherches sur le besoin de sommeil consecutif à une veille prolongée, *Z. Allg. Physiol.,* 14, 235, 1913.

36. **Legendre, R. and Piéron, H.,** Le rapports entre les conditions physiologiques et les modifications histologiques des cellules cérébrales dans l'insomnie expérimentale, *C. R. Soc. Biol.*, 62, 312, 1907.
37. **Legendre, R. and Piéron, H.,** Le problème des facteurs du sommeil. Résultats d'injections vasculaires et intra-cérébrales de liquides insomniques, *C. R. Soc. Biol.*, 68, 1077, 1910.
38. **Legendre, R. and Piéron, H.,** Des résultas histophysiologiques de l'injection intra-occipito-atlantoidienne de liquides insomniques, *C. R. Soc. Biol.*, 68, 1108, 1911.
39. **Legendre, R. and Piéron, H.,** Du développment, au cours de l'insomnie expérimentale, de propriétés hypnotoxiques des humeurs en relation avec le besoin croissant de someil, *C. R. Soc. Biol.*, 70, 190, 1911.
40. **Legendre, R. and Piéron, H.,** De la propriété hypnotoxiques des humeurs développée au cours d'une veille prolongée, *C. R. Soc. Biol.*, 72, 210, 1912.
41. **Economo, C. v.,** Sleep as a problem of localization, *J. Nerv. Ment. Dis.*, 71, 249, 1930.
42. **Pavlov, I.,** Die normale Tätigkeit und allgemeine Konstitution der Großhirnrinde, *Skand. Arch. Physiol.*, 44, 32, 1923.
43. **Berger, H.,** Über das Elektrenkephalogramm des Menschen, *J. Psychol. Neurol.*, 40, 160, 1930.
44. **Hess, W. R.,** *Hypothalamus und Thalamus*, 2. Aufl., Georg Thieme Verlag, Stuttgart, 1968.
45. **Bremer, F.,** Cerveau "isolé" et physiologie du sommeil, *C. R. Soc. Biol.*, 118, 1235, 1935.
46. **Nauta, W. J. H.,** Hypothalamic regulation of sleep in rats. An experimental study, *J. Neurophysiol.*, 9, 285, 1946.
47. **Moruzzi, G. and Magoun, H. W.,** Brain stem reticular formation and activation of the EEG, *Electroencephalogr. Clin. Neurophysiol.*, 1, 455, 1949.
48. **Ivy, A. C. and Schnedorf, J. G.,** On the hypnotoxin theory of sleep, *Am. J. Physiol.*, 119, 342, 1937.
49. **Schnedorf, J. G. and Ivy, A. C.,** An examination of the hypnotoxin theory of sleep, *Am. J. Physiol.*, 125, 491, 1939.
50. **Forsgren, E.,** Über die Beziehungen zwischen Schlaf und Leberfunktion, *Skand. Arch. Physiol.*, 60, 299, 1930.
51. **Holmgren, H.,** Über die hypnotische Wirkung von Wasserextrakt, aus der Leber schlafender Tiere bereitet, *Skand. Arch. Physiol.*, 63, 75, 1931.
52. **Zondek, H. and Bier, A.,** Brom im Blute bei manisch-depressivem Irresein, *Klin. Wochenschr.*, 11, 633, 1932.
53. **Zondek, H. and Bier, A.,** Der Bromgehalt der Hypophyse und seine Beziehungen zum Lebensalter, *Klin. Wochenschr.*, 11, 759, 1932.
54. **Zondek, H. and Bier, A.,** Hypophyse und Schlaf, *Klin. Worchenschr.*, 11, 760, 1932.
55. **Kroll, F.-W.,** Über das Vorkommen von übertragbaren schlaferzeugenden Stoffen im Hirn schlafender Tiere, *Z. Gesamte Neurol. Psychiatr.*, 146, 208, 1933.
56. **Kroll, F.-W.,** Gibt es einem humoralen Schlafstoff im Schlafhirn?, *Dtsch. Med. Wochenschr.*, 77, 879, 1952.
57. **Garcia, J. A.,** Régulation hormonale du sommeil, *Ann. Méd. Psychol.*, 108, 452, 1950.
58. **Dikshit, B. B.,** Action of acetylcholine on the "sleep center", *J. Physiol. (London)*, 83, 42P, 1934.
59. **Kalter, S. and Katzenstein, C.,** Über die Bedeutung des intermediären Alkohols für Schlaf und Narkose, *Mench. Med. Wochenschr.*, 79, 793, 1932.
60. **Schütz, F.,** Induction of sleep by simultaneous administration of posterior pituitary extracts and water, *Nature*, 153, 432, 1944.
61. **Faltz, P., Kayser, C., and Rouillard, J.,** Sommeil et hydrémie, *C. R. Soc. Biol.*, 140, 301, 1946.
62. **Demole, V.,** Pharmalologisch-anatomische Untersuchungen zum Problem des Schlafs, *Arch. Exp. Pathol. Pharmakol.*, 120, 229, 1927.
63. **Dubois, R.,** Sommeil naturel par autonarcose carbonique provoqué expérimentalement, *C. R. Soc. Biol.*, 53, 231, 1901.
64. **Brissmoret, A. and Joanin, A.,** Sur le properiété pharmacodynamiques de la cholestéerine, *C. R. Soc. Biol.*, 72, 824, 1912.
65. **Economo, C. v.,** Théorie du sommeil, *J. Neurol. Psychiatry*, 28, 437, 1928.
66. **Salmon, A.,** Le role des corrélations cortico-diencéphaliques et dienphalo-hypophysaires dans la régulation de la veille et du sommeil, *Presse Med.*, 27, 509, 1937.
67. **Errera, L.,** Note sur théorie toxique du sommeil, *C. R. Soc. Biol.*, 43, 5, 1891.
68. **Friede, R.,** Über zentral-humorale Schlafsteuerung, *Acta Neuroveg. (Wien)*, 2, 270, 1951.
69. **Bouchard, C.,** Sur la théorie toxique du sommeil et de la veille, *C. R. Acad. Sci.*, 152, 564, 1911.
70. **Bancroft, W. D. and Rutzler, J. E., Jr.,** The agglomeration theory of sleep, *Proc. Natl. Acad. Sci. U.S.A.*, 19, 73, 1933.
71. **Alekseyeva, T. T.,** Correlation of nervous and humoral factors in the development of sleep in non-disjointed twins, *Zh. Vyssh. Nervn. Deyat. im. I. P. Pavlova*, 8, 835, 1958.
72. **Scharrer, E. and Scharrer, B.,** Secretory cells within hypothalamus, *Res. Publ. Assoc. Res. Nerv. Ment. Dis.*, 20, 170, 1940.

73. **Kornmüller, A. E., Lux, H. D., Winkel, K., and Klee, M.,** Neurohumoral ausgelöste Schlafzustände an Tieren mit gekreuztem Kreislauf unter der Kontrolle von EEG-Ableitungen, *Naturwissenschaften,* 48, 503, 1961.
74. **Monnier, M., Koller, T., and Graver, S.,** Humoral influences of induced sleep and arousal upon electrical brain activity of animals with crossed circulation, *Exp. Neurol.,* 8, 264, 1963.
75. **Monnier, M. and Schoenenberger, G. A.,** Erzeuzung, Isolierung und Characterisierung eines physiologischen Schlaffaktor "delta", *Schweiz. Med. Wochenschr.,* 103, 1733, 1973.
76. **Hösli, L., Monnier, M., and Koller, T.,** Humoral transmission of sleep and wakefulness. I. Method for dialysing psychotropic humors from the cerebral blood, *Pfluegers Arch.,* 282, 54, 1965.
77. **Monnier, M. and Hösli, L.,** Humoral transmission of sleep and wakefulness. II. Hemodialysis of sleep inducing humor during stimulation of the thalamic somnogenic area, *Pfluegers Arch.,* 282, 60, 1965.
78. **Schoenenberger, G. A., Cueni, L. B., Hatt, A. M., and Monnier, M.,** Caractérisation physico-chimique du facteur hypnogène delta, *Rev. Neurol., Paris,* 126, 427, 1972.
79. **Schneidermann, N., Monnier, M., and Hösli, L.,** Humoral transmission of sleep. IV. Cerebral and visceral effects of sleep dialysate, *Pfluegers Arch.,* 288, 65, 1966.
80. **Monnier, M. and Hatt, A. M.,** Humoral transmission of sleep. V. New evidence from production of pure sleep hemodialysates, *Pfluegers Arch.,* 329, 231, 1971.
81. **Monnier, M., Hatt, A. M., Cueni, L. B., and Schoenenberger, G. A.,** Humoral transmission of sleep. VI. Purification and assessment of a hypnogenic fraction of "sleep dialysate" (factor delta), *Pfluegers Arch.,* 331, 257, 1972.
82. **Schoenenberger, G. A., Cueni, L. B., Monnier, M., and Hatt, A. M.,** Humoral transmission of sleep. VII. Isolation and physical-chemical characterization of the "sleep inducing factor delta", *Pfluegers Arch.,* 338, 1, 1972.
83. **Schoenenberger, G. A., Cueni, L. B., Hatt, A. M., and Monnier, M.,** Isolation and physical-chemical characterization of a humoral, sleep inducing substance in rabbits (factor 'delta'), *Experientia,* 28, 919, 1972.
84. **Monnier, M., Dudler, L., and Schoenenberger, G. A.,** Humoral transmission of sleep. VIII. Effects of the "sleep factor delta" on cerebral, motor and visceral activities, *Pfluegers Arch.,* 345, 23, 1973.
85. **Monnier, M., Dudler, L., Gächter, R., and Schoenenberger, G. A.,** Humoral transmission of sleep. IX. Activity and concentration of the sleep peptide delta in cerebral and systemic blood fractions, *Pfluegers Arch.,* 360, 225, 1975.
86. **Schoenenberger, G. A. and Monnier, M.,** A naturally occurring delta-EEG enhancing nonapeptide in rabbits. X. Final isolation, characterization and activity test, *Pfluegers Arch.,* 369, 99, 1977.
87. **Schoenenberger, G. A. and Monnier, M.,** Characterization of a delta-electroencephalogram(-sleep)-EEG-inducing peptide, *Proc. Natl. Acad. Sci. U.S.A.,* 74, 1282, 1977.
88. **Sachs, J., Ungar, J., Waser, P. G., and Borbély, A. A.,** Factors in cerebrospinal fluid affecting motor activity in the rat, *Neurosci. Lett.,* 2, 83, 1976.
89. **Pappenheimer, J. R.,** The sleep factor, *Sci. Am.,* 235(2), 24, 1976.
90. **Pappenheimer, J. R., Miller, T. B., and Goodrich, C. A.,** Sleep-promoting effects of cerebrospinal fluid from sleep-deprived goats, *Proc. Natl. Acad. Sci. U.S.A.,* 58, 513, 1967.
91. **Fencl, V., Koski, G., and Pappenheimer, J. R.,** Factors in cerebrospinal fluid from sleep-deprived goats that affect sleep and activity in rats, *J. Physiol. (London),* 216, 565, 1971.
92. **Pappenheimer, J. R., Fencl, V., Karnovsky, M. L., and Koski, G.,** Peptides in cerebrospinal fluid and their relation to sleep and activity, *Res. Publ. Assoc. Res. Nerv. Ment. Dis.,* 53, 201, 1974.
93. **Pappenheimer, J. R., Koski, G., Fencl, V., Karnovsky, M. L., and Krueger, J.,** Extraction of sleep-promoting factor S from cerebrospinal fluid and from brains of sleep-deprived animals, *J. Neurophysiol.,* 38, 1299, 1975.
94. **Krueger, J. M., Pappenheimer, J. R., and Karnovsky, M. L.,** Sleep-promoting factor S: purification and properties, *Proc. Natl. Acad. Sci. U.S.A.,* 75, 5235, 1978.
95. **Ringle, D. A. and Herndon, B. L.,** Plasma dialysates from sleep-deprived rabbits and their effect on the electrocorticogram of rats, *Pfluegers Arch.,* 303, 344, 1968.
96. **Ringle, D. A. and Herndon, B. L.,** Effects on rats of CSF from sleep-deprived rabbits, *Pfluegers Arch.,* 306, 320, 1969.
97. **Moruzzi, G.,** The sleep-waking cycle, *Ergeb. Physiol. Biol. Chem. Exp. Pharmakol.,* 64, 1, 1972.
98. **Nagasaki, H., Iriki, M., Inoué, S., and Uchizono, K.,** The presence of a sleep-promoting material in the brain of sleep-deprived rats, *Proc. Jpn. Acad.,* 50, 241, 1974.
99. **Honda, K., Ichikawa, H., and Inoué, S.,** Special rat cages for the assay of the sleep substance, *Rep. Inst. Med. Dent. Eng.,* 8, 149, 1974.
100. **Honda, K. and Inoué, S.,** Establishment of a bioassay method for the sleep-promoting substance, *Rep. Inst. Med. Dent. Eng.,* 12, 81, 1978.
101. **Honda, K. and Inoué, S.,** Effects of sleep-promoting substance on sleep-waking patterns of male rats, *Rep. Inst. Med. Dent. Eng.,* 15, 115, 1981.

102. **Uchizono, K., Inoué, S., Iriki, M., Ishikawa, M., Komoda, Y., and Nagasaki, H.,** Purification of the sleep-promoting substances from sleep-deprived rat brain, in *Peptides: Chemistry, Structure and Biology,* Walter, R. and Meienhofer, J., Eds., Ann Arbor Science Publishers, Ann Arbor, MI, 1975, 667.
103. **Nagasaki, H., Iriki, M., and Uchizono, K.,** Inhibitory effect of the brain extract from sleep-deprived rats (BE-SDR) on the spontaneous discharges of crayfish abdominal ganglion, *Brain Res.,* 109, 202, 1976.
104. **Uchizono, K., Nagasaki, H., Iriki, M., Inoué, S., Ishikawa, M., and Komoda, Y.,** Humoral control of sleep, in *Neurohumoral Correlate of Behaviour,* Subrahmanyam, S., Ed., Thompson Press (India), Madras, 1977, 35.
105. **Uchizono, K., Higashi, A., Iriki, M., Nagasaki, H., Ishikawa, M., Komoda, Y., Inoué, S., and Honda, K.,** Sleep-promoting fractions obtained from the brain-stem of sleep-deprived rats, in *Integrative Control Functions of the Brain,* Vol. 1, Ito, M., Tsukahara, N., Kubota, K., and Yagi, K., Eds., Kodansha, Tokyo, 1978, 392.
106. **Nagasaki, H., Kitahama, K., Valatx, J.-L., and Jouvet, M.,** Sleep-promoting effect of the sleep-promoting substance (SPS) and delta-sleep-inducing peptide (DSIP) in the mouse, *Brain Res.,* 192, 276, 1980.
107. **Inoué, S., Honda, K., Komoda, Y., Ishikawa, M., Nagasaki, H., Iriki, M., Higashi, A., and Uchizono, K.,** Purified sleep-promoting substance: its effect on the rat circadian sleep-waking rhythm, *Sleep Res.,* 8, 80, 1979.
108. **Uchizono, K., Ishikawa, M., Iriki, M., Inoué, S., Komoda, Y., Nagasaki, H., Higashi, A., Honda, K., and McRae-Degueurce, A.,** Purification of sleep-promoting substances (SPS), in *Advances in Pharmacology and Therapeutics II, Vol. 1,* Yoshida, H., Hagiwara, Y., and Ebashi, S., Eds., Pergamon Press, Oxford, 1982, 217.
109. **Komoda, Y., Ishikawa, M., Nagasaki, H., Iriki, M., Honda, K., Inoué, S., Higashi, A., and Uchizono, K.,** Uridine, a sleep-promoting substance from brainstems of sleep-deprived rats, *Biomed. Res.,* 4 (Suppl.), 223, 1983.
110. **Mendelson, W. B., Gillin, J. C., and Wyatt, R. J.,** The search for circulating sleep-promoting factors, in *Advances in Pharmacology and Therapeutics II,* Vol. 1, Yoshida, H., Hagiwara, Y., and Ebashi, S., Eds., Pergamon Press, Oxford, 1982, 227.
111. **Drucker-Colín, R., Rojas-Ramírez, J. A., Vera-Trueba, J., Monroy-Ayala, G., and Hernández-Peón, R.,** Effect of crossed-perfusion of the midbrain reticular formation upon sleep, *Brain Res.,* 23, 269, 1970.
112. **Drucker-Colín, R.,** Crossed perfusion of a sleep inducing brain tissue substance in conscious cats, *Brain Res.,* 56, 123, 1973.
113. **Drucker-Colín, R., Spanis, C. W., Cotman, C. W., and McGaugh, J. L.,** Changes in protein levels in perfusates of freely moving cats: relation to behavioral state, *Science,* 187, 963, 1975.
114. **Drucker-Colín, R., Zamora, J., Bernal-Pedraza, J., and Sosa, B.,** Modification of REM sleep and associated phasic activities by protein synthesis inhibitors, *Exp. Neurol.,* 63, 458, 1979.
115. **Drucker-Colín, R., Tuena de Gómez-Puyou, M., Gutiérrez, M. C., and Dreyfus-Cortés, G.,** Immunological approach to the study of neurohumoral sleep factors: effects on REM sleep of antibodies to brain stem proteins, *Exp. Neurol.,* 69, 563, 1980.
116. **Spanis, C. W., Gutiérrez, M. C., and Drucker-Colín, R. R.,** Neurohumoral correlates of sleep: further biochemical and physiological characterization of sleep perfusates, *Pharmacol. Biochem. Behav.,* 5, 165, 1976.
117. **Prospéro-Garcia, O., Morales, M., Arankowsky-Sandoval, G. and Drucker-Colín, R.,** Vasoactive intestinal polypeptide (VIP) and cerebrospinal fluid (CSF) of sleep-deprived cats restores REM sleep in insomniac recipients, *Brain Res.,* 385, 169, 1986.
118. **Yanagisawa, I. and Yoshikawa, H.,** Micro-determination of bromine in cerebrospinal fluid, *Clin. Chim. Acta,* 21, 217, 1968.
119. **Yanagisawa, I. and Yoshikawa, H.,** A bromine compound isolated from human cerebrospinal fluid, *Biochim. Biophys. Acta,* 329, 283, 1973.
120. **Torii, S., Mitsumori, K., Inubushi, S., and Yanagisawa, I.,** The REM sleep-inducing action of a naturally occurring organic bromine compound in the encéphale isolé cat, *Psychopharmacologia,* 29, 65, 1973.
121. **Krueger, J. M., Bascik, J., and García-Arraras, J.,** Sleep-promoting material from human urine and its relation to factor S from brain, *Am. J. Physiol.,* 238, E116, 1980.
122. **García-Arraras, J.,** Effects of sleep-promoting factor from human urine on sleep cycles of cats, *Am. J. Physiol.,* 214, E269, 1981.
123. **Krueger, J., Pappenheimer, J. R., and Karnovsky, M. L.,** The composition of sleep-promoting factor isolated from human urine, *J. Biol. Chem.,* 257, 1664, 1982.
124. **Martin, S. A., Karnovsky, M. L., Krueger, J. M., Pappenheimer, J. R., and Biemann, K.,** Peptide-glycans as promoters of slow-wave sleep. I. Structure of the sleep-promoting factor isolated from human urine, *J. Biol. Chem.,* 259, 12652, 1984.

125. **Krueger, J., Karnovsky, M. L., Martin, S. A., Pappenheimer, J. R., Walter, J., and Biemann, K.,** Peptide glycans as promoters of slow-wave sleep. II. Somnogenic and pyrogenic activities of some naturally occurring muramyl peptides; correlations with mass spectrometric structure determination, *J. Biol. Chem.,* 259, 12659, 1984.
126. **Kalter, V. G. and Folk, G. E., Jr.,** Humoral control of mammalian hibernation, *Comp. Biochem. Physiol.,* 63A, 7, 1979.
127. **Oeltgen, P. R., Walsh, J. W., Hamann, S. R., Randall, D. C., Spurrier, W. A., and Myers, R. D.,** Hibernation "trigger": opioid-like inhibitory action on brain function of the monkey, *Pharmacol. Biochem. Behav.,* 17, 1271, 1982.
128. **Spurrier, W. A., Oeltgen, P. R., and Myers, R. D.,** Hibernation "trigger" from hibernating woodchucks *(Marmota monax)* induces physiological alternations and opiate-like responses in the primate *(Macacca mulatta), J. Therm. Biol.,* 12, 139, 1987.
129. **Ruit, K. A., Bruce, D. S., Chiang, P. P., Oeltgen, P. R., Welborn, J. R., and Nilekani, S. P.,** Summer hibernation in ground squirrels *(Citellus tridecemlineatus)* induced by injection of whole or fractionated plasma from hibernating black bears *(Ursus americanus), J. Therm. Biol.,* 12, 135, 1987.
130. **Bruce, D. S., Cope, G. W., Elam, T. R., Ruit, K. A., Oeltgen, P. R., and Su, T.-P.,** Opioid and hibernation. I. Effects of naloxone on bear HIT's depression of guinea pig ileum contractility and on induction of summer hibernation in the ground squirrel, *Life Sci.,* 41, 2107, 1987.
131. **Oeltgen, P. R., Welborn, J. R., Nuchols, P. A., Spurrier, W. A., Bruce, D. S., and Su, T.-P.,** Opioid and hibernation. II. Effects of kappa opioid U69593 on induction of hibernation in summer-active ground squirrels by "hibernation induction trigger" (HIT), *Life Sci.,* 41, 2115, 1987.
132. **Stanton, T. L., Daley, J. C., III, and Salzman, S. K.,** Prolongation of hibernation bout duration by continuous intracerebroventricular infusion of melatonin in hibernating ground squirrels, *Brain Res.,* 413, 350, 1987.
133. **Lyman, C. P.,** Recent theories of hibernation, in *Hibernation and Torpor in Mammals and Birds,* Lyman, C. P., Willis, J. S., Malan, A., and Wang, L. C. H., Eds., Academic Press, New York, 1982, 283.
134. **Walker, J. M., Haskell, E. H., Berger, R. J., and Heller, H. C.,** Hibernation at moderate temperatures: a continuation of slow wave sleep, *Experientia,* 37, 726, 1981.
135. **Walker, L. E., Walker, J. M., Palca, J. W., and Berger, R. J.,** A continuum of sleep and shallow torpor in fasting doves, *Science,* 221, 194, 1983.
136. **Geiser, F.,** Hibernation and daily torpor in two pygmy possums (*Cercartetus* spp., *Marsupulia), Physiol. Zool.,* 60, 93, 1986.
137. **Geiser, F. and Baudinette, R. V.,** Seasonality of torpor and thermoregulation in three dasyurid marsupials, *J. Comp. Physiol.,* 157, 335, 1987.
138. **Purpura, D. P.,** A neurohumoral mechanism of reticulo-coritcal activation, *Am. J. Physiol.,* 186, 250, 1956.
139. **Monnier, M. and Hösli, L.,** Dialysis of sleep and waking factors in blood of the rabbit, *Science,* 146, 796, 1964.
140. **Toh, C. C.,** Preparation of a nerve-stimulating phosphopeptide (substance B, nerveside) in brain extracts and an enzyme in brain which inactivates it, *J. Physiol. (London),* 173, 420, 1964.
141. **Toh, C. C.,** The regional distribution of the nerve-stimulating phosphopeptide (nerveside, substance B) in the central nervous system, *J. Physiol. (London),* 188, 451, 1967.
142. **Scherschlicht, R., Bonetti, E. P., and Toh, C. C.,** Effects of intracerebroventricularly injected "nerveside" on free behaviour and EEG activity of rats. A pilot study, *Arch. Int. Pharmacodyn.,* 239, 221, 1979.
142a. **Komoda, Y.,** unpublished.
143. **Riou, F., Cespuglio, R., and Jouvet, M.,** Endogenous peptides and sleep in the rats. I. Peptides decreasing paradoxical sleep, *Neuropeptides,* 2, 243, 1982.
144. **Takahashi, Y., Kipnis, D. M., and Daughaday, W. H.,** Growth hormone secretion during sleep, *J. Clin. Invest.,* 47, 2079, 1968.
145. **Stern, W. C., Jalowiec, J. E., Shabshelowitz, H., and Morgane, P. J.,** Effects of growth hormone on sleep-waking patterns in cats, *Horm. Behav.,* 6, 189, 1975.
146. **Drucker-Colín, R. R., Spanis, C. W., Hunyadi, J., Sassin, J. F., and McGaugh, J. L.,** Growth hormone effects on sleep and wakefulness in the rat, *Neuroendocrinology,* 18, 1, 1975.
147. **Pavel, S., Psatta, D., and Goldstein, R.,** Slow-wave sleep induced in cats by extremely small amounts of synthetic and pineal vasotocin injected into the third ventricle of the brain, *Brain Res. Bull.,* 2, 251, 1977.
148. **Pavel, S.,** Pineal arginine vasotocin: an extremely potent sleep-inducing nonapeptide hormone. Evidence for the involvement of 5-hydroxytryptamine containing neurons in its mechanism of action, in *Sleep 1978,* Popoviciu, L., Asigian, B., and Badiu, G., Eds., S. Karger, Basel, 1980, 381.
149. **Coculescu, M., Simionescu, L., Cristoveanu, A., Servanescu, A., Matulevicius, V., Grigorescu, F., and Temeli, E.,** Administration of arginine-vasotocin (AVT) to human subjects, *Progr. Brain Res.,* 52, 535, 1979.

150. **Coculescu, M., Servanescu, A., and Temeli, E.,** Influence of arginine vasotocin administration on nocturnal sleep of human subjects, *Waking Sleeping,* 3, 273, 1979.
151. **Mendelson, W. B., Gillin, J. C., Pisner, G., and Wyatt, R. J.,** Arginine vasotocin and sleep in the rat, *Brain Res.,* 182, 246, 1980.
152. **Ueno, R., Ishikawa, Y., Nakayama, T., and Hayaishi, O.,** Prostaglandin D_2 induces sleep when microinjected into the preoptic area of conscious rats, *Biochem. Biophys. Res. Commun.,* 109, 576, 1982.
153. **Guilleminault, C. and Baker, T. L.,** Sleep and electroencephalography: points of interest and points of controversy, *J. Clin. Neurophysiol.,* 1, 275, 1984.
154. **Akimoto, H.,** The past and present of sleep study, *Seishin Igaku,* 27, 237, 1985.
155. **Graf, M. V. and Kastin, A. J.,** Delta-sleep-inducing peptide (DSIP): a review, *Neurosci. Biobehav. Rev.,* 8, 83, 1984.
156. **Graf, M. V. and Kastin, A. J.,** Delta-sleep-inducing peptide (DSIP): an update, *Peptides,* 7, 1165, 1986.
157. **Monnier, M. and Schoenenberger, G. A.,** The peptidergic modulation of sleep with the delta sleep inducing peptide as a prototype, in *Functions of the Nervous System, Vol. 4,* Monnier, M. and Meulders, M., Eds., Elsevier, Amsterdam, 1983, 161.
158. **Schoenenberger, G. A.,** Characterization, properties and multivariate functions of delta-sleep-inducing peptide (DSIP), *Eur. Neurol.,* 23, 321, 1984.
159. **Borbély, A. A.,** A two process model of sleep regulation, *Human Neurobiol.,* 1, 195, 1982.
160. **Borbély, A. A.,** Pharmacological approaches to sleep regulation, in *Sleep Mechanisms and Functions,* Mayes, A., Ed., Van Nostrand Reinhold (UK), Wokingham, 1983, 232.
161. **Daan, S., Beersma, D. G. M., and Borbély, A. A.,** Timing of human sleep: recovery process gated by a circadian pacemaker, *Am. J. Physiol.,* 246, R161, 1984.
162. **Kobayashi, T.,** The role of a sleep substance as analyzed by a sleep structure model, *Annu. Rep. Natl. Inst. Physiol. Sci., Okazaki,* 8, 452, 1987.
163. **Koella, W. P.,** The organization and regulation of sleep. A review of the experimental evidence and a novel integrated model of the organizing and regulating apparatus, *Experientia,* 40, 309, 1984.
164. **Koella, W. P.,** A partial theory of sleep, *Eur. Neurol.,* 25, (Suppl. 2), 9, 1986.
165. **Koella, W. P.,** Neurotransmitters and sleep — a synthesis, in *Sleep '84.* Koella, W. P., Rüther, E., and Schulz, H., Eds., Gustav Fischer Verlag, Stuttgart, 1985, 28.
166. **Jouvet, M.,** Biogenic amines and the states of sleep, *Science,* 163, 32, 1969.
167. **Jouvet, M.,** The role of monoamines and acetylcholine-containing neurons in the regulation of the sleep-waking cycle, *Ergeb. Physiol. Biol. Chem. Exp. Pharmakol.,* 64, 166, 1972.
168. **Danguir, J.,** Internal milieu and sleep homeostasis, in *Sleep Peptides: Basic and Clinical Approaches,* Inoué, S. and Schneider-Helmert, D., Eds., Japan Scientific Societies Press, Tokyo/Springer-Verlag, Berlin, 1988, 53.
169. **Chastrette, N., Lin, Y. L., Faradji, H., and Cespuglio, R.,** Hypnogenic properties of desacetyl-α-MSH and CLIP (ACTH 18-39), in *Sleep '86,* Koella, W. P., Obál, F., Schulz, H., and Visser, P., Eds., Gustav Fischer Verlag, Stuttgart, 1988, 165.
170. **Chastrette, N., Clement, H. W., Prevautel, H., and Cespuglio, R.,** Proopiomelanocortin components: differential sleep-waking regulation?, in *Sleep Peptides: Basic and Clinical Approaches,* Inoué, S. and Schneider-Helmert, D., Eds., Japan Scientific Societies Press, Tokyo/Springer-Verlag, Berlin, 1988, 27.
171. **Drucker-Colín, R., Aguilar-Roblero, R., and Arankowsky-Sandoval, G.,** Re-evaluation of the hypnogeic factor notion, in *Sleep: Neurotransmitters and Neuromodulators,* Wauquier, A., Gaillard, J. M., Monti, J. M., and Radulovacki, M., Eds., Raven Press, New York, 1985, 291.
172. **Krueger, J. M., Walter, J., and Levin, C.,** Factor S and related somnogens: an immune theory for slow-wave sleep, in *Brain Mechanisms of Sleep,* McGinty, D. J., Drucker-Colín, R., Morrison, A., and Parmeggiani, P. L., Eds., Raven Press, New York, 1987, 253.
173. **Hartmann, E. L.,** *The Functions of Sleep,* Yale University Press, New Haven, CT, 1973.
174. **Oswald, I.,** *Sleep,* Penguin Books Ltd., Harmondworth, 1980.
175. **Webb, W. B.,** Theories in modern sleep research, in *Sleep Mechanisms and Functions,* Mayes, A., Ed., Van Nostrand Reinhold (UK), Wokingham, 1983, 1.
176. **Inoué, S.,** Sleep and sleep substances, *Brain Dev.,* 8, 469, 1986.
177. **Inoué, S.,** Sleep substances, *Clin. Neurosci.,* 5, 22, 1987.
178. **Inoué, S.,** Interference among sleep substances simultaneously infused in unrestrained rats, in *Sleep '86,* Koella, W. P., Obál, F., Schulz, H., and Visser, P., Eds., Gustav Fischer Verlag, Stuttgart, 1988, 175.
179. **Inoué, S., Kimura, M., Honda, K., and Komoda, Y.,** Sleep peptides: general and comparative aspects, in *Sleep Peptides: Basic and Clinical Approaches,* Inoué, S. and Schneider-Helmert, D., Eds., Japan Scientific Societies Press, Tokyo/Springer-Verlag, Berlin, 1988, 1.
180. **Ehlers, C. L. and Kupfer, D. J.,** Hypothalamic peptide modulation of EEG sleep in depression: a further application of the S-process hypothesis, *Biol. Psychiatry,* 22, 513, 1987.

181. **Ernst, A. and Schoenenberger, G. A.,** DSIP: basic findings in human beings, in *Sleep Peptides: Basic and Clinical Approaches,* Inoué, S. and Schneider-Helmert, D., Eds., Japan Scientific Societies Press, Tokyo/Springer-Verlag, Berlin, 1988, 131.
182. **Matsumura, H., Honda, K., Goh, Y., Ueno, R., Sakai, T., Inoué, S., and Hayaishi, O.,** Awaking effect of prostaglandin E_2 in freely moving rats, *Brain Res.,* in press.
183. **Horne, J. A.,** *Why We Sleep: Perspectives on the Function of Sleep in Humans and Other Animals,* Oxford University Press, Oxford, 1988.
184. **Turek, F. W.** Pharmacological probes of the mammalian circadian clock: use of the phase response curve approach, *Trends Pharmacol. Sci.,* 8, 212, 1987.
185. **Rhyner, T. A., Biguet, N. F., Berrard, S., Borbély, A. A., and Mallet, J.,** An efficient approach for the selective isolation of specific transcripts from complex brain mRNA populations, *J. Neurosci. Res.,* 16, 181, 1986.
186. **Honda, K., Yanagawa, M., Fujii, K., and Inoué, S.,** Time series analyses of sleep-waking cycles in rats, *Proc. JSMEBE Prof. Group Meet. Biol. Time Ser. Signals Fluct.,* 1, 77, 1988.
187. **Inoué, S. Honda, K., Fujii, H., Yanagawa, M., and Yamamoto, M.,** Effects of sleep substances on time series characteristics of rat sleep-waking states, *Sleep Res.,* 17, 336, 1988.
188. **Nakahama, H., Yamamoto, M., Ishii, N., Fujii, H., and Aya, K.,** Dependency as a measure to estimate the order and the values of Markov processes, *Biol. Cybern.,* 25, 207, 1977.
189. **Adrien, J.,** Endogenous sleep factors: a critical approach, in *Sleep '86,* Koella, W. P., Obál, F., Schulz, H., and Visser, P., Eds., Gustav Fischer Verlag, Stuttgart, 1988, 180.
190. **Borbély, A.,** Endogenous sleep-promoting substances, *Trends Pharmacol. Sci.,* 3, 350, 1982.
191. **Borbély, A.,** Endogenous sleep-substances and sleep regulation, *J. Neural Transm.,* Suppl. 21, 243, 1986.
192. **Drucker-Colín, R. and Velverde-R., C.,** Endocrine and peptide functions in the sleep-waking cycle, in *Sleep — Clinical and Experimental Aspects,* Ganten, D. and Pfaff, D., Eds., Springer-Verlag, Berlin, 1982, 37.
193. **Inoué, S.,** *Nemuri no Sei o Motomete,* Dobutsu-sha, Tokyo, 1986.
194. **Inoué, S.,** Sleep and sleep substances, *Taisha,* 25 (Suppl. 327), 551, 1988.
195. **Inoué, S. and Borbély, A. A., Eds.,** *Endogenous Sleep Substances and Sleep Regulation,* Japan Scientific Societies Press, Tokyo/VNU Science Press BV, Utrecht, 1985.
196. **Inoué, S., Nagasaki, H., and Uchizono, K.,** Endogenous sleep-promoting factors, *Trends Neurosci.,* 5, 218, 1982.
197. **Komoda, Y.,** Sleep substance, *Farumashia,* 21, 423, 1985.
198. **Kovalzon, V. M.,** Search of a "sleep hormone", *Priroda,* 4, 13, 1983.
199. **Pappenheimer, J. R.,** Sleep factor in CSF, brain and urine, *Front. Horm. Res.,* 9, 173, 1982.
200. **Takahashi, Y. and Kato, N.,** Toward clinical application of endogenous sleep-promoting factors, *Jpn. J. Neuropsychopharmcol.,* 6, 235, 1984.
201. **Uchizono, K.,** Studies on sleep substances, *Nihon Rinsho,* 42, 989, 1984.
202. **Ursin, R. and Borbély, A. A.,** Endogenous sleep factors, in *Sleep 1982,* Koella, W. P., Ed., S. Karger, Basel, 1983, 106.

Chapter 2

TECHNIQUES FOR ISOLATION AND EVALUATION

I. SITES OF PRESENCE

It is absolutely required that an endogenous sleep substance be present in the organism. It seems unlikely, however, that a specifically differentiated glandular organ exists for the production and release of a certain sleep substance. It is rather likely that, as mentioned in the previous chapter, a large number of neuromodulators occurring in the body may exert a more or less sleep-enhancing action when timely and properly liberated. The brain seems to be a most probable site of production, since sleep is essentially regulated by the brain. However, the production site(s) may not necessarily be restricted to cerebral tissues, since sleep substances can be humorally transported to the brain. Hence it is possible to extract sleep substance(s) either from the brain or from the body fluid, i.e., blood, lymphatic fluid, and cerebrospinal fluid (CSF).

The production of sleep substances may increase in accordance with an elevation of sleep pressure. Sleep deprivation is apparently one of the best ways to artificially elevate the demand for sleep. Indeed, the earliest pioneer works by Ishimori[1] and Legendre and Piéron[2] demonstrated the presence of a hypnogenic factor in the brain tissue of long-term sleep-deprived dogs. A majority of candidate substances pursued in later studies were also extracted directly from the cerebral tissue, either from the whole brain or from the brainstem, of different animals (Factor S[3] in cattle, goats, rabbits, and sheep, interleukin-1 [IL1][4] in mice, rapid-eye-movement [REM] sleep proteins[5] in cats, sleep-promoting substance [SPS][6] in rats, etc.). Many sleep-related substances are known to be present in the brain, i.e., cholecystokinin (CCK),[7] corticotropin-releasing factor (CRF),[8] delta-sleep-inducing peptide (DSIP),[9] growth hormone-releasing factor (GRF),[8] insulin,[10] prostaglandin D_2 (PGD_2),[11] piperidine,[12] somatostatin (SRIF),[8] vasoactive intestinal polypeptide (VIP),[13] etc. Benzodiazepine receptor agonists are present in both peripheral organs and the central nervous system.[14]

Pavel[15] demonstrated sleep-modulatory effects of AVT purified from bovine pineal glands. Melatonin, another pineal hormone, exhibits a sleep-enhancing activity in many species.[16] A high concentration of DSIP[9] and PGD_2[11] is radioimmunologically detected in the rat pineal gland. Pituitary gland (hypophysis) is also regarded as one of the production sites of sleep substances. Namely, in addition to the known specific tropic actions, hypophyseal hormones from the anterior lobe (adrenocorticotropic hormone [ACTH],[17] growth hormone [GH],[18] and prolactin[19]), from the intermediate lobe (melanophore-stimulating hormone and corticotropin-like intermediate lobe peptide [CLIP],[20,21] and from the posterior lobe (arginine vasopressin and an unknown factor in crude extracts[19]) are reported to be either sleep-enhancing or sleep-suppressive, although their physiological significance is still unclear. Gastrointestinal hormones such as insulin,[22] CCK,[23] and VIP[24] also exert sleep modification. Steroid hormones derived from the gonad[25] and the adrenal[26] may also affect sleep (see Chapter 8).

Legendre and Piéron[2] reported that their "hypnotoxin" was found also in the serum and CSF of sleep-deprived dogs. This fact indicates that a sleep substance produced in the brain may be liberated into CSF and that it may enter the systemic blood circulation. Then it finally may be excreted into urine. Actually, for example, DSIP was extracted from the venous blood of sleeping rabbits;[27] Factor S was originally extracted from the CSF of sleep-deprived goats[28] and finally from the morning urine of human beings;[29] 1-methylheptyl-γ-bromoacetoacetate (γ-Br or MHBAA) was extracted from human CSF.[30] DSIP-like im-

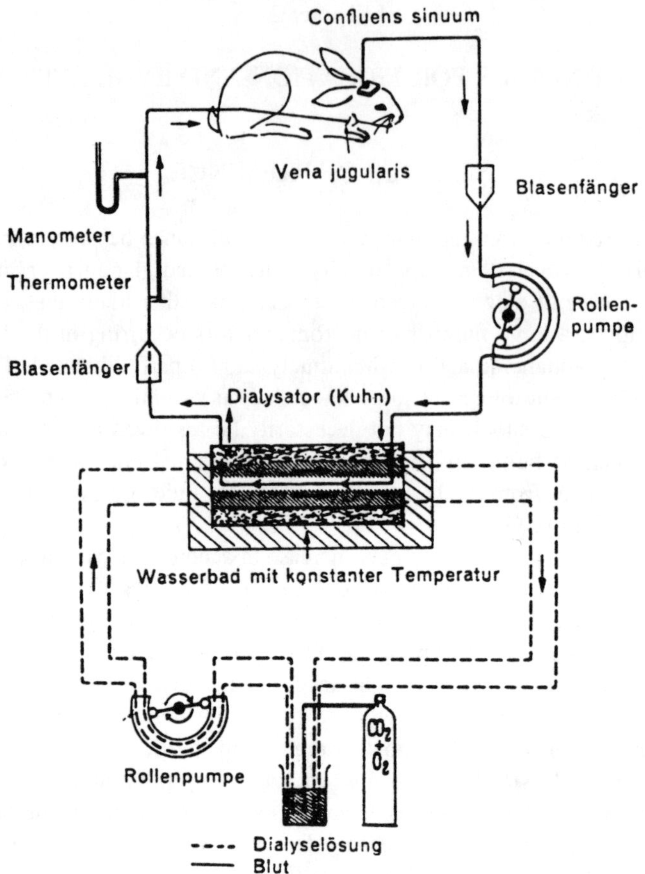

FIGURE 1. Extracorporal dialysis of the sleep factor DSIP in the cerebral venous blood of sleeping rabbits electrically stimulated in the thalamus. (From Monnier, M. and Schoenenberger, G. A., *Schweiz, Med. Wochenschr.*, 103, 1733, 1973. With permission.)

munoreactivity was detected in human blood,[31] and human and cow milk.[32-34] Peptide-like sleep-promoting factors seem to be present in the urine of sleep-deprived humans.[35] Recently, it has been suggested that muramyl peptides (MP) are liberated into blood as a result of the digestion of bacteria by macrophages.[36]

II. EXTRACTION AND PURIFICATION PROCEDURES

A. DSIP from Rabbit Venous Blood

The dialyzate containing endogenously produced DSIP was obtained originally by filtration from the cerebral venous blood of sleeping rabbits.[37-39] The donor rabbits were intermittently stimulated in the ventromedial intralaminary thalamus ("the somnogenic area") by electric pulses of 7 to 12 ms, 0.6 to 0.7 V, and 5 to 6 Hz, and submitted for 50 min to extracorporal *in vivo* hemodialysis. In the inner circuit (see Figure 1), blood from the occipital cranial sinus (confluens sinuum) was withdrawn by a roller pump through an artificial kidney (dialyzer) and returned to the jugular vein. In the dialyzer, blood components with a molecular weight (mol wt) below 40,000 diffused into the dialyzing fluid of the outer circuit. The dialyzate (70 ml/50 min) was collected in the container and further purified. The hemodialyzate prepared by the same method in animals that were not stimulated served as a control.

Table 1
STEPS OF ISOLATION AND CHARACTERIZATION OF DSIP FROM THE HEMODIALYZATES OF SLEEPING RABBITS[42]

Step	Main line fractions	Methods and analysis
1	Cerebral venous blood ↓ Dialyzate	Electrical thalamic stimulation of donors; hemodialysis; bioassay
2	↓ Dry powder	Freeze-drying; bioassay
3	↓ Supernatant (H_2O soluble)	300 mg H_2O + centrifugation (4°C; $2000 \times g$/10 min); bioassay
4	↓ Ultrafiltrate (mol wt <1000)	Ultrafiltration: UM-05 membrane; bioassay
5	↓ Dry powder	Freeze-drying; bioassay
6	↓ Desalted (G-15) Peak 1 + 2 (ninhydrin) Peak 1 (280 nm) 355< mol wt <1165	Desalting by G-15 Sephadex®; 1 g/2 ml H_2O (500 ml column − H_2O ninhydrin + 280 nm absorption; bioassay
7	↓ 5 Ninhydrin-positive bands	Freeze-drying; preparative thin-layer chromatography; bioassay
8	↓ Supernatant	Elution with H_2O; centrifugation (4°C; $2000 \times g$/10 min); bioassay
9	↓ 5 Ninhydrin-positive bands	Freeze-drying; preparative high-voltage paper electrophoresis; bioassay
10	↓ Eluted material	Elution with H_2O; freeze-drying; bioassay
11	↓ Eluted material	G-15 column: elution with H_2O; UV-spectroscopy at 280 nm; bioassay
12	↓ Higher mol wt material (ninhydrin positive)	Preparative thin-layer electrophoresis on silica-gel; elution with H_2O; bioassay
13	↓ Supernatant	Centrifugation (4°C; $2000 \times g$/10 min); bioassay
14	↓ Ultrafiltrate (mol wt = ca. 800)	Ultrafiltration; bioassay
15	↓ Dry powder	Analytical high-voltage paper electrophoresis; UV-spectroscopy; freeze-drying; bioassay
16	↓ Amino acid composition	Bioassay; amino acid analysis and hydrolysis
17	↓ DSIP	Sequence analysis
18	↓ Synthetic DSIP	Peptide synthesis

The hemodialyzates, which were freeze-dried and then dissolved in metal-free, distilled water, were submitted to 7 successive fractionations composed of 15 steps, i.e., gel-filtrations on column chromatographies of Sephadex® G-10 and G-15 and electrophoresis, analytical thin-layer chromatographies, and preparative high-voltage paper electrophoresis (see Table 1). At each isolation step, hypnogenic and behavior-modulating activities of every fraction diluted 1:10 and 1:20 were tested under double-blind conditions.[41,42] The amino acid sequence analysis specified the chemical structure of the hypnogenic factor as a novel nonapeptide with a molecular weight of 849:

Trp-Ala-Gly-Gly-Asp-Ala-Ser-Gly-Glu

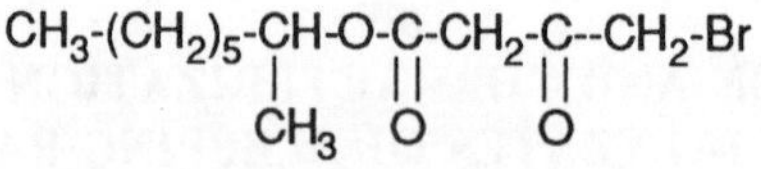

$$CH_3\text{-}(CH_2)_5\text{-}\underset{\displaystyle CH_3}{\underset{|}{CH}}\text{-}O\text{-}\underset{\displaystyle O}{\underset{||}{C}}\text{-}CH_2\text{-}\underset{\displaystyle O}{\underset{||}{C}}\text{-}CH_2\text{-}Br$$

FIGURE 2. Chemical structure of 1-methylheptyl-γ-bromoacetoacetate (γ-Br or MHBAA).

Tryptophan is located as the amino terminal and glutamic acid as the carboxy terminal.[43] This peptide, DSIP, and its eight analogs were synthesized in the final step.

B. γ-Br from Human CSF

The novel organic bromine compound, γ-Br, was originally isolated from CSF.[30] A large amount (23 l) of human CSF, pooled with an equal volume of ethanol, was filtered through filter paper. The filtrate was concentrated to about 15 l under reduced pressure at 20°C, further condensed to 0.5 l using an apparatus of lyophilization, and lyophilized. The material was then extracted by chloroform. The collected extract was evaporated *in vacuo* to yield a few milliliters of oily, brown-yellow residue. The ether solution of this residue was washed with 5% $NaHCO_3$, with 1 *M* H_2SO_4 and with water, and dried with anhydrous Na_2SO_4. The oily, yellow residue thus obtained was mixed with chloroform and subjected to silicic acid-column chromatography with a developing solution of methanol, acetone, ethyl acetate, and chloroform. The eluate of the first band was evaporated and subjected to rechromatography on a silicic acid column with carbon tetrachloride. The eluates of two reddish-yellow and yellow bands, interconvertible in chloroform, were fractionized by thin-layer chromatography to yield about 40 mg of a bromine-containing substance. After the elemental analysis, alkaline hydrolysis, and comparison with synthetic substances, this substance was finally identified as γ-Br. The chemical structure is shown in Figure 2.

C. SPS from Rat Brainstems

SPS was extracted from the brainstems of 24-h sleep-deprived rats.[6,44] Approximately 5000 male rats, weighing 200 to 300 g, were deprived of total sleep for 24 h in specially designed cages[45] (see Figure 3). Immediately after the termination of sleep deprivation between 09.30 and 10.30 h, the rats were decapitated. Their brainstems, including the medulla oblongata, pons, mesencephalon, and hypothalamus, were removed under ice cooling and homogenized with and dialyzed against distilled water. The dialyzates were lyophilized and the residue of the brainstem extracts (designated as BSE) — (see Figure 4) was stored at −74°C until use. For control, BSE was similarly prepared from normal untreated rats. However, this preparation exhibited little sleep-enhancing effect.[6]

Further purification of BSE was done by five different steps of column and preparative high-performance liquid chromatographies (HPLC). At each isolation step, hypnogenic and behavior-modulating activities of every fraction were tested by the crayfish, mouse, and rat bioassays (see below) under double-blind conditions. The first step was performed by cation-exchange column chromatography on SP Sephadex® C-25 eluted by a linear gradient from water to 0.25 *M* ammonium formate. The active material eluted just before γ-aminobutyric acid (GABA), which was used as a marker substance, was fractioned in the second step by gel filtration over Sephadex® G-10 eluted with 0.02 *M* ammonium formate.

Two different fractions designated as SPS-A and SPS-B were separated from BSE. SPS-A was further purified in the following steps. Step 3: elution over Sephadex® G-10 with 0.02 *M* acetic acid; step 4: elution over SP Sephadex® C-25 by a linear gradient from water to 0.25 *M* ammonium formate; step 5: elution over preparative HPLC. Two different fractions, SPS-A1 and SPS-A2, were hypnogenic. An amount of 2.9 μg SPS-A1 was obtained from BSE corresponding with 1000 sleep-deprived rats. The various chromatographic be-

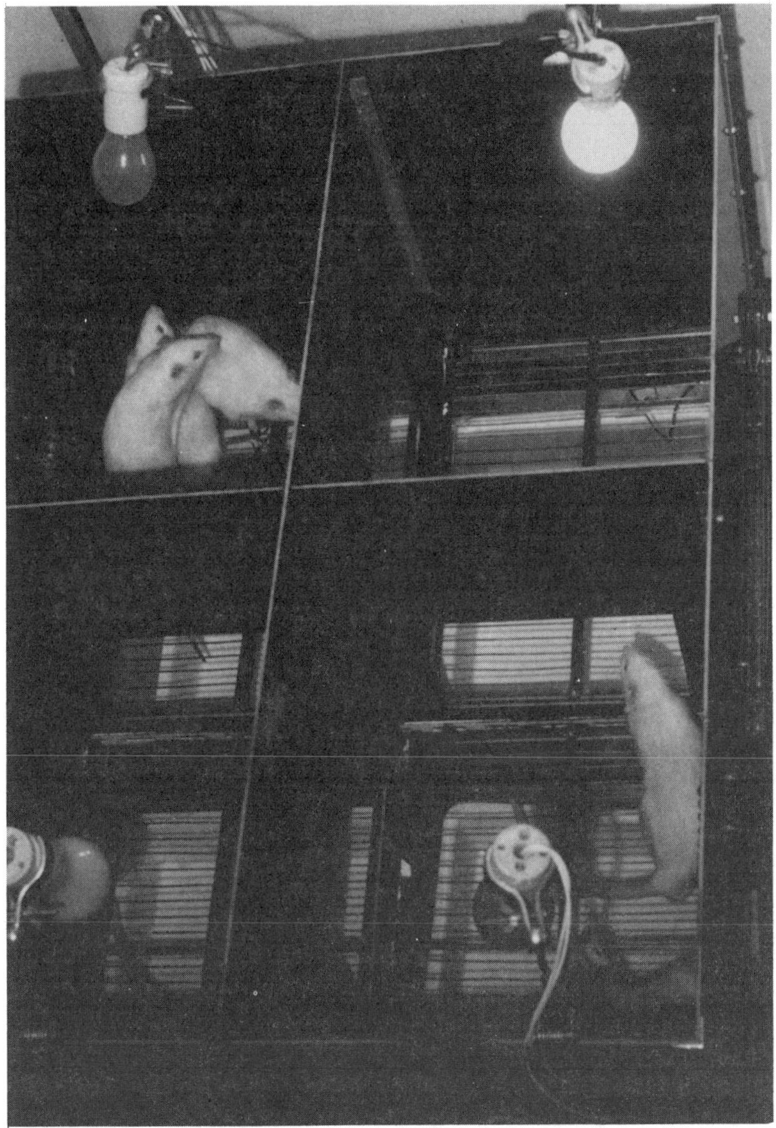

FIGURE 3. Sleep-deprivation cage composed of four rooms connected by gates. Electric current (3.5 mA, 50 Hz, 30 to 40 V) is automatically supplied to one of the room floors for 60 s at 4-min intervals, sequentially circulating from one floor to another. A ceiling light above each room indicates the current passage, turning on and off 5 s prior to the initiation and the termination, respectively, of the floor shock. Rats can easily learn how to avoid the shock and thus they are forced to move without sleeping. (From Inoué, S. and Honda, K., *Kagaku,* 49, 91, 1979. With permission.)

haviors and the relative absorbance at 254 and 280 nm on the chromatograms of HPLC suggested that SPS-A1 might be uridine. Consequently, SPS-A1 was compared with the authentic sample of uridine by mass and UV spectra, by HPLC, and by bioassay. Thus, it was finally identified as uridine.[44]

In further steps, SPS-B was separated into two active subfractions, "purified SPS-B" and SPS-X. Subsequent purification of the former subfraction has yielded at least two active components. These constituents have not yet been chemically identified and the purification procedures are still in progress (see Figure 5).

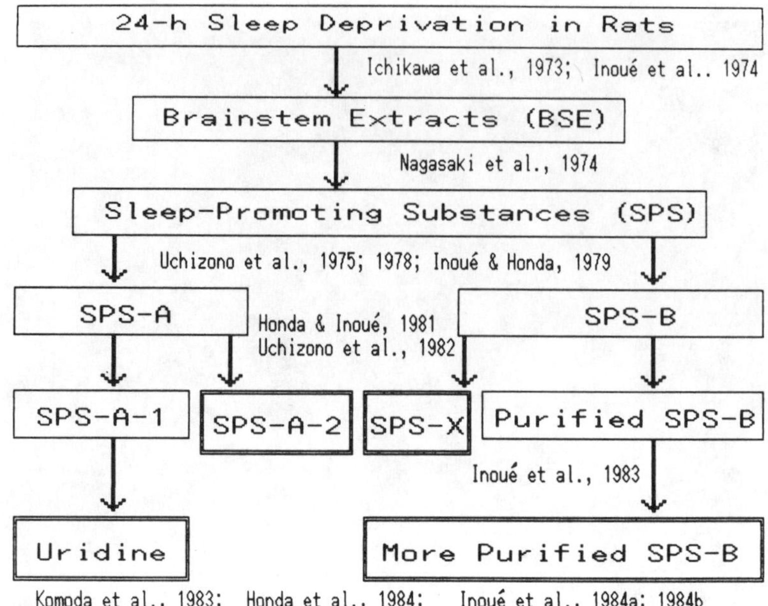

FIGURE 4. Steps of purification of SPS. (From Inoué, S., Honda, K., and Komoda, Y., in *Sleep: Neurotransmitters and Neuromodulators*, Wauquier, A., et al., Eds., Raven Press, New York, 1985, 305. With permission.)

FIGURE 5. The SPS research procedures illustrated by Haruo Nagai (appeared in *The Asahi Simbun*, Tokyo, on March 12, 1975).

Table 2
STEPS OF PURIFICATION OF FACTOR S FROM HUMAN URINE[29,48,49]

Step	Methods and analysis
1	Collection of urine
	↓ Filter through filter paper
2	Cation-exchange chromatography on CM-Sephadex®
	↓ Batch method; Elute Factor S with pH 1.9 buffer; Volume reduction
3	Gel-filtration on Sephadex® G-10
	↓
4	Anion-exchange chromatography on DEAE-Sephadex®
	↓ Elute Factor S with 1 *M* NaCl; Volume reduction
5	Gel-filtration on Sephadex® G-10
	↓
6	Cation-exchange chromatography on SP-Sephadex®
	↓ Elute Factor S with pH 2.83 buffer
7	Gel-filtration on Sephadex® G-10
	↓
8	Ascending paper chromatography
	↓ Elute section: R_F 0.1—0.3
9	React product with Fluram®
	↓
10	Gel-filtration on Sephadex® G-10
	↓
11	Amino acid analysis
	↓
12	Fast atom bombardment/mass spectrometry

D. Factor S from Human Urine

The attempt to extract and purify Factor S was done originally by using CSF withdrawn from 48-h sleep-deprived goats chronically implanted with cisternal guide tubes.[28] However, as described in Chapter 1, CSF was inadequate for obtaining a sufficient quantity. Human urine was then finally chosen to extract Factor S.[29,48,49]

It was supposed that Factor S may be found at highest concentrations in the brain at the onset of sleep and present in urine on waking. In order to avoid diurnal and/or menstrual variations in Factor S in the brain and CSF, urine from first morning micturition was obtained from healthy male adults, aged 23 to 33 years. For small-scale preparation (less than 20 l), the urine was collected, frozen within 1 h, and stored at −20°C. The frozen material was thawed and filtered through filter paper just prior to extraction. For large-scale preparation (up to 400 l), the urine was treated with carboxymethyl- (CM) Sephadex® resin on the day of collection, without freezing. The materials were treated by 12 steps of purification and identification procedures, as summarized in Table 2. Between purification steps, samples were stored at −20°C. Bioassays in rabbits were used to detect sleep-promoting activity in various fractions after each purification step (see below).

After chromatography on SP-Sephadex® (steps 6 and 7), several adjacent eluates were found to be active. Each was taken separately through the remainder of the purification program. Acid hydrolysis of the final purified fractions released glutamic acid (Glu), alanine (Ala), diaminopimelic acid (Dap), and glycine. The apparent molar ratios of Glu/Ala/Dap were 2:2:1. Since Dap is a constituent of bacterial peptideglycans, Krueger et al.[48] supposed that their purified fractions might contain amino sugars. The sample was then subjected to mild acid hydrolysis, followed by amino acid analysis. Thus, it was determined that muramic

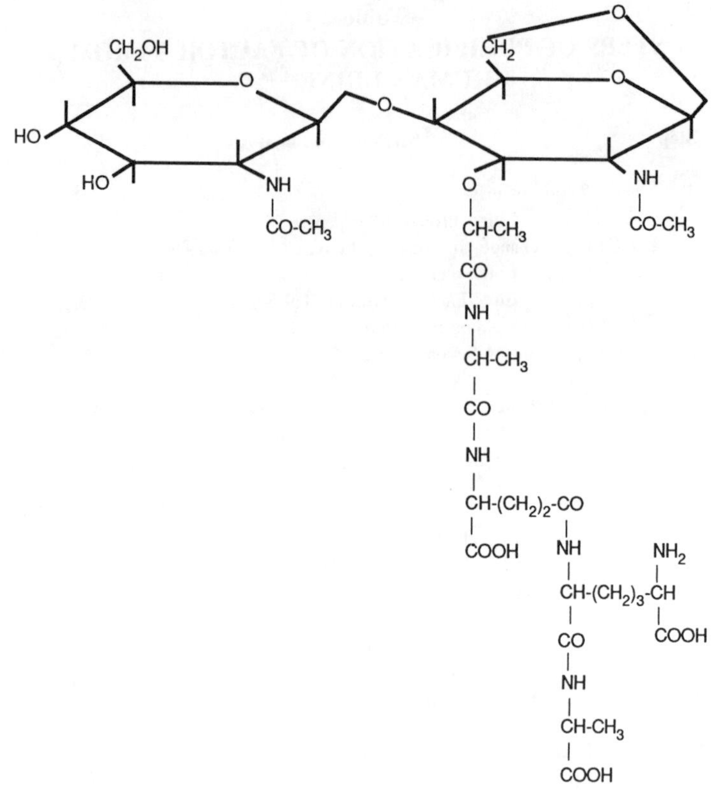

FIGURE 6. Chemical structure of urinary Factor S.

acid was equimolar with Dap, whereas glucosamine was equimolar with glycine. Hence, urinary Factor S was proven to be a small glycopeptide.

At the final step, fast atom bombardment/mass spectrometry was used to determine the structure of urinary Factor S. The major somnogenic constituent was a peptideglycan of a molecular weight of 921 with the structure of *N*-acetylglucosamyl-*N*-acetylanhydromuramylalanylglutamyldiaminopimelylalanine (see Figure 6).[49] The anhydro linkage is between C-1 and C-6 of the muramyl entity. Two additional substances accompanied this compound: the hydrated form with the muramyl entity of a free reducing end and a free hydroxyl on C-6 (molecular weight = 939), and an anhydro analog lacking the terminal alanine (molecular weight = 850).

III. BIOASSAY PROCEDURES

A. Standardized Short-Term (Within 1 d) Sleep Bioassay

1. 90-min Delta-Sleep Assay for DSIP in Rabbits

The assessment of the DSIP-containing dialyzate and its purified fractions was done by electroencephalography (EEG) and kinesigraphy in mildly restrained and/or freely moving rabbits of both sexes weighing 2.7 to 3 kg[50-52] (see Figure 7). EEG was recorded from the motor cortex. A thin cannula was chronically implanted for the intracerebroventricular (i.c.v.) infusion in the meso-diencephalic ventricle. After 2 d or later, the rabbits were used for bioassay.

The test materials from the sleep dialyzate were dissolved in 50 μl of a CSF-like solution, and i.c.v. infused by an automatic infusion pump at a rate of 50 μl/25 min (see Figure 8).

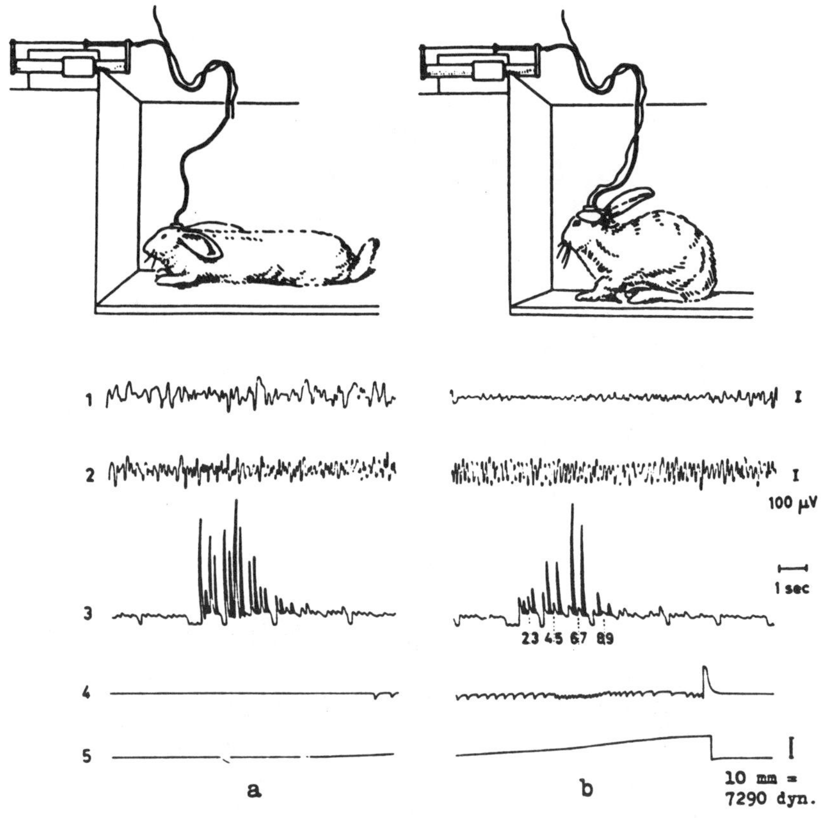

FIGURE 7. Sleep bioassay for DSIP in rabbits. (From Monnier, M. and Schoenenberger, G. A., *Schweiz. Med. Wochenschr.*, 103, 1733, 1973. With permission.)

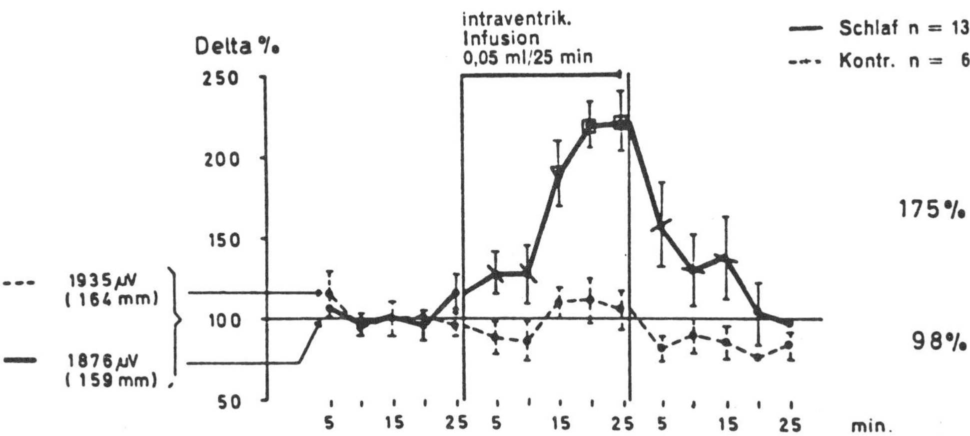

FIGURE 8. Effect of a purified fraction of DSIP on EEG delta activity. (From Monnier, M. and Schoenenberger, G. A., *Schweiz. Med. Wochenschr.*, 103, 1733, 1973. With permission.)

In later studies,[42,43] a shorter infusion period (50 µl/3.5 min) was adopted. Recordings of EEG, behavioral kinesigram, and other physiological correlates (blood and CSF pressure, and heart and respiration rate) were done during a 20-min preinfusion reference period, a 25-min infusion reference period (or 20-min period = 3.5-min infusion + 16.5-min post-adaptation, in the later bioassays), and a 25- to 50-min postinfusion test period. Hence an

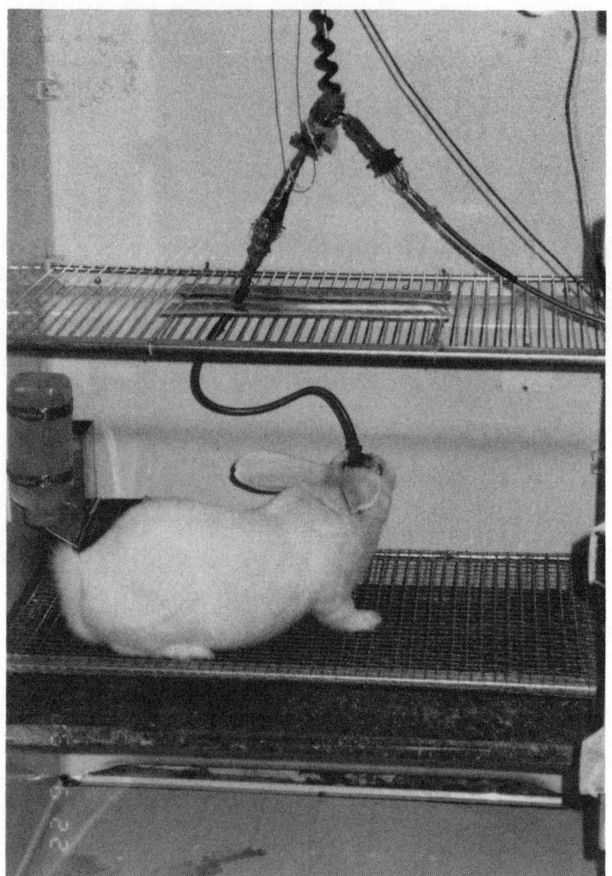

FIGURE 9. Experimental cage for the assay of Factor S.

experiment required 70 to 95 min. Each experiment was carried out on a fresh rabbit and the animal was sacrificed thereafter.

The hypnogenic effect was revealed by initial spindle bursts (13 Hz) on the cortical EEG, followed by a progressive increase of slow delta waves (2 to 3 Hz). It was also revealed by a sleep behavior with decreased motor activity developed in the kinesigram. The slow EEG delta activity during the postinfusion period was compared with that of the preinfusion reference period. For this purpose, amplitude and duration of all delta waves were integrated by a wave analyzer and averaged for each 5-min period.

2. 7 1/2-h SWS Assay for Factor S in Rabbits

According to Krueger et al.[29] male New Zealand white rabbits, weighing 3 to 4 kg, were deprived of food and water for 24 h and then anesthetized with pentobarbital. They were implanted with stainless steel screw electrodes for EEG recording over the frontal, parietal, and occipital cortex and with a guide tube in the cerebral lateral ventricle. At least 1 week was allowed for recovery from the operation. Rabbits were housed in a room with a 12-h light-dark cycle. Before each experiment, they were brought to the experimental cages (see Figure 9) for an overnight acclimation period. Infusions and recordings were performed during the light hours.

Experimental samples were dissolved in artificial CSF (3 m*M* KCl, 1.15 m*M* $CaCl_2$, and 0.96 m*M* $MgCl_2$) in nonpyrogenic sterile saline (155 m*M* NaCl) and infused i.c.v. over a

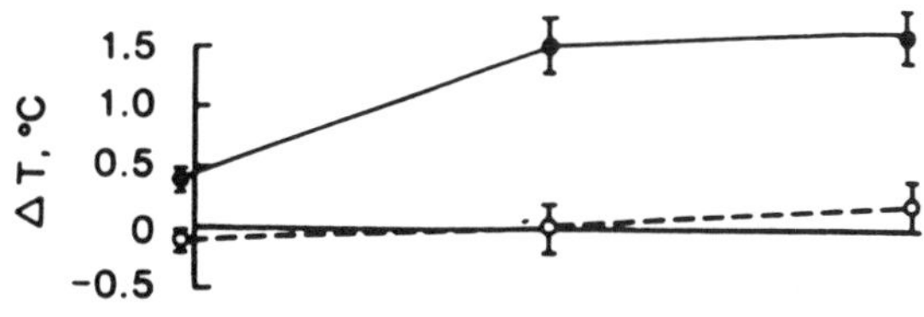

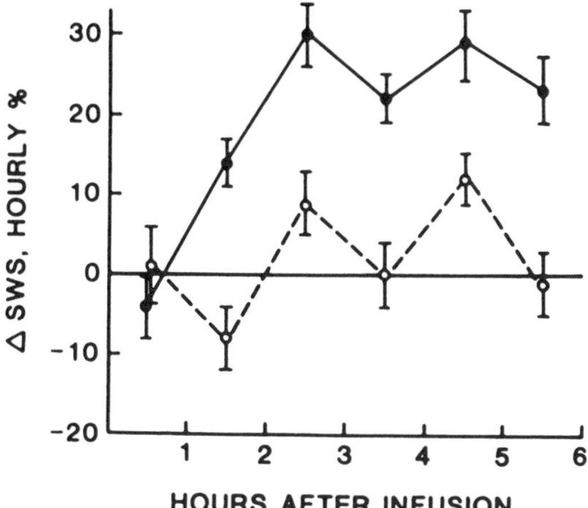

FIGURE 10. Effects of a 45-min i.c.v. infusion of urinary Factor S (solid line, 1 to 5 pmol, n = 11) and an inactive muramyl peptide (PGM_a, 100 to 1000 pmol) on rabbit SWS and rectal temperature. (From Krueger, J. M., *Sleep: Neurotransmitters and Neuromodulators*, Wauquier, A., Gaillard, J. M., Monti, J. M., and Radulovacki, M., Eds., Raven Press, New York, 1985, 319. With permission.)

45- or 90-min period at a rate of 3 μl/min. EEG, rectified slow-wave EEG (0.5 to 4 Hz), and bodily movements were recorded during the infusion period and for the next 6 h. Bodily movements were measured by a pressure transducer as pressure changed inside a water-filled tube attached to the EEG cable. Measurement of rectal temperature was done intermittently. Control recordings of 6 h were obtained from uninfused rabbits. A rotary commutator allowed unrestrained movement of the rabbits.

Recordings were analyzed in two ways, i.e., duration of SWS determined by visual scoring of the EEG and of the filtered, rectified EEG recording, and amplitude of slow waves (mean rectified slow-wave voltages) averaged over 2 min. The duration of SWS episodes was defined as that period of time during which EEG slow waves (0.5 to 4 Hz) were not interrupted for more than 10 s. A result on the effect of urinary Factor S and an inactive muramyl peptide is shown in Figure 10.

A similar bioassay technique was applied to male Sprague-Dawley rats kept under a 10-h light and 14-h dark cycle. Experimental samples were dissolved in artificial CSF and i.c.v. infused over a 30-min period at a rate of 3 μl/min up to the end of the light period. EEG was recorded during the infusion and for the next 18 h. However, the rat bioassay was not routinely adopted as the standardized assay for Factor S.

3. 8-h Sleep Assay for Endogenous Peptides in Rats

Riou et al.[54] tested sleep-modulatory activity of various endogenous peptides. Under chloral-hydrate anesthesia, screw electrodes were inserted into the skulls of adult male OFA rats weighing 200 to 350 g for recording the fronto-occipital EEG, and electromyographic (EMG) electrodes were inserted into the neck and temporal muscles. Two other electrodes were placed on both sides of the ocular globe to enable recording of the electro-oculogram. A stainless steel guide cannula was implanted in the lateral ventricle. To avoid CSF flow back a stylet was inserted into the cannula.

After surgery, the rats were placed under a 12-h light-dark schedule (lights on: 06.00 to 18.00 h) at a controlled temperature (24 ± 1°C). Control polygraphic recordings began 2 weeks after surgery between 10.00 and 18.00 h. For at least 5 d prior to the experiments, the rats were injected with 0.9% saline either i.c.v. (2 μl at 37°C) or intraperitoneally (i.p.; 0.5 ml) at 10.00 h. During the subsequent experimental sessions, peptides were either i.c.v. or i.p. injected under the same conditions. The variations induced by the treatments were compared to control values of the same rat. The i.c.v. injection of peptide solutions were made by inserting a Hamilton needle into the guide cannula. The injection lasted for 1 min, and to avoid liquid flow back, the needle was then left in the cannula for 30 s. The 8-h polygraphic recordings were scored visually according to the classical criteria of the sleep stages. Total time of SWS, PS and wakefulness, number and mean duration of episodes of SWS and PS, and latency of the first SWS and paradoxical sleep (PS) were analyzed.

Riou et al.[24] also tested the somnogenic effects of VIP i.c.v. administered at the onset of the dark period. In this case, the EEG recordings were done throughout the 12-h dark period (18.00 to 08.00 h).

4. 6-h Sleep Assay for Adenosine in Rats

For the assessment of sleep-modulatory effects of adenosine and its analogs, Radulovacki and colleagues[55] and Yanik et al.[56] categorized "S_1" (or slow-wave sleep-1, SWS_1) and "S_2" (or slow-wave sleep-2, SWS_2) from SWS in rats. S_1 was defined by the appearance of spindles and less than 50% low frequency, high-amplitude EEG delta slow waves, while S_2 was defined as having greater than 50% delta slow waves. According to their criteria, wakefulness was characterized by a high-frequency, low-amplitude EEG and relatively high EMG tone. PS was characterized by a high-frequency, low-amplitude EEG and, except for occasional muscle twitches, an absence of EMG tone. "S_2 latency" was defined as the time from the injection of saline or drugs to the appearance of the first 2-min S_2 sleep episode, and "PS latency" was defined as the time between the injection and the appearance of the first full 1-min episode of PS.

Adult male Sprague-Dawley rats weighing 300 to 350 g were maintained on a 12-h light-dark cycle, with the light period from 08.00 to 20.00 h. EEG electrodes consisted of 3 to 4 stainless steel screws (used for either monopolarly or bipolarly) which were threaded into the skull above the parietal, occipital, and/or frontal cortex. EMG electrodes consisted of two ball-shaped wires which were inserted into the bilateral neck musculature. After surgery, the animals were individually housed and allowed a 1-week recovery period. All rats were habituated to the cable system 12 or 48 h before the sleep evaluation, and handled so as to minimize the stress involved in the actual experimental procedures.

On the day of the experiments, the rats were placed under a beam equipped with a series of mercury slip-ring commutators, from which shielded recording cables hung into their cages. Thus, each animal could be recorded with relatively free movements. The rats received i.p. injections of either control (vehicle) or drug solutions (dissolved in 0.9% saline) 15 min before the start of the recordings. All recordings started in the morning and lasted for 6 h. A closed-circuit television camera permitted observation of the animals during the experiments. The hypnogenic effects were evaluated by comparing the amounts of wakefulness, S_1, S_2, PS, and total sleep time, and sleep latencies between the saline-injected and the drug-injected rats during the 0- to 3-h, 3- to 6-h, and 0- to 6-h postinjection periods.

The same procedure was applied to the i.c.v. injection experiments of some nucleosides.[57]

In this study, rats were chronically implanted with a stainless steel guide cannula in the right lateral cerebral ventricle. On the day of the experiments, drugs dissolved in 5 μl of 0.9% saline were i.c.v. injected over a 30-s period by means of a thinner injection cannula, which passed through and projected 1 mm beyond the tip of the guide cannula.

5. 1-h Assay in Rats

Obál[23] routinely used an acute i.c.v. injection method in rats. Samples were dissolved in 1 to 4 μl of artificial CSF and injected into the left lateral cerebral ventricle 10 to 15 min before the onset of the dark period under 12-h light and 12-h dark cycles. Prior to the experiment, the rats were habituated to the experimental conditions for 7 to 10 d, including injection of the solvent for 5 d. The amount of SWS and PS for the first postinjection hour were calculated from the registered polygram of EEG and locomotor activity.

B. Standardized Long-Term Sleep Bioassay Techniques

1. 4-d Assay for SPS in Rats

The rat is a night-active animal and a polyphasic sleeper. Although a short-term sleep assay in rats may exhibit a considerable fluctuation in experimental results which depend largely on the time of day, young adult male rats kept in a controlled environment show little variation in the daily amount of sleep,[58] whereas female rats exhibit 4- or 5-d estrous cycles with significant reductions in the amount of nocturnal SWS and PS at the time of proestrus.[59]

In freely behaving male rats of the Sprague-Dawley strain, which were uninterruptedly infused with 0.9% saline in the third cerebral ventricle, the cumulative amount of SWS and PS attained about 400 min and about 80 min, respectively, in the 12-h light period and about 220 min and about 30 min, respectively, in the 12-h dark period.[60] On the basis of this stability, we developed a long-term i.c.v. infusion technique for the bioassay of sleep substances in freely behaving rats.[46,58,61] This technique, thereafter designated as the long-term i.c.v. infusion technique, has enabled us to quantitatively compare the somnogenic property of putative sleep substances.

Male Sprague-Dawley rats, raised in a closed colony on a 12-h light-dark schedule (lights on: 08.00 to 20.00 h) under a constant air-conditioned environment of 25 ± 1°C and 60 ± 6% relative humidity, were used. At the age of 60 to 70 d, animals weighing 300 to 350 g were anesthetized with pentobarbital, placed on a stereotaxic apparatus and implanted with three electrodes of gold-plated screws over the frontal and occipital cortex for EEG recording, two nuchal hook electrodes for EMG recording, and a stainless steel cannula in the third ventricle for continuous infusion (see Figure 11).[61] A silver plate was fixed on the skull and served as an indifferent electrode. The locus of the cannular tip was ascertained by instantaneous X-ray photography. For the purpose of monitoring brain temperature (T_{br}), a cupper-constantan thermocouple, embedded in a stainless-steel tube, or a thermistor electrode, was implanted into the thalamus.

The rats were individually housed in a special cage,[62] which enabled the continuous monitoring of locomotor activity by means of an attached vibration sensor. A cannular feed-through slip ring fixed above the cage guaranteed the free movement of the rats under continued i.c.v. infusion and successive recordings of EEG, EMG, and T_{br}. Each cage was placed in a sound-proof, electromagnetically shielded chamber under the same environmental conditions as above. The rats were continuously infused with 0.9% saline at a rate of 10 or 20 μl/h 1 week after surgery. Simultaneously, their EEG, EMG, and locomotion were polygraphically recorded.

After observing the establishment of circadian rhythms in sleep-waking activity under the steady infusion of saline, the rats were subjected to the experimental treatments. A bioassay was done in 4 consecutive days: a baseline day for control recordings when the rats were

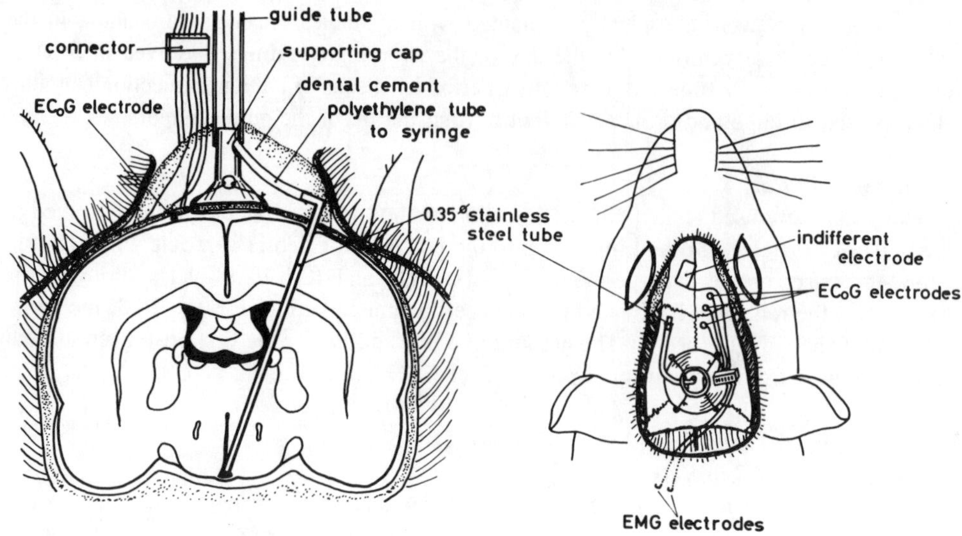

FIGURE 11. Implantation of electrodes and a cannula in the recipient rat for the long-term sleep assay. (From Honda, K. and Inoué, S., *Rep. Inst. Med. Dent. Eng.*, 12, 81, 1978. With permission.)

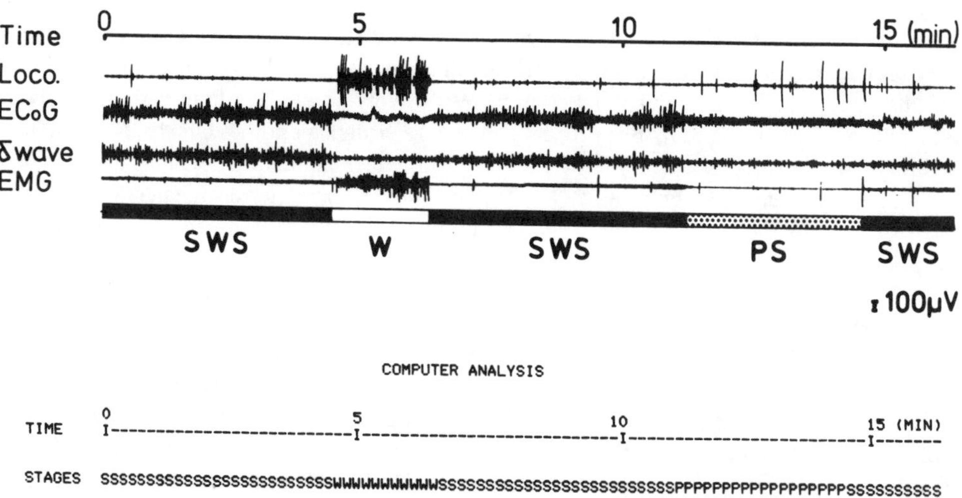

FIGURE 12. Top: typical polysomnographic record with visually scored episode(s) of wakefulness (W), slow-wave sleep (SWS), and paradoxical sleep (PS). Bottom: computer-stored record corresponding with the above. SWS and PS are expressed by S and P, respectively. (After Honda, K. and Inoué, S., *Rep. Inst. Med. Dent. Eng.*, 15, 115, 1981. With permission.)

continuously infused with saline; an experimental day when the steady saline infusion was replaced by a 10-h infusion of a sample solution dissolved in 100 or 200 μl saline beginning at 1 h before the onset of the dark period (between 19.00 and 05.00 h); and 2 recovery days when the saline solution was continuously infused.

Combining polygraphic records of locomotion, EEG, delta EEG waves, and EMG (see Figure 12), sleep-waking states were visually scored on a large-scale digitizer. The successive states of SWS, PS, and wakefulness (W) were directly fed into and numerically processed by a computer-aided device system[58,63-65] (see Figure 13). The minimal scoring interval was 12 s of recording time or 6 mm of polygraphic sheet. The following criteria were used for

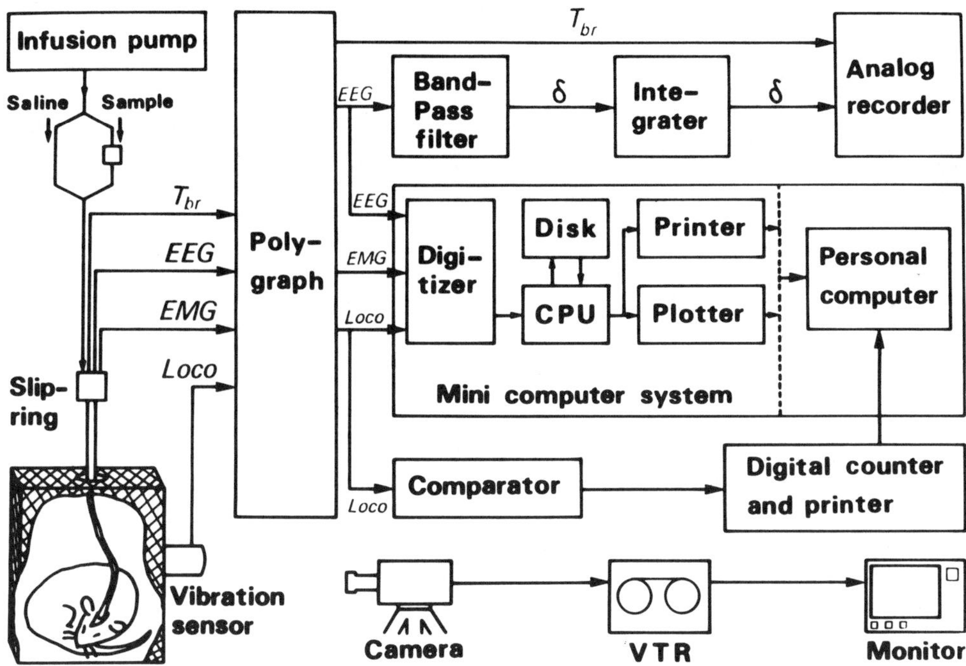

FIGURE 13.. Total sleep assay system of the long-term i.c.v. infusion technique. (From Inoué, S., Honda, K., and Komoda, Y., in *Sleep: Neurotransmitters and Neuromodulators,* Wauquier, A., et al., Eds., Raven Press, New York, 1985, 305. With permission.)

the scoring. Wakefulness was characterized by a low-amplitude (up to 100 μV) and high-frequency (more than 12 Hz) EEG, a high-amplitude EMG, the presence of locomotor activity, and the absence of EEG delta waves. SWS was characterized by a high-amplitude (more than 100 μV) and low-frequency (up to 12 Hz) EEG, a low-amplitude EMG, the absence of locomotor activity, and the presence of EEG delta waves occupying more than half of the minimal scoring interval. PS was characterized by a low-amplitude and high-frequency EEG, almost no trace of EMG except for occasional muscle twitches, the presence of intermittent, slight locomotor activity, and the absence of EEG delta waves. PS was also characterized by succeeding SWS. Any episode that lasted for less than 12 s was added to the preceding episode and not regarded as an independent episode. Examples of the experiment records are illustrated in Figures 14 and 15.

In parallel with the above bioassay, two other simplified screening techniques were applied to partially purified SPS-A fractions.[44] One was to detect an inhibitory effect on the neuronal firing activity of the abdominal ganglion[66] and the stretch receptor[67] in crayfish *in vitro* perfused with sample solutions. The other was to detect a sleep-promoting effect by measuring daily amounts of SWS and/or PS in mice before and after i.p. injection of sample solutions.[68]

2. 11-d Assay for Gastrointestinal Hormones in Rats

Danguir and Nicolaidis[22] performed very long-term i.c.v. infusion experiments in rats for assessing the somnogenicity of insulin. Prior to the infusion, male Wistar rats weighing 300 to 350 g were implanted with cortical electrodes for EEG recording and a stainless steel guide cannula, which was placed 1 mm over the lateral ventricle for i.c.v. infusion. After surgery, the rats were housed individually in a room illuminated from 08.00 to 20.00 h under regulated ambient temperature of 24 ± 1°C. The cages were Plexiglas® cylinders open at the top so as to allow chronic EEG recordings and continuous infusion in the freely moving rats. After a 1-week recovery period, EEG was monitored for 11 consecutive days:

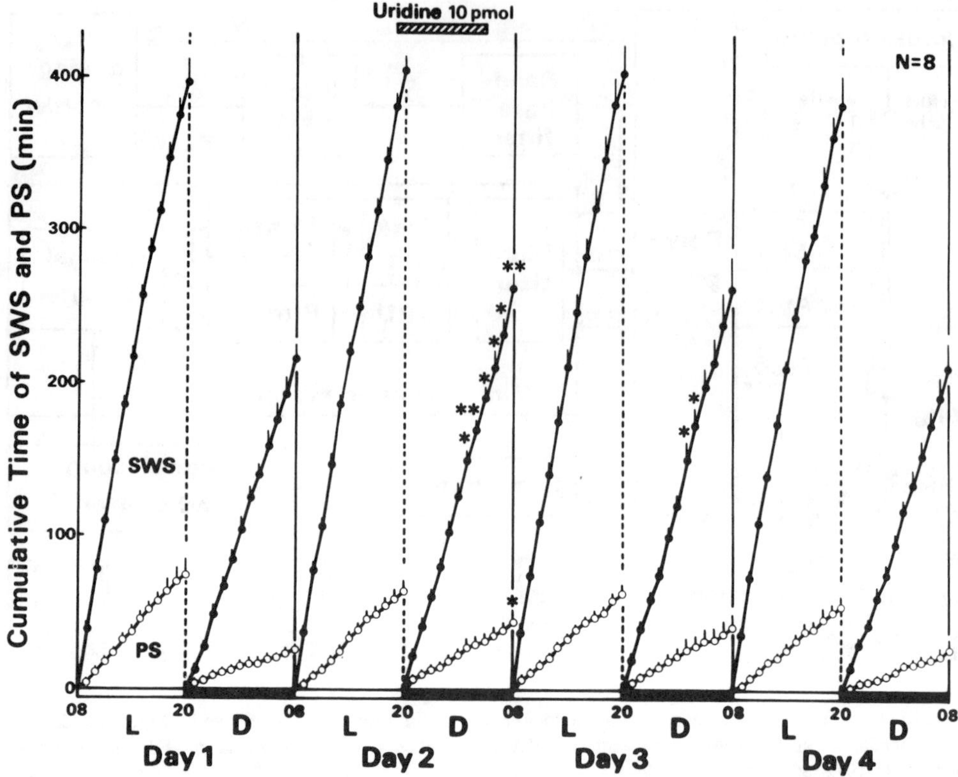

FIGURE 14. A 4-d analysis of the effect of 10-h i.c.v.-infused uridine (10 pmol) on day 2 at 19.00 through 05.00 h (indicated by a bar) on the time-course changes in cumulative sleep amounts in otherwise saline-infused, freely behaving rats. Vertical lines on each hourly value indicate standard error of mean. L and D represent the environmental light and dark period, respectively. $*p < 0.05$; $**p < 0.01$: Significantly different from the values of day 1. (From Honda, K., Komoda, Y., Nishida, S., Nagasaki, H., Higashi, A., Uchizono, K., and Inoué, S., *Neurosci. Res.*, 1, 243, 1984. With permission.)

the first 3 d for control recordings, the second 3 d for continuous i.c.v. infusion of artificial CSF (50 μl/d), the third 3 d for continuous i.c.v. infusion of insulin, and the final 2 d for insulin-withdrawal recordings. The i.c.v. infusions were done by inserting a needle, 1 mm longer than the implanted guide cannula and connected to an infusion pump through a special swivel, into the guide cannula and fixing it on the latter. The hourly and daily amounts of SWS and PS were averaged and compared among the treatments.

Similar but shorter (7-d) experiments were done for bioassay of SRIF by Danguir.[69] The difference from the above was as follows: 2 d for control recordings, 1 d for CSF infusion, 2 d for SRIF infusion, and 2 d for withdrawal recordings.

3. 2-d Assay for Neuropeptides in Rats

Obál et al.[70] routinely used 2-d assays for evaluating the somnogenic activity of several neuropeptides. Samples were dissolved in 1 to 4 μl of artificial CSF and acutely injected into the left lateral cerebral ventricle of rats 10 to 15 min before the onset of the dark period. On the day of the experiment and the previous day (baseline), 24-h recordings of EEG, locomotor activity, and cortical temperature were done. Prior to the experiment, the rats were habituated to the experimental conditions for 7 to 10 d, including i.c.v. injection of the solvent for 5 d. In addition, the functional circulating of CSF was verified by means of the drinking response to i.c.v. injection of angiotensin (100 ng in 1 μl), since this substance elicited drinking in about 2 min by stimulating preoptic structures after reaching the third cerebral ventricle. The amount of SWS and PS was calculated at 3-h intervals from the registered polygram of EEG and locomotor activity.

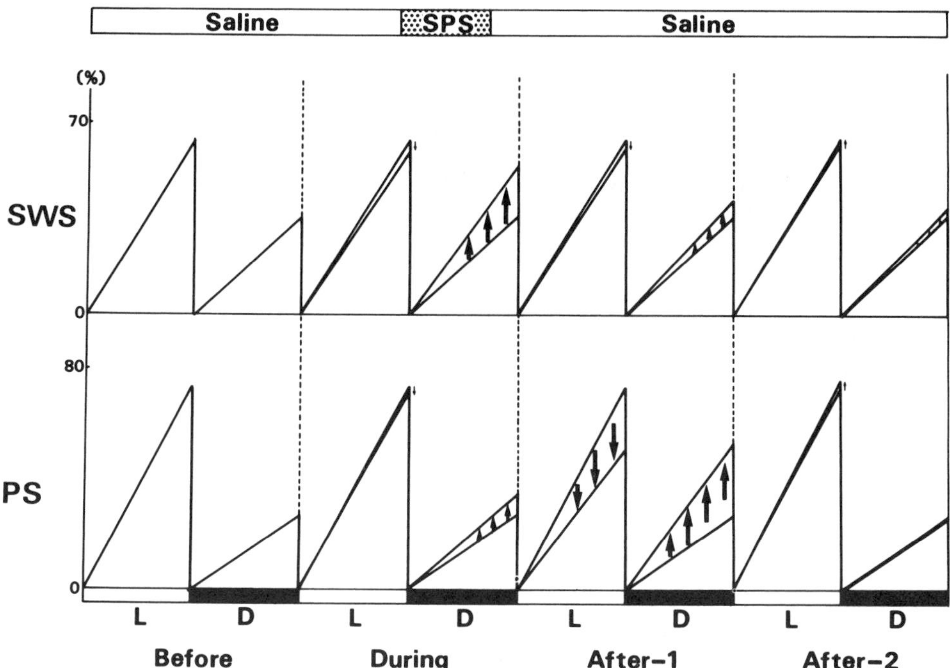

FIGURE 15. Effects of 10-h i.c.v. infusion of two units (equivalent to the two brainstems of sleep-deprived donor rats) of SPS-B at 19.00 to 05.00 h (indicated by a hatched bar on the top) on the time-course changes in the amount of sleep in otherwise saline-infused, freely behaving rats. The graph shows the cumulative amount of SWS (above) and PS (below) in the light (L) and the dark (D) period during the 4-d observation period. The arrows indicate the deviation from the baseline. For other explanations, see the Figure 14 caption. (From Inoué, S., Nagasaki, H., and Uchizono, K., *Trends Neurosci.*, 5, 218, 1982. With permission.)

REFERENCES

1. **Ishimori, K.,** True cause of sleep — A hypnogenic substance as evidenced in the brain of sleep-deprived animals, *Tokyo Igakkai Zasshi,* 23, 429, 1909.
2. **Legendre, R. and Piéron, H.,** Recherches sur le besoin de sommeil consecutif à une veille prolongée, *Z. Allg. Physiol.,* 14, 235, 1913.
3. **Pappenheimer, J. R., Koski, G., Fencl, V., Karnovsky, M. L., and Krueger, J.,** Extraction of sleep-promoting factor S from cerebrospinal fluid and from brains of sleep-deprived animals, *J. Neurophysiol.,* 38, 1299, 1975.
4. **Tobler, I., Borbély, A. A., Schwyzer, M., and Fontana, A.,** Interleukin-1 derived from astrocytes enhances slow wave activity in sleep, *Eur. J. Pharmacol.,* 104, 191, 1984.
5. **Drucker-Colín, R., Rojas-Ramírez, J. A., Vera-Trueba, J., Monroy-Ayala, G., and Hernández-Peón, R.,** Effect of crossed-perfusion of the midbrain reticular formation upon sleep, *Brain Res.,* 23, 269, 1970.
6. **Nagasaki, H., Irki, M., Inoué, S., and Uchizono, K.,** The presence of a sleep-promoting material in the brain of sleep-deprived rats, *Proc. Jpn. Acad.,* 50, 241, 1974.
7. **Straus, E., Ryder, S. W., Eng, J., and Yalow, R. S.,** Immunochemical studies relating to cholecystokinin in brain and gut, *Rec. Progr. Horm. Res.,* 37, 447, 1981.
8. **Sandow, J. and König, W.,** Chemistry of the hypothalamic hormones, in *The Endocrine Hypothalamus,* Jeffcoate, S. L. and Hutchinson, J. S. M., Eds., Academic Press, London, 1978, 149.
9. **Kastin, A. J., Nissen, C., Schally, A. V., and Coy, D. H.,** Radioimmunoassay of DSIP-like material in rat brain, *Brain Res. Bull.,* 3, 691, 1978.
10. **Van Houten, M., Posner, B. I., Kopriwa, B. M., and Brawer, J. R.,** Insulin-binding sites in the rat brain: *in vivo* localization to the circumventricular organs by quantitative radioautography, *Endocrinology,* 105, 666, 1979.

11. **Narumiya, S., Ogorochi, T., Nakao, K., and Hayaishi, O.,** Prostaglandin D_2 in rat brain, spinal cord, and pituitary: basal levels and regional distribution, *Life Sci.*, 31, 2093, 1982.
12. **Giacobini, E.,** Piperidine: a new neuromodulator or a hypnogenic substance?, *Adv. Biochem. Psychopharmacol.*, 15, 17, 1976.
13. **Fahrenkrug, J. and Schaltalizky de Muckadell, O. B.,** Distribution of vasoactive intestinal polypeptide in the porcine central nervous system, *J. Neurochem.*, 31, 1445, 1978.
14. **Skolnick, P., Mendelson, W. B., and Paul, S. M.,** Benzodiazepine receptors in the central nervous system, in *Psychobiology of Sleep*, Wheatley, D., Ed., Raven Press, New York, 1981, 117.
15. **Pavel, S.,** Pineal arginine vasotocin: an extremely potent sleep-inducing nonapeptide hormone. Evidence for the involvement of 5-hydroxytryptamine containing neurons in its mechanism of action, in *Sleep 1978*, Popoviciu, L., Asigian, B., and Badiu, G., Eds., S. Karger, Basel, 1980, 381.
16. **Lieberman, H. R.,** Behavior, sleep and melatonin, *J. Neural Transm.*, Suppl. 21, 233, 1986.
17. **Chastrette, N., Clement, H. W., Prevautel, H., and Cespuglio, R.,** Proopiomelanocortin components: differential sleep-waking regulation?, in *Sleep Peptides: Basic and Clinical Approaches*, Inoué, S. and Schneider-Helmert, D., Eds., Japan Scientific Societies Press, Tokyo/Springer-Verlag, Berlin, 1988, 27.
18. **Stern, W. C., Jalowiec, J. E., Shabshelowitz, H., and Morgane, P. J.,** Effects of growth hormone on sleep-waking patterns in cats, *Horm. Behav.*, 6, 189, 1975.
19. **Jouvet, M., Buda, C., Cespuglio, R., Chastrette, N., Denoyer, M., Sallanon, M., and Sastre, J. P.,** Hypnogenic effects of some hypothalamo-pituitary peptides, *Clin. Neuropharmacol.*, 9(Suppl. 4), 465, 1986.
20. **Chastrette, N. and Cespuglio, R.,** Effets hypnogènes de la des-acetyl-α-MSH et du CLIP (ACTH 18-39) chez le rat, *C.R. Acad. Sci. Paris*, 301, 527, 1985.
21. **Chastrette, N. and Cespuglio, R.,** Influence of proopiomelanocortin-derived peptides on the sleep-waking cycle of the rat, *Neurosci. Lett.*, 62, 365, 1985.
22. **Danguir, J. and Nicolaidis, S.,** Chronic intracerebroventricular infusion of insulin causes increase of slow wave sleep in rats, *Brain Res.*, 306, 97, 1984.
23. **Obál, F., Jr.,** Effects of peptides (DSIP, DSIP analogues, VIP, GRF and CCK) on sleep in the rat, *Clin. Neuropharmacol.*, 9(Suppl. 4), 459, 1986.
24. **Riou, F. R., Cespuglio, R., and Jouvet, M.,** Endogenous peptides and sleep in the rat. III. The hypnogenic properties of vasoactive intestinal polypeptide, *Neuropeptides*, 2, 265, 1982.
25. **Branchey, M., Branchey, L., and Nadler, R. D.,** Effects of estrogen and progesterone on sleep patterns of female rats, *Physiol. Behav.*, 6, 743, 1971.
26. **Mendelson, W. B., Martin, J. V., Wagner, R. R., Perlis, M., Majewska, M., and Paul, S. M.,** Hypnotic properties of an endogenous corticosteroid, *Sleep Res.*, 16, 108, 1987.
27. **Schoenenberger, G. A.,** Characterization, properties and multivariate functions of delta-sleep-inducing peptide (DSIP), *Eur. Neurol.*, 23, 321, 1984.
28. **Fencl, V., Koski, G., and Pappenheimer, J. R.,** Factors in cerebrospinal fluid from sleep-deprived goats that affect sleep and activity in rats, *J. Physiol. (London)*, 216, 565, 1971.
29. **Krueger, J. M., Bacsik, J., and García-Arrarás, J.,** Sleep-promoting material from human urine and its relation to factor S from brain, *Am. J. Physiol.*, 238, E116, 1980.
30. **Yanagisawa, I. and Yoshikawa, H.,** A bromine compound isolated from human cerebrospinal fluid, *Biochim. Biophys. Acta*, 329, 283, 1973.
31. **Kastin, A. J., Castellanos, P. F., Banks, W. A., and Coy, D. H.,** Radioimmunoassay of DSIP-like material in human blood: possible protein binding, *Pharmacol. Biochem. Behav.*, 15, 969, 1981.
32. **Graf, M. V., Hunter, C. A., and Kastin, A. J.,** Presence of delta-sleep-inducing peptide-like material in human milk, *J. Clin. Endocrinol. Metab.*, 59, 127, 1984.
33. **Ernst, A., Monti, J. C., and Schoenenberger, G. A.,** Delta-sleep-inducing-peptide (DSIP) in human and cow milk, in *Sleep '84*, Koella, W. P., Rüther, E., and Schulz, H., Eds., Gustav Fischer Verlag, Stuttgart, 1985, 215.
34. **Ernst, A. and Schoenenberger, G. A.,** DSIP: basic findings in human beings, in *Sleep Peptides: Basic and Clinical Approaches*, Inoué, S. and Schneider-Helmert, D., Eds., Japan Scientific Societies Press, Tokyo/Springer-Verlag, Berlin, 1988, 131.
35. **Ursin, R.,** Endogenous sleep factors, *Exp. Brain Res.*, Suppl. 8, 118, 1984.
36. **Johannsen, L., Wecke, J., and Krueger, J. M.,** Macrophages produce a sleep-enhancing substance(s) during the digestion of bacteria, Book of Abstracts, Int. Symp. Current Trends in Slow Wave Sleep Res. (Physiological and Pathological Aspects), 1987, 42.
37. **Hösli, L., Monnier, M., and Koller, T.,** Humoral transmission of sleep and wakefulness. I. Method for dialysing psychotropic humors from the cerebral blood, *Pfluegers Arch.*, 282, 54, 1965.
38. **Monnier, M. and Hösli, L.,** Dialysis of sleep and waking factors in blood of the rabbit, *Science*, 146, 796, 1964.
39. **Monnier, M. and Hösli, L.,** Humoral transmission of sleep and wakefulness. II. Hemodialysis of sleep inducing humor during stimulation of the thalamic somnogenic area, *Pfluegers Arch.*, 282, 60, 1965.

40. **Monnier, M. and Schoenenberger, G. A.,** Erzeugung, Isolierung und Characterisierung eines physiologischen Schlaffaktor "delta", *Schweiz. Med. Wochenschr.*, 103, 1733, 1973.
41. **Schoenenberger, G. A., Cueni, L. B., Monnier, M., and Hatt, A. M.,** Humoral transmission of sleep. VII. Isolation and physical-chemical characterization of the "sleep inducing factor delta", *Pfluegers Arch.*, 338, 1, 1972.
42. **Schoenenberger, G. A., Maier, P. F., Tobler, H. J., and Monnier, M.,** A naturally occurring delta-EEG enhancing nonapeptide in rabbits. X. Final isolation, characterization and activity test, *Pfluegers Arch.*, 369, 99, 1977.
43. **Schoenenberger, G. A., Maier, P. F., Tobler, H. J., Wilson, K., and Monnier, M.,** The delta EEG (sleep)-inducing peptide (DSIP). XI. Amino-acid analysis, sequence, synthesis and activity of the nonapeptide, *Pfluegers Archiv.*, 376, 119, 1978.
44. **Komoda, Y., Ishikawa, M., Nagasaki, H., Iriki, M., Honda, K., Inoué, S., Higashi, A., and Uchizono, K.,** Uridine, a sleep-promoting substance from brainstems of sleep-deprived rats, *Biomed. Res.*, 4(Suppl.), 223, 1983.
45. **Ichikawa, H., Honda, K., and Inoué, S.,** Manufacture of sleep deprivation apparatuses for small animals, *Rep. Inst. Med. Dent. Eng.*, 7, 145, 1973.
46. **Inoué, S. and Honda, K.,** Humoral theories of sleep and searches for the sleep substance, *Kagaku,* 49, 91, 1979.
47. **Pappenheimer, J. R.,** The sleep factor, *Sci. Am.*, 235(2), 24, 1976.
48. **Krueger, J. M., Pappenheimer, J. R., and Karnovsky, M. L.,** The composition of sleep-promoting factor isolated from human urine, *J. Biol. Chem.*, 257, 1664, 1982.
49. **Martin, S. A., Karnovsky, M. L., Krueger, J. M., Pappenheimer, J. R., and Biemann, K.,** Peptideglycans as promoters of slow-wave sleep. I. Structure of the sleep-promoting factor isolated from human urine, *J. Biol. Chem.*, 259, 12652, 1984.
50. **Monnier, M. and Hatt, A. M.,** Humoral transmission of sleep. V. New evidence from production of pure sleep hemodialysates, *Pfluegers Arch.*, 329, 231, 1971.
51. **Monnier, M., Hatt, A. M., Cueni, L. B., and Schoenenberger, G. A.,** Humoral transmission of sleep. VI. Purification and assessment of a hypnogenic fraction of "sleep dialysate" (factor delta), *Pfluegers Arch.*, 331, 257, 1972.
52. **Monnier, M., Dudler, L., and Schoenenberger, G. A.,** Humoral transmission of sleep. VIII. Effects of the "sleep factor delta" on cerebral, motor and visceral activities, *Pfluegers Arch.*, 345, 23, 1973.
53. **Krueger, J. M.,** Endogenous sleep factors, in *Sleep: Neurotransmitters and Neuromodulators,* Wauquier, A., Gaillard, J. M., Monti, J. M., and Radulovacki, M., Eds., Raven Press, New York, 1985, 319.
54. **Riou, F., Cespuglio, R., and Jouvet, M.,** Endogenous peptides and sleep in the rats. I. Peptides decreasing paradoxical sleep, *Neuropeptides,* 2, 243, 1982.
55. **Radulovacki, M., Virus, R. M., Djuricic-Nedelson, M., and Green, R. D.,** Adenosine analogs and sleep in rats, *J. Pharmacol. Exp. Ther.*, 228, 268, 1984.
56. **Yanik, G., Glaum, S., and Radulovacki, M.,** The dose-response effects of caffeine on sleep in rats, *Brain Res.*, 403, 177, 1987.
57. **Radulovacki, M., Virus, R. M., Rapoza, D., and Crane, R. A.,** A comparison of the dose response effects of pyrimidine ribonucleosides and adenosine on sleep in rats, *Psychopharmacology,* 87, 136, 1985.
58. **Honda, K. and Inoué, S.,** Effects of sleep-promoting substance on sleep-waking patterns of male rats, *Rep. Inst. Med. Dent. Eng.*, 15, 115, 1981.
59. **Yamaoka, S.,** Participation of limbic-hypothalamic structures in circadian rhythm of slow wave sleep and paradoxical sleep in the rat, *Brain Res.*, 151, 255, 1978.
60. **Inoué, S., Honda, K., and Komoda, Y.,** A possible mechanism by which the sleep-promoting substance induces slow wave sleep but supresses paradoxical sleep in the rat, in *Sleep 1982,* Koella, W. P., Ed., S. Karger, Basel, 1983, 112.
61. **Honda, K. and Inoué, S.,** Establishment of a bioassay method for the sleep-promoting substances, *Rep. Inst. Med. Dent. Eng.*, 12, 81, 1978.
62. **Honda, K., Ichikawa, H., and Inoué, S.,** Special rat cages for the assay of the sleep substance, *Rep. Inst. Med. Dent. Eng.*, 8, 149, 1974.
63. **Honda, K., Komoda, Y., Nishida, S., Nagasaki, H., Higashi, A., Uchizono, K., and Inoué, S.,** Uridine as an active component of sleep-promoting substance: its effects on the nocturnal sleep in rats, *Neurosci. Res.*, 1, 243, 1984.
64. **Inoué, S., Honda, K., and Komoda, Y.,** Sleep-promoting substances, in *Sleep: Neurotransmitters and Neuromodulators,* Wauquier, A., Gaillard, J. M., Monti, J. M., and Radulovacki, M., Eds., Raven Press, New York, 1985, 305.
65. **Inoué, S., Nagasaki, H., and Uchizono, K.,** Endogenous sleep-promoting factors, *Trends Neurosci.*, 5, 218, 1982.
66. **Nagasaki, H., Iriki, M., and Uchizono, K.,** Inhibitory effect of the brain extract from sleep-deprived rats (BE-SDR) on the spontaneous discharges of crayfish abdominal ganglion, *Brain Res.*, 109, 202, 1976.

67. **Uchizono, K., Higashi, A., Iriki, M., Nagasaki, H., Ishikawa, M., Komoda, Y., Inoué, S., and Honda, K.,** Sleep-promoting fractions obtained from the brain-stem of sleep-deprived rats, in *Integrative Control Functions of the Brain,* Vol. 1, Ito, M., Tsukahara, N., Kubota, K., and Yagi, K., Eds., Kodansha, Tokyo, 1978, 392.
68. **Nagasaki, H., Kitahama, K., Valatx, J.-L., and Jouvet, M.,** Sleep-promoting effect of the sleep-promoting substance (SPS) and delta-sleep-inducing peptide (DSIP) in the mouse, *Brain Res.,* 192, 276, 1980.
69. **Danguir, J.,** Intracerebroventricular infusion of somatostatin selectively increases paradoxical sleep in rats, *Brain Res.,* 367, 26, 1984.
70. **Obál, F., Jr., Török, A., Alföldi, P., Sáry, S., Hajós, M., and Penke, B.,** Effects of intracerebroventricular injection of delta sleep-inducing peptide (DSIP) and an analogue on sleep and brain temperature in rats at night, *Pharmacol. Biochem. Behav.,* 239, 953, 1986.

Chapter 3

DELTA-SLEEP-INDUCING PEPTIDE AND ITS DERIVATIVES

I. SLEEP-PROMOTING PROPERTIES

A. Induction of Delta Sleep

1. Experiments in Rabbits

As described in Chapter 2, the search for delta-sleep-inducing peptide (DSIP) was undergone by tracing its activity to induce electroencephalographic (EEG) delta sleep[1] (see Figure 1.) EEG delta activity increases by 30 to 50% above the baseline level, if an effective dose of DSIP is intracerebroventricularly (i.c.v.), intravenously (i.v.), intraperitoneally (i.p.), or subcutaneously (s.c.) administered in rabbits. Interestingly, the dose-response curve illustrates a parabolic or bell-shaped form (see Figure 2) regardless of the administration routes. In rabbits, the optimal dose for i.c.v., i.v., and s.c. administration is 6 to 8, 30, and up to 70 nmol/kg, respectively.[2] Different research groups have confirmed the slow-wave sleep (SWS)-enhancing activity of DSIP in rabbits.[3-11] Some investigators noted that DSIP also enhanced EEG sigma (spindle) activity.[4-8,11]

Somnogenic activity of synthetic DSIP has also been reported in other animal species and in humans. According to Graf and Kastin,[12] the optimal dose of DSIP is 5 to 10 nmol/kg i.c.v., 20 to 40 nmol/kg i.v., and 100 to 200 nmol/kg i.p. or s.c. in general. Consequently doses either smaller or larger than the optimal dosage may elicit no sleep-enhancing effect.

2. Experiments in Cats

In cats, Polc et al.[13] demonstrated that i.v. injection of DSIP (30 nmol/kg) shortened sleep latency, reduced waking time, enhanced SWS, and even more markedly, paradoxical sleep (PS) over a 6-h observation period (see Figure 3). This is the first evidence that DSIP may induce PS in addition to SWS. A higher dose (300 nmol/kg) exerted little effect (see Figure 4). SWS- and PS-enhancing effect of DSIP in cats was confirmed by later studies of Scherschlicht[9] and Scherschlicht et al.[10] They found that DSIP administered either i.v. (25 μg/kg) or s.c. (100 μg/kg) augmented sleep time during a 6-h postinjection period. Again, PS increased more markedly (171 to 177% of control) than SWS (116 to 119%). Karmanova et al.[14,15] reported that suboccipitally administered DSIP (10 to 20 μg/kg) induced an increase of EEG delta sleep, a marked prolongation of latency to sleep onset and PS, and of SWS episodes, without changing the total amount of sleep in a 2-h observation period. From these results, the Russian investigators suggested that the DSIP-induced sleep is not identical to natural sleep.

Susic and co-workers[16-18] extensively investigated the sleep-inducing activity of DSIP. An acute injection of DSIP (7 nmol/kg) into the lateral ventricle of PS-deprived cats induced a significant increase in deep SWS (characterized by continuous high-voltage EEG slow waves) and reduced wakefulness and light SWS during an 8-h postinjection period. However, no change was observed in the amount of total sleep, total SWS, and total PS.[16] In similarly treated normal cats, DSIP i.c.v. injected at 8 h significantly reduced sleep latency, and significantly increased total sleep and total deep SWS during an 8-h period. This was due mainly to a prolongation of deep SWS episodes. No change occurred in PS parameters.[17] A single s.c. injection of DSIP (120 nmol/kg) brought about almost the same results, although the effects were less pronounced.[18]

3. Experiments in Rats

In rats, many investigators from different laboratories have reported sleep-enhancing effects of DSIP. Kafi et al.[19] demonstrated a significant increase in the amount of total sleep

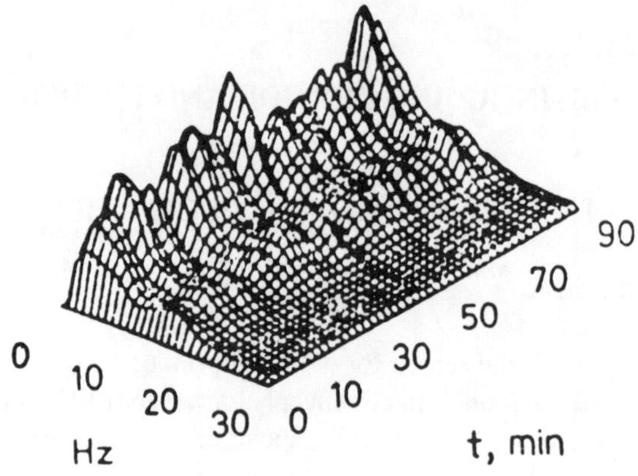

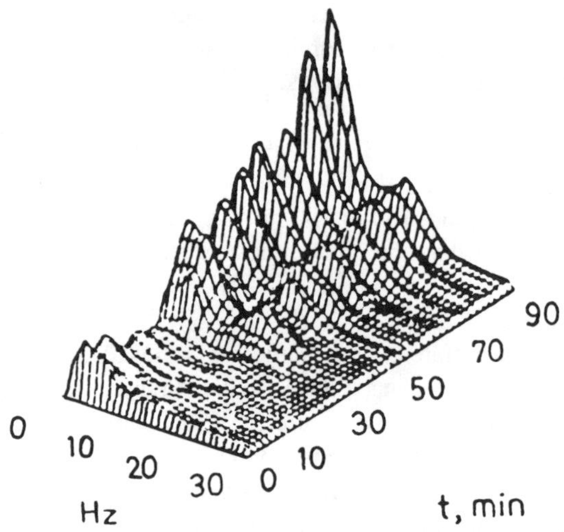

FIGURE 1. Tridimensional power spectra of EEG from frontal cortex of rabbits infused with synthetic DSIP and with artificial CSF. (From Monnier, M. and Schoenenberger, G. A., *Sleep 1976*, Koella, W. P. and Levin, P., Eds., S. Karger, Basel, 1977, 257. With permission.)

and total SWS during a 1- to 4.5-h period following i.v. injection of DSIP (30 to 80 nmol/kg) at 09.00 h. Miller et al.[20,21] observed an increase in EEG delta activity after i.p. administration of DSIP (80 μg/kg). Karmanova et al.[14,15] reported that suboccipitally administered DSIP (10 to 20 μg/kg) induced an increase of EEG delta sleep at the expense of wakefulness and PS and a marked prolongation of latency to sleep onset and PS in a 2-h observation period. Ursin and Larsen[22] and Ursin[23] found about a 20% increase of total sleep and total SWS during a 5-h diurnal period following i.c.v. injection (into the lateral cerebral ventricle) of DSIP (7 nmol/kg). Young and Key[24] injected rats with 5 mg of DSIP into the lateral cerebral ventricle four times at 30-min intervals during the dark period. SWS and PS were significantly increased in the following 3-h period.

Inoué et al.[25-32] analyzed the sleep-modulatory effects of DSIP by their long-term i.c.v. infusion technique (into the third cerebral ventricle). The earliest samples of DSIP (presented by Drs. G. A. Schoenenberger and M. Graf) exhibited a marked, prompt, but short-lasting SWS-inducing activity at a dose of 2.5 nmol/10 h per animal during a nocturnal period[24-26]

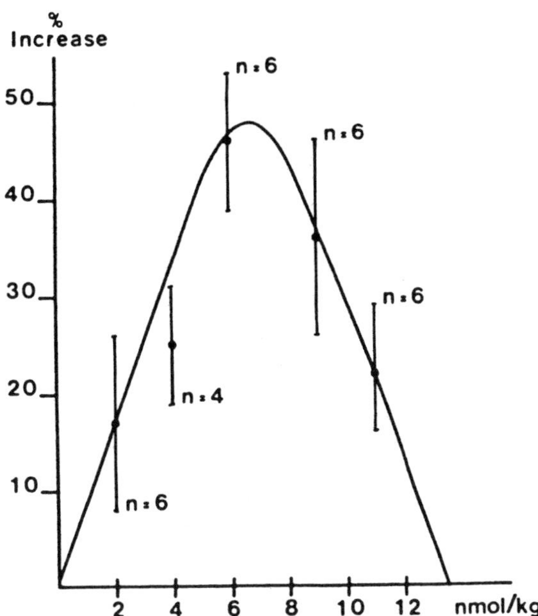

FIGURE 2. Bell-shaped or inverted U-shaped dose response curve obtained from rabbits i.c.v. infused with DSIP. (From Schoenenberger, G. A., *Eur. Neurol.*, 23, 321, 1984. With permission.)

(see Figure 5). However, the later samples synthesized by three different laboratories were less effective or almost ineffective[27-31] (see Figure 6). Since nuclear magnetic resonance analysis indicated that three of the samples were chemically identical to DSIP itself,[26a] the difference in the bioassay results could be attributed to unknown differences in experimental conditions. Presumably some degradation of the samples might be involved, since it has been suggested that degraded DSIP may contain a fragment that increases wakefulness instead of sleep.[33]

Wachtel et al.[34] have recently reported that i.p. injections of DSIP (300 μg/kg/d) in the early morning for 4 consecutive days partially restored the deep stage of SWS and PS in stress-induced insomniac rats. Interestingly, such sleep-restoring effects of DSIP were still observed 1 month after the termination of the DSIP treatment.

4. Experiments in Other Animals

In the preliminary study using a beagle dog, Takahashi et al.[35] observed that infusions of DSIP (20 μg/kg in 0.5 ml saline) or saline alone into the lateral cerebral ventricle over a period of 2 min at 13.00 h on 6 consecutive days resulted in a significant reduction in the amount of total sleep as compared to baseline (mean: 723 min/d). However, the amount of sleep after DSIP infusions (653 min/d) was significantly larger than that after saline infusions (613 min/d). Thus, DSIP partially blocked sleep disturbance caused by the treatment.

In mice, Nagasaki et al.[36] detected that DSIP i.p. injected at a dose of 30 nmol/kg at the dark period onset significantly increased total sleep and total SWS during the following 24 h.

5. Experiments in Healthy Humans

Blois et al.[37] injected ten subjects with DSIP (25 nmol/kg i.v.) at 09.00 h and polygraphically monitored sleep for the subsequent daytime hours. DSIP significantly increased total

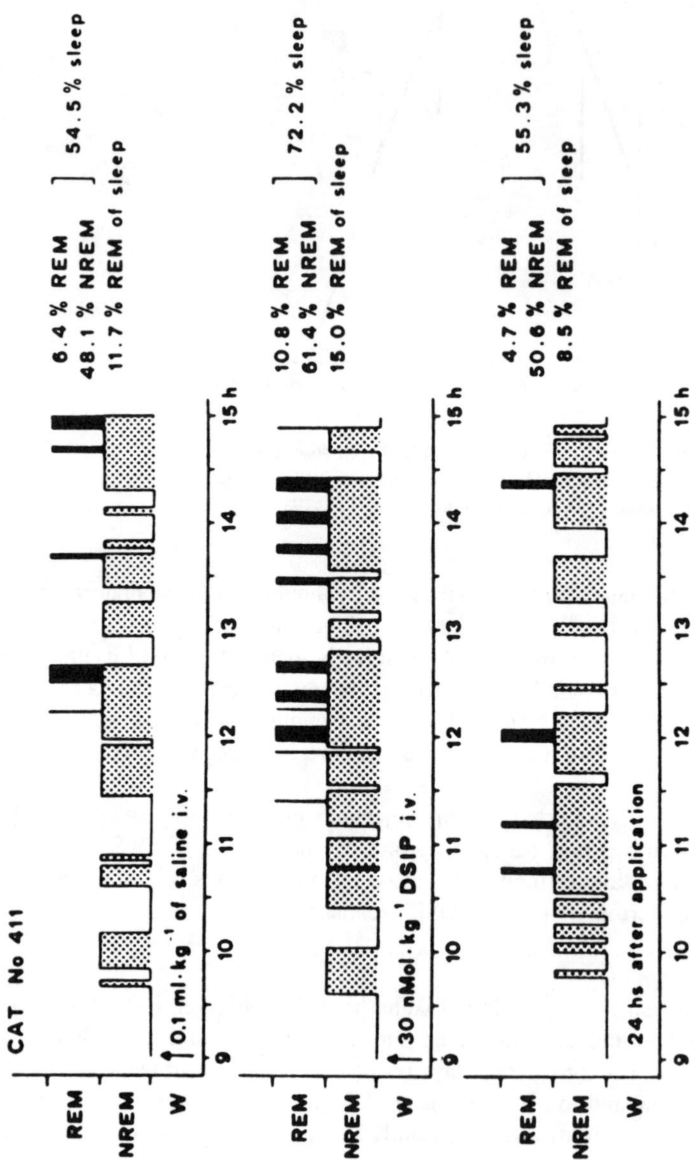

FIGURE 3. 6-h hypnograms in a cat injected with 0.1 ml/kg of saline (top), 30 nmol/kg of DSIP (middle), and none (bottom) on 3 consecutive days. REM: rapid-eye-movement sleep or paradoxical sleep; NREM: non-REM sleep or slow-wave sleep; W: wakefulness. (From Polc, P., Schneeberger, J. and Haefely, W., *Neurosci. Lett.*, 9, 33, 1978. With permission.)

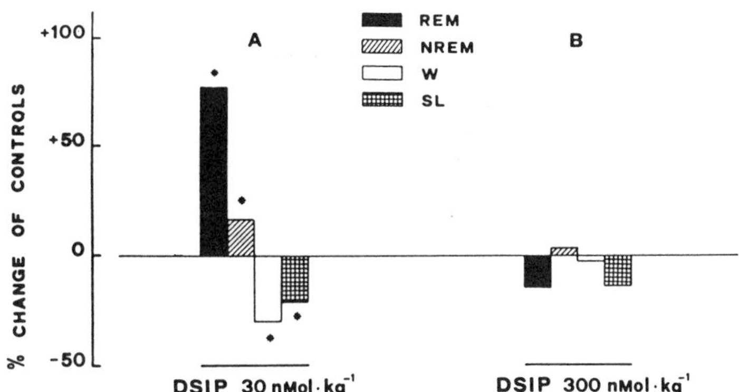

FIGURE 4. Effects of two doses of DSIP as compared to the control level (0%). Values are the means of five cats. SL: sleep latency. See Figure 3 for the other abbreviations. $*p < 0.05$. (From Polc, P., Schneeberger, J. and Haefely, W., *Neurosci. Lett.*, 9, 33, 1978. With permission.)

sleep time during the 120 to 160-min postinjection period. The EEG activities of spindles, theta, delta, and K potentials obviously increased as compared to the placebo condition. The Swiss investigators also i.v. injected healthy volunteers with DSIP (30 nmol/kg) before retiring and monitored sleep for 24 h. No difference of sleep parameters was found between the previous placebo night and the DSIP night. However, in the following night of no treatment, a significant enhancement of total sleep time occurred.

Schneider-Helmert and associates[38-41] i.v. infused six healthy subjects with DSIP at a dose of 25 nmol/kg during the daytime under twilight conditions. The infusion rate greatly influenced the effect of DSIP: the infusion of DSIP over 20, 7.5, 2.5, and 1-min periods brought about increases in total sleep time within 2 h, 1 h, 30 min, and no effect, respectively. The optimal effects seem to be obtained by 4- to 8-min infusions, provided a parabolic infusion rate-response curve might be applicable.[39]

B. Induction of Paradoxical Sleep

In cats, DSIP (30 nmol/kg i.v., 120 nmol/kg s.c.) markedly induced PS[9,10,13] (see above). In rats, as mentioned before, PS together with SWS increased after i.c.v. injections of DSIP (5 mg 4 times).[24] In insomniac human patients, DSIP treatments elevated the amount of REM sleep to the normal level.[38-42]

C. No Somnogenic Activity or Enhancement of Wakefulness

Several papers deal with the "no sleep-enhancing activity of DSIP". Medvedjev and Bakharev[43] and Karmanova et al.[44] reported the i.c.v.-administered DSIP at doses of 20 nmol/kg or more induced narcosis-like effects in cats, dogs, rabbits, and rats, whereas smaller doses were ineffective. Krueger et al.[45] reported that DSIP (18 nmol) was ineffective when i.c.v. infused for 90 min in rabbits. Tobler and Borbély[46] demonstrated that in rats neither an acute injection of DSIP into the lateral cerebral ventricle (8 nmol per animal) or the third cerebral ventricle (24 nmol), nor a 1.5- to 2-h infusion into the third cerebral ventricle (7 nmol) resulted in a change in the amount of total sleep, total SWS, and total PS during a 1- to 2-h diurnal postadministration period, except for a significant reduction in EEG delta activity.

Mendelson et al.[47-49] observed that DSIP (6 and 30 nmol/kg i.c.v.) increased sleep latency and intermittent wakefulness and decreased SWS during the first 2 h in rats. Yehuda and Mostofsky[50] found no somnogenic effect of DSIP i.p. injected in rats at 3 doses (0.1, 1, or 3 mg/kg). According to Obál et al.,[33,51,52] DSIP (7 nmol/kg) acutely injected into the lateral

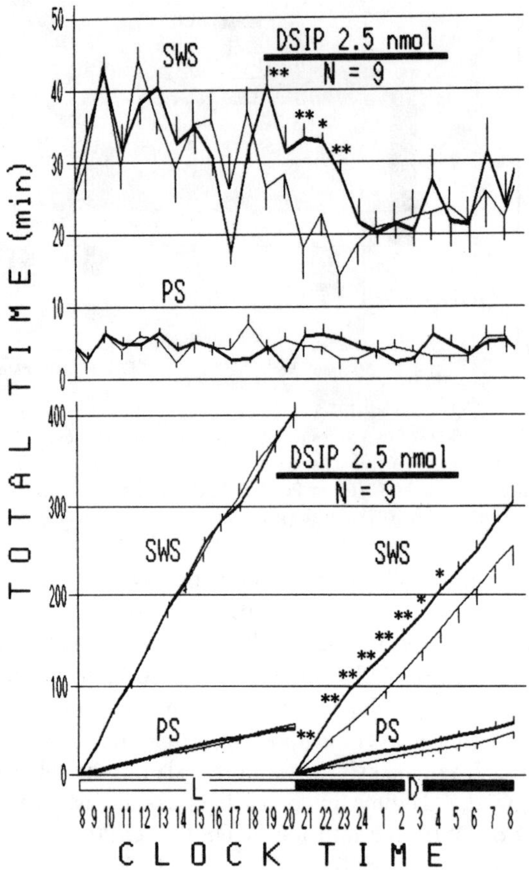

FIGURE 5. Effects of 10-h i.c.v. infusion of 2.5 nmol DSIP at 19.00 to 05.00 h (indicated by a solid bar) on the time-course changes in sleep amounts in otherwise saline-infused, freely behaving rats. The graph at the top shows the hourly summated SWS and PS amounts, while the one on the bottom illustrates their cumulative values in the environmental light (L) and dark (D) periods. Thin and thick curves stand for the baseline and the experimental day, respectively. The vertical lines on each hourly value indicate standard error of mean. $*p < 0.05$. (From Inoué, S., Honda, K., Komoda, Y., Uchizono, K., Ueno, R., and Hayaishi, O., *Proc. Natl. Acad. Sci. U.S.A.*, 81, 6240, 1984. With permission.)

cerebral ventricle of rats at the dark period onset failed to increase nocturnal sleep. Both SWS and PS decreased, while wakefulness increased during a 6- to 9-h postinjection period. Kimura et al.[28-31] and Inoué et al.[32] compared sleep-modulatory activity of synthetic DSIP (2.5 nmol per rat) from four different sources by their long-term i.c.v. infusion technique and found little activity in two samples (see Figure 6).

Sommerfelt[53] observed an elevation of wakefulness at the expense of light SWS and PS in cats after i.p. injection of DSIP (30 nmol/kg) during a 10-h postinjection period of daytime. This result seems to be attributable to the effect of DSIP in reducing ''redundancy sleep''.

It is rather complicated to elucidate the discrepancy between the marked positive and negative somnogenic effects of DSIP. The controversial results may be caused by several reasons. Graf and Kastin[54] pointed out that the following factors might be involved in variable

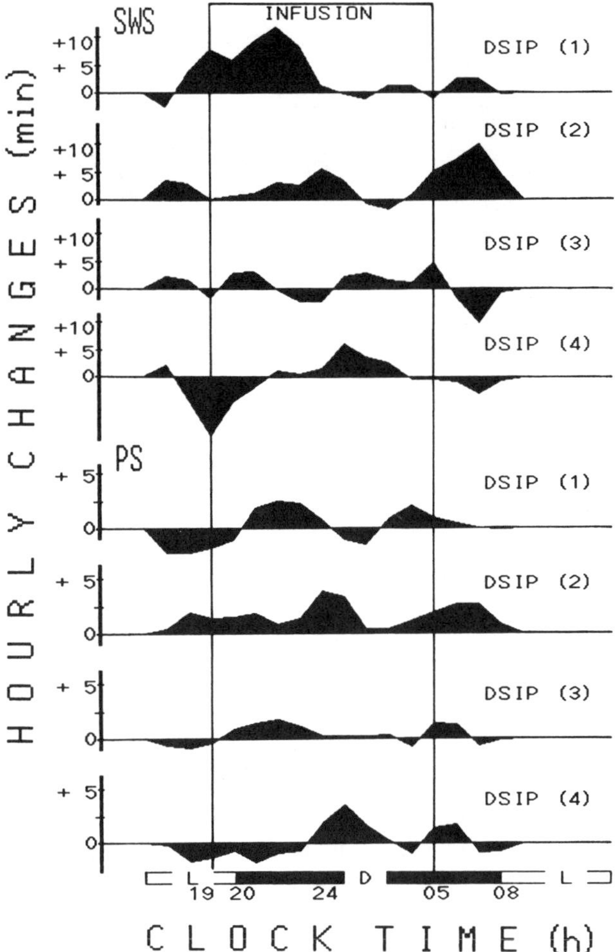

FIGURE 6. Comparison of the somnogenic effects of four different samples of DSIP (each 2.5 nmol/10 h) i.c.v. infused in freely moving rats. Curves show hourly increments expressed as the difference from the baseline value. The four top and four bottom curves represent SWS and PS, respectively. L and D stand for the environmental light and dark periods, respectively. DSIP (1), (2), (3), and (4) indicate the samples presented by Drs. G. A. Schoenenberger and M. Graf, purchased from Peptide Institute, Inc., Osaka, presented by Drs. E. Herzog and D. Gillessen, and presented by Dr. V. M. Kovalzon, respectively. Note that the scales for SWS and PS are different. (From Inoué, S., Kimura, M., Honda, K., and Komoda, Y., in *Sleep Peptides: Basic and Clinical Approaches,* Inoué, S. and Schneider-Helmert, D., Eds., Japan Scientific Societies Press, Tokyo/Springer-Verlag, Berlin, 1988, 1. With permission.)

responses to exogenously administered DSIP: an optimal dosage, an optimal infusion time, an adequate time of day, the content of endogenous DSIP dependent on the circadian and/or seasonal rhythm, and the existence of an indirect triggering mechanism.

D. Circadian Variations in Effectiveness

We observed that DSIP, at the same dose that induced a profound SWS increase during a 12-h nocturnal infusion period, was entirely ineffective if the infusion period was shifted by 12 h[27,55] (i.e., between 07.00 and 17.00 h). Thus, during the circadian resting phase,

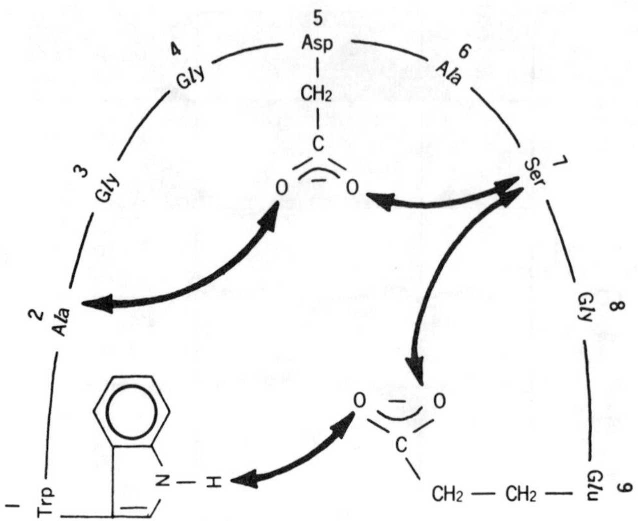

FIGURE 7. Presumed structural conformation of DSIP in aqueous solutions. (From Schoenenberger, G. A., *Eur. Neurol.*, 23, 321, 1984. With permission.)

rats did not react to exogenously supplied DSIP. Similar results were also obtained after the diurnal infusion of uridine and prostaglandin D_2 (PGD_2), as will be mentioned in later chapters. In this connection, it should be noted that DSIP (100 μg/kg i.p.) did not alter motor activity of rats under daylight conditions, and no anxiolytic effect was found with the same dose of i.v.-injected DSIP in a conflict procedure.[9] Also, DSIP (30 nmol/kg i.v.) administered four times within 24 h differentially affected the levels of several cerebral neurotransmitters and plasma proteins in rats.[56]

II. CHEMICAL AND BIOCHEMICAL ASPECTS

A. Conformation of DSIP and Its Synthetic Analogs

DSIP in an aqueous environment seems to exhibit a pseudo-cyclic structure, as predicted by theoretical considerations[57,58] (see Figure 7). This structural conformation is stable due to the interactions between the side chains of the amino terminal Trp and the carboxy terminal Glu. Somnogenic activity of the peptide seems to be closely related to the structure.

Obál et al.[51] compared the sleep-modulatory activity of DSIP and its three different analogs, D-Trp^1-DSIP, D-Tyr^1-DSIP, and D-Trp^1-DSIP$_{1\text{-}6}$. Each substance (7 nmol/kg) was injected into the lateral cerebral ventricle of rats at the dark period onset and the sleep-waking activity was monitored during the 12-h dark period. DSIP did not increase sleep but increased wakefulness, whereas both D-Trp^1-DSIP and D-Tyr^1-DSIP promoted sleep in the first part of the night. In contrast, D-Tyr^1-DSIP$_{1\text{-}6}$, in which the carboxy terminal is removed, exerted a prompt arousing effect. From the results, the Hungarian and Russian investigators have suggested that, although the active structure of DSIP is quickly degraded by the enzymic splitting of Trp from the original molecule, that of D-Tyr^1-DSIP and D-Tyr^1-DSIP might be preserved and able to elicit a sleep-promoting effect. The enhancement of wakefulness by D-Tyr^1-DSIP$_{1\text{-}6}$ may support the idea that DSIP presumably contains a fragment that increases wakefulness.[34]

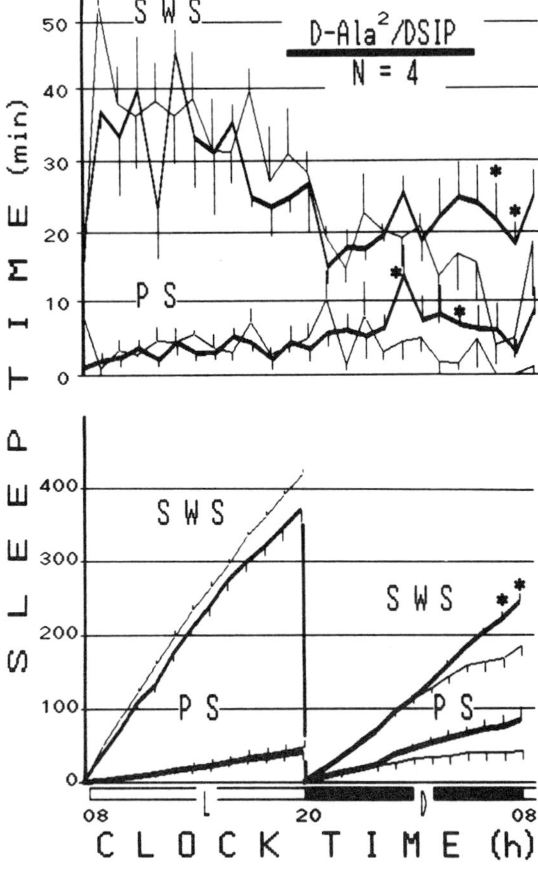

FIGURE 8. Effects of 10-h i.c.v. infusion of 2.5 nmol D-Ala²-DSIP at 19.00 to 05.00 h (indicated by a solid bar) on the time-course changes in sleep amounts in otherwise saline-infused, freely behaving rats. See Figure 5 for other explanations. (From Kimura, M., Honda, K., Kovalzon, V. M., and Inoué, S., *Rep. Med. Dent. Eng.*, 20, 47, 1986. With permission.)

Kovalzon et al.[59] examined hypnogenic activity of D-Tyr¹-DSIP, which is known to be stable against aminopeptidases. An acute injection in doses of 0.7 to 70 μg/kg into the lateral cerebral ventricle of rabbits induced a significant increase in SWS during a 3-h diurnal postinjection period. I.v. injections were also effective in a narrow dose range from 30 to 70 μg/kg. Two other DSIP analogs, D-Tyr¹-DSIP and D-Ala²-DSIP, exhibited a similar hypnogenic activity.[60] Kimura et al.[28-31] examined the sleep-enhancing activity of DSIP and its analogs, presented by Dr. V. M. Kovalzon, using their long-term i.c.v. infusion technique, and demonstrated positive results for D-Tyr¹-DSIP and D-Ala²-DSIP (see Figure 8), but not for the parent DSIP and D-Ala²-$DSIP_{1-6}$.

Kastin and colleagues[61] reported that a DSIP analog, D-Ala⁴-DSIP-NH_2, entered the brain of rats after i.p. injection more readily than DSIP itself. They also reported that D-Ala⁴-DSIP-NH_2 significantly increased EEG delta and theta activity after i.p. injection and that the effects were more marked than the parent DSIP.[62]

B. Other Active Analogs

Since the identification and synthesis of DSIP by Schoenenberger et al.,[63] a number of DSIP analogs have been synthesized by different laboratories, mainly in Switzerland, China, the U.S.S.R., and the U.S. The original method of DSIP synthesis yielded 2 isomers in the ratio 1:4. The product contained only 20% of the authentic DSIP of the natural sequence with Asp linked through the α-carboxyl group, whereas the remainder contained the isomer with Asp linked through the β-carboxyl group.[63] The latter β-DSIP has no somnogenic activity. Ji and colleagues[4,8] succeeded in synthesizing only α-DSIP by the classical solution method, as well as by the solid-phase method.[5-7] The Chinese investigators also synthesized an active analog, Phe^5-DSIP.[5,6]

Dobrynin et al.[64] applied a gene-engineering technique to artificially construct DSIP by ligations of deoxyribonucleotides. Sargsyan et al.[65] and Mikhaleva et al.[66] synthesized 11 different analogs of DSIP, in which 6 were active in suppressing the neuronal firing activity of snail cerebral ganglia by 30 to 40% and elevating their resting potential by 1 to 4 mV. These substances are Tyr^5-DSIP, Tyr^7-DSIP, Tyr^8-DSIP, Tyr^9-DSIP, $Ava^{3,4}$-DSIP, and $Glu(NH_2)$-DSIP.

Taking the important role of phosphorylation and dephosphorylation events in the functional processing of peptides on a membrane and signal-transmitting level into account, Schoenenberger et al.[67] synthesized the phosphorylated analog, $(Ser^7\text{-}PO_4)$DSIP, or P-DSIP. The sleep-related activity of this substance is dealt with in the following section.

Furthermore, Obál and associates[33,51] reported that ω-amino-caprilyl-DSIP (C-DSIP) exerted wakefulness-enhancing and sleep-suppressive effects in rats.

C. Phosphorylated DSIP (P-DSIP)

Early studies on the biological activities of P-DSIP revealed that its potency of increasing SWS and PS in cats was six times greater than that of the parent compounds.[9] Graf et al.[68-70] and Schoenenberger and Graf[71] found that repeated evening injections of P-DSIP (0.1 nmol/kg i.v.) in rats under 12-h light-dark cycles produced alternations of circadian locomotor behavior. High locomotor activity typical of the dark period was reduced, while the animals became relatively more active during the light period. If the animals were kept under continuous illumination, a similar behavioral activation occurred during the subjective daytime, while a relatively reduced activity took place during the subjective nighttime. Interestingly, a comparable behavioral modification was induced by the same schedule of injections of DSIP at a dose of 30 nmol/kg, which is 300 times greater than that of P-DSIP.

The sleep-promoting potency of P-DSIP was five times greater than that of DSIP when tested by our long-term i.c.v. nocturnal infusion assay in rats.[30,32,72] We have found a dose-dependent somnogenic effect of P-DSIP at a range of 0.025 to 25 nmol. A significant SWS-enhancing effect, comparable to that induced by 2.5 nmol of DSIP of the earliest preparation,[26] was induced by the administration of 0.5 nmol of P-DSIP (see Figure 9). Lower and higher doses were ineffective. The dose response curve shows a bell shape for both SWS and PS (see Figure 10). Nakagaki et al.[73,74] and Nakagaki and Takahashi[75] acutely injected P-DSIP at doses of 0.02 or 0.2 nmol/kg into the third cerebral ventricle of rats shortly before the dark period onset. The higher dose induced a significant increase in the total amount of SWS and PS during the 12-h dark period, while the lower dose exerted no somnogenic effect.

Interestingly, recent studies have revealed that, at least in human beings, this substance exists as an endogenous sleep-regulating factor,[77] although P-DSIP was originally synthesized after DSIP on a theoretical basis.[67] Functional significance of P-DSIP as an endogenous sleep regulator is dealt with in a later section.

D. Degradation

DSIP seems to degrade rapidly. Quantitative amino acid analysis revealed that degradation of DSIP occurred within 15 min in brain extracts[78] and homogenates.[79] An even faster disappearance of exogenous DSIP was observed in plasma. Kato et al.[80] observed that

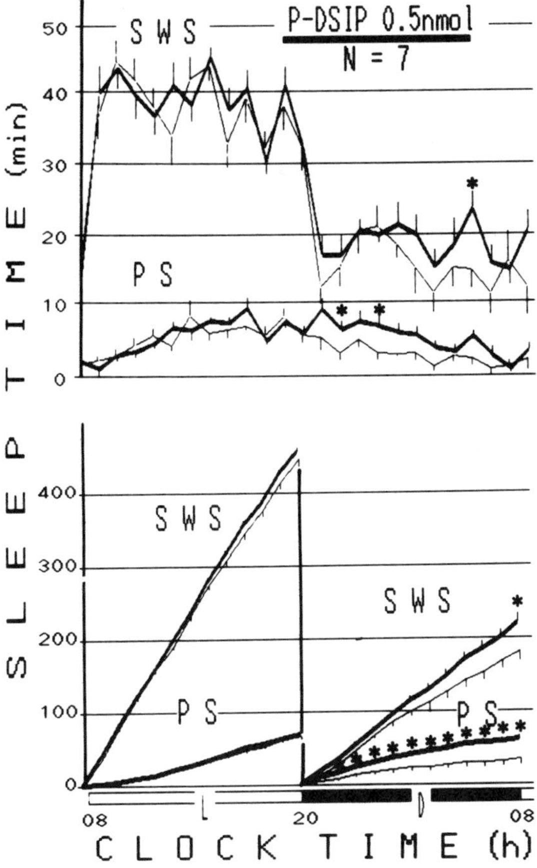

FIGURE 9. Effects of 10-h i.c.v. infusion of 0.5 nmol P-DSIP at 19.00 to 05.00 h (indicated by a solid bar) on the time-course changes in sleep amounts in otherwise saline-infused rats. See Figure 5 for other explanations. (From Inoué, S., Kimura, M., Honda, K., and Komoda, Y., in *Sleep Peptides: Basic and Clinical Approaches,* Inoué, S. and Schneider-Helmert, D., Eds., Japan Scientific Societies Press, Tokyo/Springer-Verlag, Berlin, 1988, 1. With permission.)

exogenously supplied DSIP disappeared with half-lives of 2 to 4 min in dogs, monkeys, and rats. The first and most important step of metabolism of DSIP is the cleavage of the N-terminal Trp. Graf et al.[81] found that the incubation of ^{3}H-DSIP labeled at Tyr^1 also cleaved the labeled Trp rapidly. Mean half-life was 3 to 4 min in rat plasma and 7 to 8 min in human plasma. P-DSIP is more potent than DSIP itself, since its half-life was about 11 h in human milk and no degradation at all was found for several h in cow milk.[82] Hence, it is speculated that P-DSIP is the precursor molecule of DSIP; dephosphorylation of P-DSIP produces DSIP, which is in turn rapidly cleaved to Trp and the 2-9 fragment.[12]

E. Receptors

In cultures of the rat lower brainstem (the pons and the medulla oblongata), Hösli et al.[83,84] observed binding sites of ^{3}H-DSIP on small, medium, and large neurons but not on glial cells. The binding sites were localized predominantly on the cell bodies and the primary dendrites, except on some neurons where binding sites were also detected on secondary and tertiary dendrites. Van Dijk and Schoenenberger[85] studied the DSIP membrane binding site in rat brain homogenates and concluded that the DSIP and Trp binding sites are not identical.

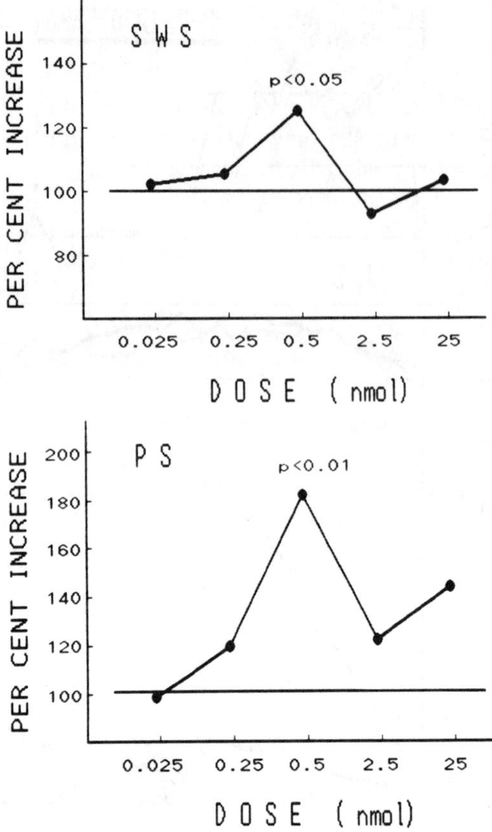

FIGURE 10. Bell-shaped dose response curve obtained from rats i.c.v. infused with P-DSIP. (From Inoué, S., Kimura, M., Honda, K., and Komoda, Y., in *Sleep Peptides: Basic and Clinical Approaches,* Inoué, S. and Schneider-Helmert, D., Eds., Japan Scientific Societies Press, Tokyo/Springer-Verlag, Berlin, 1988, 1. With permission.)

III. BODILY DISTRIBUTIONS

A. Brain and Peripheral Organs

Kastin et al.[86] developed a radioimmunoassay for DSIP and detected DSIP-like immunoreactivity (DSIP-LI) widespread in the rat brain. The highest brain value (12 ng/g tissue) was found in the thalamus, whereas the hippocampus and the hypothalamus contained slightly less DSIP-LI (9 and 7 ng/g, respectively). The pituitary gland also exhibited a high immunoreactivity, although the presence of DSIP was dubious. Similar results were obtained by Rozhanets et al.[87] who reported no great difference among the different regions of the rat brain (from 1 ng/g tissue in the thalamus to 3 ng/g in the hippocampus). Kato and associates[80,88,89] developed an enzyme immunoassay for DSIP and observed a similar distribution of DSIP-LI in the rat and the human brain. However, the regional distribution of DSIP-LI varied fivefold from 3.2 ng/g tissue in the nucleus accumbens to 0.7 ng/g in the cortex or the pons-medulla in rats. No change in brain DSIP-LI occurred after 24-h total sleep deprivation.

According to Kastin et al.,[90] DSIP-LI was present in the fetal brain at increased levels shortly before birth. The content of DSIP-LI in the whole brain rapidly increased during the postnatal development toward 20 d of age, decreasing slightly thereafter. Interestingly, DSIP-LI in the thalamus positively correlated with the time used for running at maze-learning in the rat.[91]

Kramarova et al.[92] found seasonal changes in brain content of DSIP-like material in ground squirrels. During the active period of June and July and the torpid period of December, the content of DSIP-like material was relatively low, whereas it increased pronouncedly during the hibernating and aroused periods of January and February. From these results, the Russian researchers suggested that DSIP does not participate in sleep induction, but in regulating the annual rhythms of activity.

Constantidinis et al.[93] immunohistologically visualized the presence of DSIP in the rat brain using an anti-DSIP antiserum. Broad areas of neocortex and subcortical structures contained specific DSIP-LI. They also demonstrated the presence of DSIP in the rat hypophysis. Feldman and Kastin[94,95] studied the distribution of immunoreactive DSIP neurons in the rat brain after increasing the number by colchicine treatment. Their survey suggested that DSIP was related to the sensory and visceral system in the brain.

Using radioimmunoassay and immunocytochemistry, Ekman et al.[96] demonstrated that DSIP-LI material was present in a subpopulation of the cells storing adrenocorticotropic hormone (ACTH) and α-melanotropin. They also demonstrated that in the adrenal medulla, DSIP-LI material was detected in a major population of the noradrenalin-storing cells. Hence, the anterior and intermediate lobes of the hypophysis and the adrenal medulla seem to be potential sources of DSIP-LI peptides.

Graf and Kastin[97] examined 12 peripheral organs of the rat by radioimmunoassay. The occurrence of DSIP-LI materials were detected and found to be relatively high in the gastrointestinal tract, kidney, spleen, and pancreas, and relatively low in the adrenal, thymus, liver, and muscle. Graf et al.[98] detected DSIP-LI also in the placenta (16 to 25 pg/g).

B. Body Fluids

DSIP was originally extracted from the venous blood of sleeping rabbits, as already described (see Chapter 2). The natural occurrence of DSIP in body fluids of different animals has been evidenced by many investigators.[12] The structure of DSIP is bound in large protein molecules, since radioimmunoreactivity could be detected in both free and bound forms of DSIP-like materials in human plasma.[99] Banks et al.[100] found that in dogs, exogenously administered DSIP increased the bound form of DSIP in the plasma without changing the level of the free form. Takahashi et al.[101] noted an increase in the plasma level of DSIP after 12 h sleep deprivation in dogs. Kato and colleagues[88] demonstrated the presence of DSIP-LI in human plasma and cerebrospinal fluid (CSF) by enzyme immunoassay. However, the values of DSIP-LI varied considerably according to the investigators and methods (25 pg/ml to 4 ng/ml). This may be dependent on the fact that the different antibodies can recognize only the free form of DSIP or all forms of DSIP.

A renewed approach by Graf et al.[102] revealed that the plasma level of DSIP-LI was 105 pg/ml in dogs, 260 pg/ml in humans, 155 pg/ml in rabbits, and 291 pg/ml in rats. They also found DSIP-LI at a level of 33 pg/ml in human CSF and 108 pg/ml in human urine. Thus, it seems likely that the free form of DSIP really exists in mammalian plasma. It is assumed that the freely occurring DSIP in plasma amounts to values of 50 to 100 pg/ml, whereas the total DSIP-LI material seems to be in the range of 1 to 5 ng/ml.[12]

Recently, P-DSIP-like immunoreactive material (P-DSIP-LI) has been detected by Schoenenberger and Graf[71] and Ernst and Schoenenberger.[77] The CSF and plasma levels of P-DSIP-LI and DSIP-LI dynamically change according to the time to day and the bodily states. The Swiss investigators emphasize that the mutual interactions between the two substances is crucial in the regulation of sleep (see below).

DSIP and P-DSIP occur in human milk.[71,77,98,103] Colostrum contained higher values of

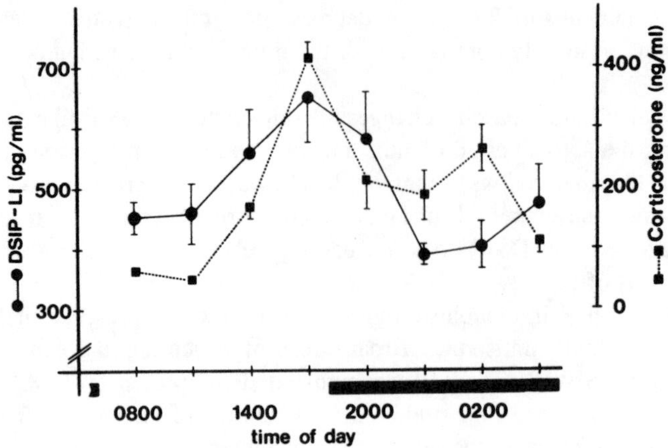

FIGURE 11. Circadian rhythm of plasma levels of DSIP-LI (solid line) and corticosterone (dotted line) in rats. (From Fischman, A. J., Kastin, A., and Graf, M., *Life Sci.*, 35, 2079, 1984. With permission.)

DSIP-LI than milk. Ernst and Schoenenberger[77] reported that the levels of DSIP-LI and P-DSIP-LI in colostrum were 12 and 2.6 ng/ml, respectively, while those of milk were 2 and 0.3 ng/ml, respectively. Since DSIP was found in plasma of neonatal rats after feeding them orally with the peptide,[104] it is possible that DSIP contained in maternal milk could influence the sleep-waking pattern in neonates.[98] Cow milk also contained both DSIP-LI and P-DSIP-LI in values of 1.1 to 1.3 and about 0.3 ng/ml, respectively.[77]

C. Circadian Rhythm of Plasma DSIP

DSIP-LI in rat plasma exhibited a circadian rhythm that paralleled the normal rhythm of corticosterone[105] (see Figure 11). Exposure to constant illumination abolished the rhythm of DSIP-LI. It was suggested that the temporal parallelism between the levels of DSIP-LI and corticosterone may represent a functional relationship between both compounds.[105] A similar rhythm of rat plasma DSIP-LI was observed by Banks et al.[106] Interestingly, DSIP-LI in human plasma also showed higher values in the evening as compared with other times of day.[88,99] No direct correlation was found between the DSIP-LI levels and sleep stages.[88] Circadian variations were also apparent in the level of DSIP-LI in human milk, high in the afternoon and low in the morning.[71,98]

Since the timing of the peak and trough in the circadian rhythm of human plasma DSIP-LI is dissociated with that of sleep-activity, it is speculated that DSIP may fulfuill a different function than that in rats.[12] Indeed, Ernst and Schoenenberger[77] have recently observed that DSIP-LI and P-DSIP-LI exhibited a biphasic pattern of circadian rhythm (see Figure 12). Statistical analyses revealed the multiple correlations with sleep parameters and mutually antagonistic effects between the two peptides. Thus, the Swiss investigators have proposed that DSIP is involved in arousal, which is reduced by an action of P-DSIP.[77] A similar circadian variation of DSIP-LI has been recently reported in human plasma[107,108] and in human urine.[108]

D. Blood-Brain Barrier

Monnier and associates[109] suggested that DSIP may pass the blood-brain barrier, since i.v.-injected DSIP (30 nmol/kg) induced a significant increase in EEG delta activity and a decrease in locomotor activity of freely behaving rabbits within 5 h. Subsequent studies also demonstrated the somnogenic effects of DSIP i.v. or i.p. administered in different animals and humans (see above), thus indirectly indicating the permeability of the blood-brain barrier

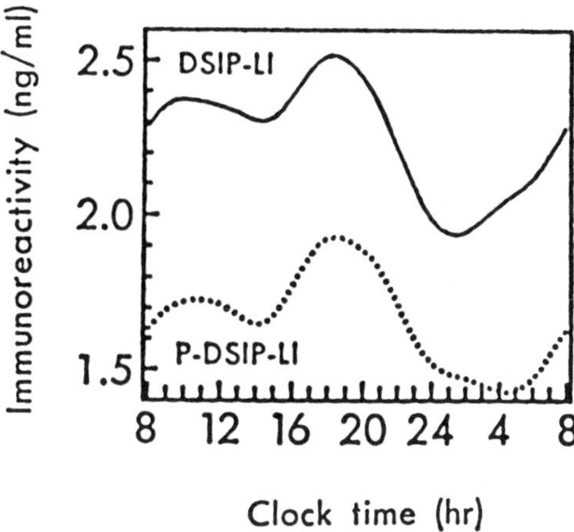

FIGURE 12. Biphasic circadian rhythm of plasma levels of DSIP-LI (solid line) and P-DSIP-LI (dotted line) in humans. (From Ernst, A. and Schoenenberger, G. A., in *Sleep Peptides: Basic and Clinical Approaches,* Inoué, S. and Schneider-Helmbert, D., Eds., Japan Scientific Societies Press, Tokyo/Springer-Verlag, Berlin, 1988, 131. With permission.)

to DSIP.[60,109,110] Olson et al.[112] observed a significant reduction of swimming activity in goldfish 3 min after intracranial injection and 6 min after i.p. injection of various peptides (80 μg/kg), and concluded that most of the peptides or their metabolites can cross the blood-brain barrier. However, DSIP did not affect the swimming activity, although D-Ala^4-DSIP did. We observed that swimming activity of cyprinid fish was modified when they were kept in DSIP and P-DSIP solutions[27,113,114] (see Chapter 10).

Banks et al.[100] reported that in dogs, exogenously administered DSIP was detected in CSF as the free form, and suggested that only the free DSIP may pass the blood-brain barrier into the CSF. They also demonstrated evidence of the transfer of DSIP-LI from lactating rat mothers to the plasma of their pups, and from the gastrointestinal tract of infant rats to their brain via the blood-brain barrier.[103] DSIP and its analog D-Ala^4-DSIP-NH_2 significantly increased delta EEG activity in rats after i.p. injection.[115] Thus peripherally administered DSIP can reach the brain and increase sleep activity.

IV. EXTRA-SLEEP EFFECTS

A. Multiple Functions of DSIP

Since synthetic DSIP has become available, a variety of studies have been conducted to analyze the physiological and pharmacological effects of DSIP and unveil its properties which are not directly related to sleep.[2,12,54,67,116] Such sleep-unrelated properties are called "extra-sleep" effects.[82] Since the multivariate "programming" effect of DSIP is proposed by taking all these properties into consideration (see below), the extra-sleep effects of DSIP are briefly summarized here.

B. Effects on Temperature

Yehuda et al.[117] reported that a low dose (0.1 mg/kg) of DSIP i.p. injected caused hypothermia in rats maintained at 4°C, but larger doses (1 or 3 mg/kg) did not. In contrast, at 22°C all three doses caused hyperthermia. However, DSIP reversed the usual D-amphetamine-induced hyperthermia to hypothermia. A similar effect was obtained in D-amphetamine-pretreated mice.[118] According to Obál et al.,[33] DSIP and C-DSIP (7 nmol/kg each) injected into the lateral cerebral ventricle of rats at the dark period onset did not affect brain temperature. Kovalzon et al.[59] reported that D-Trp[1]-DSIP acutely injected into the lateral cerebral ventricle of rabbits induced a significant increase in rectal temperature.

C. Blood Pressure and Anxiolytic Action

Graf et al.[119] suggested an involvement of DSIP in the regulation of blood pressure in spontaneously hypertensive rats. They reported that spontaneously hypertensive rats exhibited about 25% higher plasma level of DSIP-LI than normal rats. Continuous infusion of DSIP (200 μg/kg/d) by osmotic minipump resulted in the maintenance of blood pressure of about 200 mmHg, significantly lower than the blood pressure of saline-infused controls (about 220 mmHg). Daily s.c. injections of a single dose of 200 μg/kg DSIP for 5 consecutive days also brought about a similar result.

Sudakov et al.[120] exposed rats to acute emotional stress by keeping them immobilized in boxes for 3 h, and giving them irregular, intermittent, short-term, episodic electric stimulations to the ventromedial hypothalamus and skin. DSIP (60 nmol/kg i.v.) increased the resistance to stress by raising the survival rate from the control level of 29 to 77%. In the stress-resistant rats, blood pressure remained relatively stable. Graf and Kastin[12] reported that exposure to ether for 1 min elicitied a rapid decrease, followed by an increase of plasma DSIP-LI in rats within 1 to 15 min, followed by a second increase over the next 50 to 80 min. They also found a significant decrease in rat plasma DSIP-LI 30 min after immobilization stress.

In morphine-induced insomniac cats, Scherschlicht[9] and Scherschlicht et al.[10] found that pretreatment with DSIP (25 μg/kg i.v. and 100 μg/kg s.c.) reversed considerably the effects of morphine. In morphine-dependent mice, withdrawal jumping frequently occurred by the injection of naloxone. DSIP (25.5 μg/kg i.v. and 85 to 2550 μg/kg s.c.) significantly suppressed the number of jumps.[9] In rabbits, DSIP (25 μ/kg i.v.) augmented SWS and prevented hyposomnia after a stressful situation arisen by confrontation with an agitated dog for 1 min.[9,10] In human opiate and alcohol addicts, the clinical symptoms and signs disappeared or markedly and rapidly improved after daily i.v. injections of DSIP, although anxiety decreased more slowly.[121]

Nakamura et al.[122] reported that DSIP (0.25 to 0.66 nmol/mouse i.c.v.) dose-dependently and time-of-day-independently produced an antinociceptive effect against tail-pinching. They also found a prominent antinociceptive effect against hot-plate test with the ED_{50} values of 1.61 nmol/mouse. Higher doses of DSIP elicited sedation and catalepsy in some mice without correlation to the analgesic activity. Shiomi and Nakamura[123] have recently suggested that DSIP may act on the perigigantocellular reticular nucleus of the medulla.

D. Neuronal Firing Activity

As described before, the neuronal firing activity of snail cerebral ganglia was suppressed by 30 to 40% and the resting potential increased by 1 to 4 mV by the presence of Tyr[5]-DSIP, Tyr[7]-DSIP, Tyr[8]-DSIP, Tyr[9]-DSIP, Ava[3,4]-DSIP, and Glu(NH_2)-DSIP.[65,66] Normanton and Wolstencroft[124] investigated the actions of DSIP on single neurons in the rat brain by iontophoretic application. DSIP excited 13 to 25% of the neurons in the cerebral and cerebellar cortex, hippocampus, and nucleus reticularis gigantocellularis in the brainstem. In the thalamus, DSIP influenced 39% of the neurons with large receptive fields, in which half of the population was excited and the rest inhibited. Normanton and Gent[125] further studied the response of neurons in the nucleus reticularis gigantocellularis to iontophoretically applied DSIP in rats and rabbits. DSIP excited and inhibited 56 and 5%, respectively, of the cells (n = 116) in rats, while 71 and 0% of the cells (n = 38) in rabbits. The effect of DSIP was short-lasting and dose-dependent and showed no significant desensitization to repeated applications.

In this regard, it should be mentioned here that Traczyk and Kosinski[126] observed that i.c.v. administration of DSIP (2.5 or 25 nmol) caused a dose-dependent increase in hippocampal electrical activity corresponding with frequency bands of 4 to 5 and 7 to 12 Hz in unrestrained rabbits, which were kept in arousal by the feedback-modulated electrical stimulation of the midbrain reticular formation.

E. Locomotor Activity

Monnier et al.[3,127,128] noted that DSIP reduced locomotor activity concomitantly with the increase in EEG delta activity in rabbits. Tobler and Borbély[46] noted that a suppressive effect on the nocturnal activity appeared after i.p. injection of DSIP (160 nmol/kg) in rats. Graf et al.[68-70] and Schoenenberger and Graf[71] reversed night-active rats into light-active animals by regular evening injections of DSIP and P-DSIP (see above). Amphetamine-induced locomotor activity in mice might be influenced largely by DSIP, depending on the dosage of the peptide, the time of day, and the level of amphetamine stimulation.[118,129]

V. INTERACTIONS WITH OTHER FACTORS

A. Other Sleep Factors

We have demonstrated the existence of complex interactions between DSIP and other co-existing sleep substances.[32,130-135] The details will be dealt with in Chapter 9.

B. Alcohol, Amphetamine, Morphine, and Naloxone

Burov et al.[136] reported that the DSIP content in different parts of the brain reduced significantly in chronically alcoholic rats. Interactions of DSIP with amphetamine are already mentioned above. Scherschlicht et al.[137] reported the antagonistic effect of DSIP to morphine insomnia in cats (for their further studies,[9,10] see above). Tissot[11] also reported antagonism between DSIP and the opiate antagonist naloxone in rabbits. Young and Key[24] observed that the somnogenic effect of DSIP was blocked by pretreatment with naloxone at a dose level considered selective for the μ-receptor, suggesting an interaction between DSIP and opioid-controlled systems in the brain. Shiomi and Nakamura[123] have recently demonstrated no direct action of DSIP on the opioid receptors.

C. Hypothalamo-Hypophyseal Hormones

Takahashi et al.[35] reported that DSIP induced growth hormone (GH) release in sleep-deprived dogs. However, Graf and Kastin[54] found no change in plasma GH concentration after DSIP administration in rats. They noted that prolactin release was suppressed over the next 20 h after either evening or morning injection of DSIP in rats. The concentration of corticosterone was also reduced. Graf et al.[138] further demonstrated that DSIP inhibited corticosterone release induced by CRF but not by ACTH in rats. Okajima and Hertting[139] analyzed the inhibitory action of DSIP on corticosterone using rat anterior pituitary tissues *in vitro*. They concluded that DSIP inhibits CRF-induced ACTH release at the pituitary level through the inhibition of the cyclic AMP system in corticotrophs.

Sun et al.[140] reported that the hypothalamic immunoreactive thyrotropin-releasing hormone (TRH or TRF) content in rats increased significantly after i.v. injection of DSIP (250 μg/kg), whereas the plasma TRH concentration tended to decrease. Plasma thyrotropin (TSH) levels decreased significantly in a dose-dependent manner after i.v. injection of DSIP at a dosage range from 75 to 400 μg/kg. Plasma TRH and TSH responses to cold as well as plasma TSH response to TRH were inhibited by DSIP administration (250 μg/kg). It is suggested that DSIP acts on both the hypothalamus and the pituitary to inhibit TRH and TSH release.

Sahu and Kalra[141] found that DSIP (2 to 30 μg i.c.v.) promoted luteinizing hormone (LH) release in estrogen-progesterone-primed ovariectomized rats but not in ovariectomized rats. DSIP had no effect on either basal or LH-releasing hormone (LHRH)-induced LH release *in vitro*. Thus, it is likely that DSIP may activate the hypothalamic neural circuitry responsible for the stimulation of LHRH and LH release.

On the other hand, Nakagaki et al.[73,142] concluded that endogenous DSIP in the CSF and/or the periventricular tissues is not responsible for sleep, since rabbit antiserum specific for the C-terminal of DSIP, which was i.c.v.infused over 10 min shortly before the dark period or the light period onset, exerted no significant change in sleep parameters in rats.

D. Antagonism Between DSIP and P-DSIP

As already explained, Ernst and Schoenenberger[77] have recently noted mutually antagonistic effects between DSIP and P-DSIP, eventually proposing a hypothesis that DSIP is involved in arousal, whereas P-DSIP is related to the induction of sleep.

E. Neurotransmitters

There are several papers dealing with the effects of DSIP on the level of neurotransmitters. Ashmarian and Dovedova[143] reported that DSIP increased monoamine oxidase A activity in brain mitochondria to 200 to 280% of the normal value, although its effect on the serotonergic system was unclear. Acetylcholinesterase activity was not affected by DSIP. Yehuda and Mostofsky,[144] however, proposed that the action of DSIP is best understood in the context of mediation via serotonergic mechanisms, since the effect of DSIP on temperature was potentiated by α-methyl-apra-tyrosine and/or L-tryptophan. Graf et al.[56] observed time-of-day-dependent responses of neurotransmitter concentrations to DSIP in rat brains: dopamine and noradrenalin variably increased and/or decreased while serotonin was suppressed. Sun et al.[140] suggested that the inhibitory action of DSIP on TRH and TSH secretion may be modified by amines of the central nervous system, since the pretreatment by pimozide or parachlorophenylalanine, but not 5-hydroxytryptophan or L-DOPA, prevented the effect of DSIP. On the other hand, Graf and Kastin[12] and Graf and Schoenenberger[145] have recently demonstrated a direct interaction of DSIP with adrenergic transmission in rat pineal glands.

F. Lymphokines

Yehuda et al.[146] have recently reported a modulatory effect of DSIP on the lymphokine system. DSIP added in the culture medium caused an initial decrease to be followed by an increase in interleukin-1 levels in rat spleen cells *in vitro*. It is interesting that, as will be described in Chapter 5, lymphokines are closely related to an SWS enhancement.

VI. CLINICAL APPLICATIONS

Schneider-Helmert and associates[39,41,42,147,148] i.v. infused six insomniacs aged 36 to 54 years with DSIP at a dose of 25 nmol/kg over a 4-min period prior to bedtime, and found that DSIP induced longer and more efficient sleep, more REM sleep, and better sleep stability as compared to the placebo treatment. There was no indication of a hangover effect after repeated injections of DSIP.

Subsequent applications of DSIP to therapeutic treatments for insomniacs,[40-42,149] a narcoleptic,[150] opiate addicts and alcoholics,[121] and patients with chronic, pronounced pain episodes[151] reportedly resulted in an improvement of the diseases. These results are dealt with in a special issue of the journal *European Neurology* (see Figure 13) and partly summarized by Schoenenberger and Schneider-Helmert.[152,153]

Recently, Schneider-Helmert[154] reported that DSIP administration 1 h before retiring for 6 consecutive days (each time: 30 nmol/kg i.v.) resulted in an improvement of sleep in 18 chronic psychophysiological insomniacs. In nine middle-aged insomniacs (29 to 59 years),

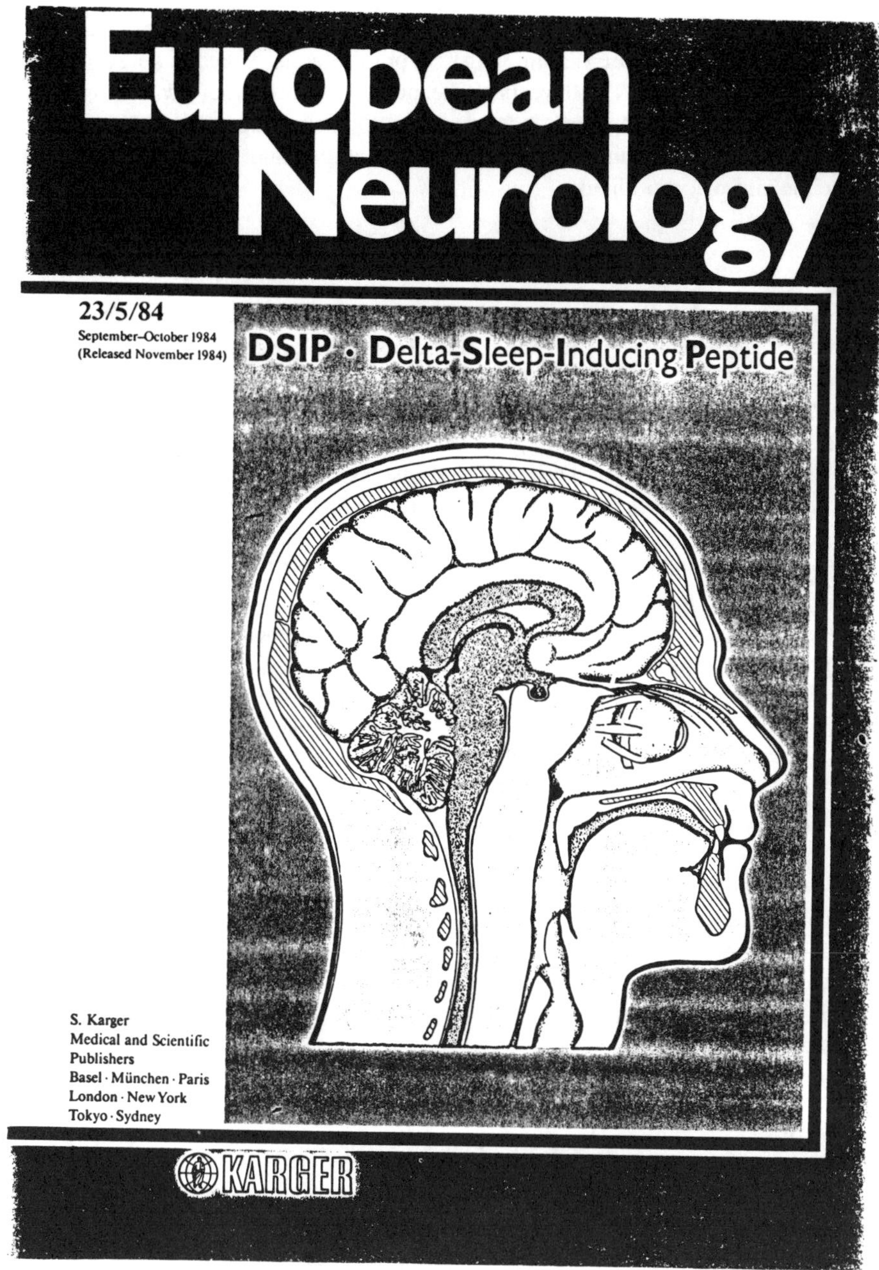

FIGURE 13. The cover page of *European Neurology*, Vol. 23, No. 5, 1984.

improvement of sleep to normal values occurred by the end of the DSIP administration and was maintained during a subsequent 1-week follow-up period. In nine similarly treated elderly insomniacs (60 to 83 years) the immediate effect was larger, but full normalization of sleep was obtained only by the end of the follow-up period. The improvement of sleep included increases in total sleep time and hence sleep efficiency, and decreases in latency to sleep onset, waking time after sleep onset, and number of awakenings. Thus, the whole population showed a complete elimination of insomnia. More recently, Schneider-Helmert and associates[155-159] have reported an improvement in sleep, daytime mood state, circadian rhythmicity of sleep, and cognitive and psychomotor performance in insomniacs.

On the contrary, Monti et al.[160] reported that little sleep-improving effect was obtained by a 4-min i.v. infusion of DSIP (25 nmol/kg) 1 h prior to going to bed in six severe chronic insomniac patients (aged 21 to 60 years). The number of nocturnal awakenings, NREM sleep latency, and total waking time decreased, and total sleep time and NREM sleep time increased during an 8-h recording period. The differences were not significant in comparison to the baseline and placebo nights.

VII. PROGRAMMING EFFECTS

As viewed above, DSIP exhibits a variety of unspecific effects for sleep regulation. Schoenenberger and Monnier[161] first called these properties "programming" effects of DSIP. Miller et al.[115] and Kastin et al.[162] pointed out that DSIP is more than a sleep peptide and that the name DSIP is misleading. Graf et al.[68,69] hypothesized that DSIP, in addition to facilitating sleep, may exert a chronopharmacological action as a natural "programming substance". According to them, DSIP-induced changes of cerebral neurotransmitters and plasma proteins may underlie the programming functions.[56] Tobler and Borbély[46] noted a delayed effect of DSIP on rat locomotor activity. Inoué et al.[26,55] observed circadian rest-activity, rhythm-dependent, sleep-modulatory effects of DSIP in rats. Sommerfelt[53] suggested that DSIP may reduce "redundancy" sleep.

On the basis of such observations, Schoenenberger and associates[2,67,77] have formulated a hypothetical concept that DSIP might be the first peptide representative of a possibly large group of "psycho-physiological programming substances" within a yet-unknown hierarchical, multidimensional network of priming molecular mechanisms. Within the theory, the following principles may be applicable.[2,77]

1. The primary target tissue is always on the same functional level, i.e., the modulation and induction of a specific metabolic molecular process or release phenomena.
2. The dose response curve of those peptides is comparable to that of a pharmacon, i.e., overdoses are harmful.
3. DSIP never seems to produce an effect over "normality". It only normalizes deviated psychobiological phenomena on one hand or counteracts or suppresses idiopathic or artificially/experimentally produced acute or chronically impaired situations on the other.
4. The optimal effect depends not only on the correct dose given, but also on the duration of the injection.
5. The DSIP effects differ with the time of the injection.

Keeping these principles in mind, Schneider-Helmert[159] concluded that the programming effect of DSIP offers an option for the optimal functions of an individual in a complex situation and that DSIP must be involved in a high-level control of sleep and waking.

REFERENCES

1. **Monnier, M. and Schoenenberger, G. A.,** Characterization, sequence, synthesis and specificity of a delta (EEG) sleep-inducing peptide, in *Sleep 1976,* Koella, W. P. and Levin, P., Eds., S. Karger, Basel, 1977, 257.
2. **Schoenenberger, G. A.,** Characterization, properties and multivariate functions of delta-sleep-inducing peptide (DSIP), *Eur. Neurol.,* 23, 321, 1984.
3. **Monnier, M., Dudler, L., Gächter, R., and Schoenenberger, G. A.,** Delta sleep-inducing peptide (DSIP): EEG and motor activity in rabbits following intravenous administration, *Neurosci. Lett.,* 6, 9, 1977.

4. **Ji, A., Li, C., Ye, Y., Xing, Q., Liu, S., Chang, W., Wang, T., and Tai, S.**, Synthesis of the delta sleep inducing peptide by the classical solution method and its physiological activity, in *Nucleic Acids and Proteins*, Shen, Z., Ed., Science Press, Beijing, 1980, 201.
5. **Liu, S. Y., Chang, W., Tai, W., and Wang, Z.**, Synthesis and physiological activity of delta-sleep-inducing-peptide (DSIP) and its analogs. II. Physiological activity, in *Nucleic Acids and Proteins*, Shen, Z., Ed., Science Press, Beijing, 1980, 209.
6. **Liu, S. Y., Zhang, W., Dai, X., and Wang, Z.**, Delta- and sigma-enhancing effects of sleep-inducing peptide synthesized by the solid phase method, *Kexue Tongbo*, 26, 753, 1981.
7. **Liu, S. Y., Zhang, W., Wang, Z., and Tai, S.**, Studies on the delta and sigma-enhancing effect of sleep-inducing peptide synthesized by the solid phase method, *Acta Biochim. Biophys. Sin.*, 13, 145, 1981.
8. **Ji, A., Li, C., Ye, Y., Lin, Y., Xing, Q., Liu, S., Zhang, W., Wang, Z., and Dai, X.**, Synthesis of delta sleep-inducing peptide (DSiP) and its physiological activity, *Sci. Sin.*, 26, 174, 1983.
9. **Scherschlicht, R.**, Pharmacological profile of delta sleep-inducing peptide (DSIP) and a phosphorylated analogue, (Ser-PO_4)DSIP, in *Sleep 1982*, Koella, W. P., Ed., S. Karger, Basel, 1983, 109.
10. **Scherschlicht, R., Aeppli, L., Polc, P., and Haefely, W.**, Some pharmacological effects of delta sleep-inducing peptide (DSIP), *Eur. Neurol.*, 23, 346, 1984.
11. **Tissot, R.**, Récepteurs de l'opium et sommeil. III. Effets de micro-injections de DSIP (delta sleep peptide) dans le thalamus médian, la substance grise centrale péri-aqueducale et le noyau du tractus solitaire du lapin, *Neuropsychobiology*, 7, 321, 1981.
12. **Graf, M. V. and Kastin, A. J.**, Delta-sleep-inducing peptide (DSIP): an update, *Peptides*, 7, 1165, 1986.
13. **Polc, P., Schneeberger, J., and Haefely, W.**, Effect of the delta sleep-inducing peptide (DSIP) on the sleep-wakefulness cycle of cats, *Neurosci. Lett.*, 9, 33, 1978.
14. **Karmanova, I. G., Maksimuk, V. F., Voronov, I. B., Bogoslovskii, M. M., Demin, N. N., Rubinskaya, N. P., and Al'bertin, S. V.**, Analysis of the effect of neuropeptide which induces sleep in cats and albino rats, *Zh. Evol. Biokhim. Fiziol.*, 15, 583, 1979.
15. **Karmanova, I. G., Maksimuk, V. F., Voronov, I. B., Bogoslovskii, M. M., Demin, N. N., Rubinskaya, N. P., and Al'bertin, S. V.**, Analysis of the actions of the neuropeptide-inducing delta-sleep in cats and white rats, *Neurosci. Behav. Physiol.*, 13, 476, 1983.
16. **Susic, V. and Masirevic, G.**, Effects of delta sleep inducing peptide on sleep cycle of cats deprived of paradoxical sleep, *Arch. Int. Physiol. Biochim.*, 93, 271, 1985.
17. **Susic, V.**, The effect of subcutaneous administration of delta sleep-inducing peptide (DSIP) on some parameters of sleep in the cat, *Physiol. Behav.*, 40, 569, 1987.
18. **Susic, V., Masirevic, G., and Totic, S.**, The effects of delta-sleep-inducing peptide (DSIP) on wakefulness and sleep patterns in the cat, *Brain Res.*, 414, 262, 1987.
19. **Kafi, S., Monnier, M., and Gaillard, J. M.**, The delta-sleep inducing peptide (DSIP) increases duration of sleep in rats, *Neurosci. Lett.*, 13, 169, 1979.
20. **Miller, L. H., Kastin, A. J., Hayes, M., Sterste, A., Garcia, J., and Coy, D. H.**, Inverse relationship between onset and duration of EEG effects of six peripherally administered peptides, *Pharamacol. Biochem. Behav.*, 15, 845, 1981.
21. **Miller, L. H., Turnbull, B. A., Kastin, A. J., and Coy, D. H.**, Sleep-wave activity of a delta sleep-inducing peptide analog correlates with its penetrance of the blood-brain barrier, *Sleep*, 9, 80, 1986.
22. **Ursin, R. and Larsen, M.**, Increased sleep following intracerebroventricular injection of the delta sleep-inducing peptide in rats, *Neurosci. Lett.*, 40, 145, 1983.
23. **Ursin, R.**, Endogenous sleep factors, *Exp. Brain Res.*, Suppl. 8, 118, 1984.
24. **Young, A. M. J. and Key, B. J.**, Antagonism of the effect of delta sleep-inducing peptide by naloxone in the rat, *Neuropharmacology*, 23, 1347, 1984.
25. **Inoué, S., Honda, K., Nishida, S., and Komoda, Y.**, Temporal changes in sleep-promoting effects of constantly infused sleep substances in rats, *Sleep Res.*, 12, 81, 1983.
26. **Inoué, S., Honda, K., Komoda, Y., Uchizono, K., Ueno, R., and Hayaishi, O.**, Differential sleep-promoting effects of five sleep substances nocturnally infused in unrestrained rats, *Proc. Natl. Acad. Sci. U.S.A.*, 81, 6240, 1984.
26a. **Kovalzon, V. M.**, personal communication.
27. **Inoué, S.**, Sleep substances: their roles and evolution, in *Endogenous Sleep Substances and Sleep Regulation*, Inoué, S. and Borbély, A. A., Eds., Japan Scientific Societies Press, Tokyo/VNU Science Press BV, Utrecht, 1985, 3.
28. **Kimura, M., Honda, K., Kovalzon, V. M., and Inoué, S.**, Effects of DSIP analogues on sleep in unrestrained rats, *Rep. Med. Dent. Eng.*, 20, 47, 1986.
29. **Kimura, M., Honda, K., and Inoué, S.**, Effects of intracerebroventricularly infused DSIP of different sources on nocturnal sleep in unrestrained rats, *Zool. Sci.*, 3, 1088, 1987.
30. **Kimura, M., Honda, K., and Inoué, S.**, Effects of DSIP analogues on nocturnal sleep in unrestrained rats, *J. Physiol., Soc. Jpn.*, 49, 444, 1987.

31. **Kimura, M., Honda, K., and Inoué, S.,** Sleep-promoting effects of several synthetic DSIP samples, *Annu. Rep. Natl. Inst. Physiol. Sci. Okazaki,* 8, 448, 1987.
32. **Inoué, S., Kimura, M., Honda, K., and Komoda, Y.,** Sleep peptides: general and comparative aspects, in *Sleep Peptides: Basic and Clinical Approaches,* Inoué, S. and Schneider-Helmert, D., Eds., Japan Scientific Societies Press, Tokyo/Springer-Verlag, Berlin, 1988, 1.
33. **Obál, F., Jr., Török, A., Alföldi, P., Sáry, S., Hajós, M., and Penke, B.,** Effects of intracerebroventricular injection of delta sleep-inducing peptide (DSIP) and an analogue on sleep and brain temperature in rats at night, *Pharmacol. Biochem. Behav.,* 23, 953, 1985.
34. **Wachtel, E., Koplik, E., Kolometsewa, I. A., Balzer, H.-U., Hecht, K., Oehme, P., and Ivanov, V. T.,** Vergleichende Untersuchungen zur Wirkung von DSIP und SP_{1-11} auf streβinduzierte chronische Schlafstörungen der Ratte, *Pharmazie,* 42, 188, 1987.
35. **Takahaski, Y., Kato, N., Nakamura, Y., Ebihara, S., Tsuji, A., and Takahashi, K.,** Delta sleep inducing peptide (DSIP): a preliminary report of effects of intraventricular infusion on sleep and GH secretion, and of plasma DSIP concentrations in the dog and rat, in *Integrative Control Functions of the Brain,* Vol. 2, Ito, M., Tsukahara, N., Kubota, K., and Yagi, K., Eds., Kodansha, Tokyo, 1980, 339.
36. **Nagasaki, H., Kitahama, K., Valatx, J.-L., and Jouvet, M.,** Sleep-promoting effect of the sleep-promoting substance (SPS) and delta-sleep-inducing peptide (DSIP) in the mouse, *Brain Res.,* 192, 276, 1980.
37. **Blois, R., Monnier, M., Tissot, R., and Gaillard, J.-M.,** Effect of DSIP on diurnal and nocturnal sleep in man, in *Sleep 1980,* Koella, W. P., Ed., S. Karger, Basel, 1981, 301.
38. **Schneider-Helmert, D., Gnirss, F., Monnier, M., Schenker, J., and Schoenenberger, G. A.,** Acute and delayed effects of DSIP (delta sleep-inducing peptide) on human sleep behavior, *Int. J. Clin. Pharmacol. Therm. Toxicol.,* 19, 341, 1981.
39. **Schneider-Helmert, D., Gnirss, F., and Schoenenberger, G. A.,** Effects of DSIP applications in healthy and insomniac adults, in *Sleep 1980,* Koella, W. P., Ed., S. Karger, Basel, 1981, 417.
40. **Schneider-Helmert, D.,** Influences of DSIP on sleep and waking behavior in man, in *Sleep 1982,* Koella, W. P., Ed., S. Karger, Basel, 1983, 117.
41. **Schneider-Helmert, D. and Schoenenberger, G. A.,** Effects of DSIP in man. Multifunctional psychophysiological properties besides induction of natural sleep, *Neuropsychobiology,* 9, 197, 1983.
42. **Schneider-Helmert, D.,** DSIP in insomnia, *Eur. Neurol.,* 23, 358, 1984.
43. **Medvedjev, V. P. and Bakharev, V. D.,** Studies on the effects of the oligopeptide, sleep neuromodulator, in rats, rabbits, and dogs, *Zh. Evol. Biokhim. Fiziol.,* 15, 379, 1979.
44. **Karmanova, I. G., Maksimuk, V. F., Voronov, I. B., Bogoslovskii, M. M., Demin, N. N., Rubinskaya, N. P., and Albertin, S. V.,** Effects of delta-sleep-inducing peptide in cats, *Zh. Evol. Biokhim. Fiziol.,* 15, 583, 1979.
45. **Krueger, J. M., Bacsik, J., and García-Arrarś, J.,** Sleep-promoting material from human urine and its relation to factor S from brain, *Am. J. Physiol.,* 238, E116, 1980.
46. **Tobler, I. and Borbély, A. A.,** Effect of delta sleep inducing peptide (DSIP) and arginine vasotocin (AVT) on sleep and locomotor activity in the rat, *Waking Sleeping,* 4, 139, 1980.
47. **Mendelson, W. B., Gillin, J. C., and Wyatt, R. J.,** Studies with the delta-sleep-inducing peptide in the rat, *Sleep Res.,* 9, 55, 1980.
48. **Mendelson, W. B., Gillin, J. C., and Wyatt, R. J.,** The search for circulating sleep-promoting factors, in *Advances in Pharmacology and Therapeutics II,* Vol. 1, Yoshida, H., Hagiwara, Y., and Ebashi, S., Eds., Pergamon Press, Oxford, 1982, 227.
49. **Mendelson, W.B., Wyatt, R. J., and Gillin, J. C.,** Whither the sleep factors?, in *Sleep Disorders: Basic and Clinical Research,* Chase, M. H. and Weitzman, E. D., Eds., Spectrum, New York, 1983, 281.
50. **Yehuda, S. and Mostofsky, D. I.,** Possible serotonergic mediation of induction of sleep signs by DSIP and by d-amphetamine, *Int. J. Neurosci.,* 16, 221, 1982.
51. **Obál, F., Jr., Kovalzon, V. M., Kalikhevich, V. N., Török, A., Alföldi, P., Sáry, S., Hajós, M., and Penke, B.,** Structure-activity relationship in the effects of delta-sleep-inducing peptide (DSIP) on rat sleep, *Pharmacol. Biochem. Behav.,* 24, 889, 1986.
52. **Obál, F., Jr.,** Effects of peptides (DSIP, DSIP analogues, VIP, GRF and CCK) on sleep in the rat, *Clin. Neuropharmacol.,* 9(Suppl. 4), 459, 1986.
53. **Sommerfelt, L.,** Reduced sleep in cats after intraperitoneal injection of δ-sleep-inducing peptide, *Neurosci. Lett.,* 58, 73, 1985.
54. **Graf, M. V. and Kastin, A. J.,** Delta-sleep-inducing peptide (DSIP): a review, *Neurosci. Biobehav. Rev.,* 8, 83, 1984.
55. **Inoué, S., Honda, K., Komoda, Y., Uchizono, K., Ueno, R., and Hayaishi, O.,** Little sleep-promoting effect of three sleep substances diurnally infused in unrestrained rats, *Neurosci. Lett.,* 49, 207, 1984.
56. **Graf, M. V., Baumann, J. B., Girard, J., Tobler, H. J., and Schoenenberger, G. A.,** DSIP-induced changes of the daily concentrations of brain neurotransmitters and plasma proteins in rats, *Pharmacol. Biochem. Behav.,* 17, 511, 1982.

57. **Akhrem, A. A., Galaktinov, S. G., Golubovich, V. P., and Kirnarsky, L. I.,** Theoretical conformational analysis of delta-sleep-inducing peptide, *Biofizika,* 27, 324, 1982.
58. **Popov, E. M., Akhmedov, N. A., Kasumov, O. K., Kasumov, J. A., and Godzhaev, N. M.,** Theoretical conformational analysis of brain peptides. I. A sleep stimulator, *Bioorg. Khim.,* 6, 1620, 1980.
59. **Kovalzon, V. M., Kalikhevich, V. N., and Churkina, S. I.,** The active analogue of the inactive "sleep peptide", *Biul. Eksp. Biol. Med.,* 101, 707, 1986.
60. **Kovalzon, V. M., Kalikhevich, V. N., Obál, F., Jr., and Kleinlogel, H.,** Hypnogenic effect of the peptide 1-D-Trp-DSIP in rabbits and rats, in *Sleep '84,* Koella, W. P., Rüther, E., and Schulz, H., Eds., Gustav Fischer Verlag, Stuttgart, 1985, 218.
61. **Kastin, A. J., Banks, W. A., Castellanos, P. F., and Coy, D. H.,** Differential penetration of DSIP peptides into rat brain, *Pharmacol. Biochem. Behav.,* 17, 1187, 1982.
62. **Miller, L. H., Turnbull, B. A., Kastin, A. J., and Coy, D. H.,** Sleep-wave activity of a delta sleep-inducing peptide analog correlates with its penetrance of the blood-brain barrier, *Sleep,* 9, 80, 1986.
63. **Schoenenberger, G. A., Maier, P. F., Tobler, H. J., Wilson, K., and Monnier, M.,** The delta EEG (sleep)-inducing peptide (DSIP). XI. Amino-acid analysis, sequence, synthesis and activity of the nona-peptide, *Pfluegers Arch.,* 376, 119, 1978.
64. **Dobrynin, V. N., Korobko, V. G., Severtsova, I. V., Vlasov, V. P., Bystrov, N. S., and Kolosv, M. N.,** Synthesis and cloning of an artificial structural gene for delta sleep inducing peptide, *Bioorg. Khim.,* 7, 1745, 1981.
65. **Sargsyan, A. S., Sumskaya, L. V., Aleksandova, I. Y., Besrukov, M. V., Mikhaleva, I. I., Ivanov, V. T., and Balaban, P. M.,** Synthesis and some biological properties of δ-sleep inducing peptide and its analogs, *Bioorg. Khim.,* 7, 1125, 1981.
66. **Mikhaleva, I., Sargsyan, A., Balashova, T., and Ivanov, V.,** Synthesis, spectral and biological properties of DSIP and its analogs, in *Chemistry of Peptides and Proteins,* Vol. 1, Voelter, W., Wünsch, E., Ovchinnikov, J., and Ivanov, V., Eds., Walter de Gruyter, Berlin, 1982, 289.
67. **Schoenenberger, G. A., Monnier, M., Graf, M., Schneider-Helmert, D., and Tobler, H. J.,** Biochemical aspects of sleep regulations and the involvement of delta-sleep-inducing peptide (=DSIP), in *Sleep: Normal and Deranged Function,* Kamphuisen, H. A. C., Bruyn, G. W., and Visser, P., Eds., Mefar BV, Leiden, 1981, 25.
68. **Graf, M. V., Christen, H., Tobler, H. J., Baumann, J. B., and Schoenenberger, G. A.,** DSIP a circadian 'programming' substance?, *Experientia,* 37, 624, 1981.
69. **Graf, M. V., Christen, H., Tobler, H. J., Maier, P. F., and Schoenenberger, G. A.,** Effects of repeated DSIP and DSIP-P administration on the circadian locomotor activity of rats, *Pharmacol. Biochem. Behav.,* 15, 717, 1981.
70. **Graf, M. V., Christen, H., and Schoenenberger, G. A.,** DSIP/DSIP-P and circadian motor activity of rats under continuous light, *Peptides,* 3, 623, 1982.
71. **Schoenenberger, G. A. and Graf, M.,** Effects of DSIP and DSIP-P on different biorhythmic parameters, in *Sleep: Neurotransmitters and Neuromodulators,* Wauquier, A., Gaillard, J. M., Monti, J. M., and Radulovacki, M., Eds., Raven Press, New York, 1985, 265.
72. **Kimura, M. and Inoué, S.,** The phosphorylated analogue of DSIP enhances slow wave sleep and paradoxical sleep in unrestrained rats, *Psychopharmacology,* in press.
73. **Nakagaki, K., Ebihara, S., Usui, S., Honda, Y., Takahashi, Y., and Kato, N.,** Effects of intraventricular injection of anti-DSIP and DSIP-P on sleep in rats, *Sleep Res.,* 16, 65, 1987.
74. **Nakagaki, K., Ebihara, S., Usui, S., Honda, Y., and Takhashi, Y.,** Sleep-promoting effect following intracerebroventricular injection of DSIP in rats, *Neurosci. Lett.,* 91, 160, 1988.
75. **Nakagaki, K. and Takahashi, Y.,** Sleep-inducing activity of DSIP-P and VIP, *Annu. Rep. Natl. Inst. Physiol. Sci. Okazaki,* 8, in press.
76. **Nakagaki, K. and Takahashi, Y.,** Sleep-promoting effect following intracerebroventricular injection of DSIP-P in rats, *Neurosci. Lett.,* in press.
77. **Ernst, A. and Schoenenberger, G. A.,** DSIP: basic findings in human beings, in *Sleep Peptides: Basic and Clinical Approaches,* Inoué, S. and Schneider-Helmert, D., Eds., Japan Scientific Societies Press, Tokyo/Springer-Verlag, Heidelberg, in press.
78. **Marks, N., Stern, F., Kastin, A. J., and Coy, D. H.,** Degradation of delta-sleep inducing peptide (DSIP) and its analogs by brain extracts, *Brain Res. Bull.,* 2, 491, 1977.
79. **Huang, J.-T. and Lajtha, A.,** The degradation of a nonapeptide, sleep inducing peptide, in rat brain: comparison with enkephalin breakdown, *Res. Commun. Chem. Pathol. Pharmacol.,* 19, 191, 1978.
80. **Kato, N., Honda, Y., Ebihara, S., Naruse, H., and Takahashi, Y.,** Development of an enzyme immunoassay for delta-sleep-inducing peptide (DSIP) and its use in the determination of the metabolic clearance rate of DSIP administered to dogs, *Neuroendocrinology,* 39, 39, 1984.
81. **Graf, M. V., Saegesser, B., and Schoenenberger, G. A.,** Separation of tryptophan from DSIP, a Trp-nonapeptide, by adsorption to aluminum oxide, *Peptides,* 157, 295, 1986.

82. **Graf, M. V., Kastin, A. J., and Fischman, A. J.,** The natural occurrence and some extra-sleep effects of DSIP, in *Endogenous Sleep Substances and Sleep Regulation,* Inoué, S. and Borbély, A. A., Eds., Japan Scientific Societies Press, Tokyo/VNU Science Press BV, Utrecht, 1985, 155.
83. **Hösli, A., Schoenenberger, G. A., and Hösli, L.,** Autoradiographic localization of binding sites for the delta sleep-inducing peptide ([^{3}H]-DSIP) on neurons of cultured rat brain, *Brain Res.,* 279, 374, 1983.
84. **Hösli, A., Schoenenberger, G. A., and Hösli, L.,** Cellular localization of binding sites for ^{3}H-DSIP on neurons of cultured rat brain stem, *Eur. Neurol.,* 23, 317, 1984.
85. **Van Dijk, A. M. A. and Schoenenberger, G. A.,** Biochemical evidence for DSIP specific binding sites in the rat brain, in *Endogenous Sleep Substances and Sleep Regulation,* Inoué, S. and Borbély, A. A., Eds., Japan Scientific Societies Press, Tokyo/VNU Science Press BV, Utrecht, 1985, 167.
86. **Kastin, A. J., Nissen, C., Schally, A. V., and Coy, D. H.,** Radioimmunoassay of DSIP-like material in rat brain, *Brain Res. Bull.,* 3, 691, 1978.
87. **Rozhanets, V. V., Yukhananov, R. Y., Chizevskaya, M. A., and Navolotskaya, E. V.,** Regional distribution of delta-sleep-inducing peptide (DSIP)-like immunoreactivity in rat brain and other organs, *Neurokhimiia,* 2, 353, 1983.
88. **Kato, N., Nagaki, S., Takahashi, Y., Namuma, I., and Saito, Y.,** DSIP-like material in rat brain, human cerebrospinal fluid, and plasma as determined by enzyme immunoassay, in *Endogenous Sleep Substances and Sleep Regulation,* Inoué, S. and Borbély, A. A., Eds., Japan Scientific Societies Press, Tokyo/VNU Science Press BV, Utrecht, 1985, 141.
89. **Nagaki, S. and Kato, N.,** Delta sleep-inducing peptide (DSIP)-like material in rat brain as determined by enzyme immunoassay: effect of sleep deprivation, *Neurosci. Lett.,* 51, 253, 1984.
90. **Kastin, A. J., Nissen, C., and Coy, D. H.,** DSIP-like immunoreactivity in the developing rat brain, *Brain Res. Bull.,* 7, 678, 1981.
91. **Kastin, A. J., Dickson, J. C., and Fischman, A. J.,** DSIP-like material in brain parts of rats running a complex maze, *Physiol. Behav.,* 33, 427, 1984.
92. **Kramarova, L. I., Kolaeva, S. H., Yukkhananov, R. Y., and Rozhanets, V. V.,** Content of DSIP, enkephalins and ACTH in some tissues of active and hibernating ground squirrels *(Citellus suslicus), Comp. Biochem. Physiol.,* 74C, 31, 1983.
93. **Constantidinis, J., Bouras, C., Guntern, R., Taban, C. H., and Tissot, R.,** Delta sleep-inducing peptide in the rat brain: an immunohistological microscopic study, *Neuropsychobiology,* 10, 94, 1983.
94. **Feldman, S. C. and Kastin, A. J.,** Localization of neurons containing delta sleep-inducing peptide in the rat brain: an immunocytochemical study, *Neuroscience,* 11, 303, 1984.
95. **Feldman, S. C. and Kastin, A. J.,** The distribution of neurons containing delta sleep-inducing peptide in the hippocampal formation, *Brain Res. Bull.,* 13, 833, 1984.
96. **Ekman, R., Bjartell, A., Ekblad, E., and Sundler, F.,** Immunoreactive delta sleep-inducing peptide in pituitary adrenocorticotropin/alpha-melanotropin cells and adrenal medullary cells of the pig, *Neuroendocrinology,* 45, 298, 1987.
97. **Graf, M. V. and Kastin, A. J.,** Delta sleep-inducing peptide (DSIP)-like material exists in peripheral organs of rats in large dissociated forms, *Proc. Soc. Exp. Biol. Med.,* 177, 197, 1984.
98. **Graf, M. V., Hunter, C. A., and Kastin, A. J.,** Presence of delta-sleep-inducing peptide-like material in human milk, *J. Clin. Endocrinol. Metab.,* 59, 127, 1984.
99. **Kastin, A. J., Castellanos, P. F., Banks, W. A., and Coy, D. H.,** Radioimmunoassay of DSIP-like material in human blood: possible protein binding, *Pharmacol. Biochem. Behav.,* 15, 969, 1981.
100. **Banks, W. A., Kastin, A. J., and Coy, D. H.,** Delta sleep-inducing peptide crosses the blood-brain-barrier in dogs: some correlations with protein binding, *Pharmacol. Biochem. Behav.,* 17, 1009, 1982.
101. **Takahashi, Y., Kato, N., Nakamura, Y., Ebihara, S., Tsuji, A., and Takahashi, K.,** Plasma concentrations of delta sleep inducing peptide (DSIP) in dogs and rats, *Sleep Res.,* 10, 71, 1981.
102. **Graf, M. V., Kastin, A. J., and Fischman, A. J.,** DSIP occurs in free form in mammalian plasma, human CSF and urine, *Pharmacol. Biochem. Behav.,* 21, 761, 1984.
103. **Ernst, A., Monti, J. C., and Schoenenberger, G. A.,** Delta-sleep-inducing-peptide (DSIP) in human and cow milk, in *Sleep '84,* Koella, W. P., Rüther, E., and Schulz, H., Eds., Gustav Fischer Verlag, Stuttgart, 1985, 215.
104. **Banks, W. A., Kastin, A. J., and Coy, D. H.,** Delta sleep-inducing peptide (DSIP)-like material is absorbed by the gastrointestinal tract of the neonatal rat, *Life Sci.,* 33, 1587, 1983.
105. **Fischman, A. J., Kastin, A., and Graf, M.,** Circadian variation of DSIP-like material in rat plasma, *Life Sci.,* 35, 2079, 1984.
106. **Banks, W. A., Kastin, A. J., and Selznik, J. K.,** Modulation of immunoactive levels of DSIP and blood-brain barrier permeability by lighting and diurnal rhythm, *J. Neurosci. Res.,* 14, 347, 1985.
107. **Ernst, A., Schulz, P., and Schoenenberger, G. A.,** Circadian variation of DSIP-like immunoreactivity in human plasma, *Sleep Res.,* 16, 609, 1987.
108. **Someya, T., Nagaki, S., Masui, A., Fujita, M., Kawada, E., and Kato, N.,** On the circadian rhythm of DSIP in body fluid, *Annu. Rep. Natl. Inst. Physiol. Okazaki,* 8, 447, 1987.

109. **Monnier, M., Dudler, L., Gächter, R., and Schoenenberger, G. A.,** Delta sleep-inducing peptide (DSIP): EEG and motor activity in rabbits following intravenous administration, *Neurosci. Lett.,* 6, 9, 1977.
110. **Kastin, A. J., Nissen, C., and Coy, D. H.,** Permeability of blood-brain barrier to DSIP peptides, *Pharmacol. Biochem. Behav.,* 15, 955, 1981.
111. **Kastin, A. J., Nissen, C., Schally, A. V., and Coy, D. H.,** Additional evidence that small amounts of a peptide can cross the blood-brain-barrier, *Pharmacol. Biochem. Behav.,* 11, 717, 1979.
112. **Olson, R. D., Kastin, A. J., Montalbano-Smith, D., Olson, G. A., Coy, D. H., and Michell, G. F.,** Neuropeptides and the blood-brain barrier in goldfish, *Pharmacol. Biochem. Behav.,* 9, 521, 1978.
113. **Inoué, S., Honda, K., and Komoda, Y.,** Effect of sleep substances on circadian swimming activity in fish, *Int. J. Biometeorol.,* 29(Suppl. 1), 118, 1985.
114. **Inoué, S., Honda, K., and Komoda, Y.,** Behavior-modulating effect of sleep substances in fish and insects, *Sleep Res.,* 14, 84, 1985.
115. **Miller, L. H., Turnbull, B. A., Kastin, A. J., and Coy, D. H.,** Sleep-wave activity of a delta sleep-inducing peptide analog correlates with its penetrance of the blood-brain barrier, *Sleep,* 9, 80, 1986.
116. **Kastin, A. J., Olson, G. A., Schally, A. V., and Coy, D. H.,** DSIP — more than a sleep peptide?, *Trends Neurosci.,* 3, 163, 1980.
117. **Yehuda, S., Kastin, A. J., and Coy, D. H.,** Thermoregulatory and locomotor effects of DSIP: paradoxical interaction with d-amphetamine, *Pharmacol. Biochem. Behav.,* 13, 895, 1980.
118. **Graf, M. V., Kastin, A. J., Coy, D. H., and Zadina, J. E.,** DSIP reduces amphetamine-induced hyperthermia in mice, *Physiol. Behav.,* 33, 291, 1984.
119. **Graf, M. V., Kastin, A. J., and Schoenenberger, G. A.,** Delta sleep-inducing peptide in spontaneously hypertensive rats, *Pharmacol. Biochem. Behav.,* 24, 1797, 1986.
120. **Sudakov, K. V., Ivanov, V. T., Koplik, E. V., Vedjaev, D. F., Mikhaleva, I. I., and Sargsjan, A. S.,** Delta-sleep-inducing peptide (DSIP) as a factor facilitating animals' resistance to acute emotional stress, *Pavlovian J. Biol. Sci.,* 18, 1, 1983.
121. **Dick, P., Costa, C., Fayolle, K., Grandjean, M. E., Khoshbeen, A., and Tissot, R.,** DSIP in the treatment of withdrawal syndromes from alcohol and opiates, *Eur. Neurol.,* 23, 364, 1984.
122. **Nakamura, A., Nakashima, M., Kanemoto, H., Sugao, T., Fukumura, Y., and Shiomi, H.,** Delta sleep-inducing peptide (DSIP) has potent analgesic activity in mice, *Eur. J. Pharmacol.,* 121, 157, 1986.
123. **Shiomi, H. and Nakamura, A.,** Analgesic activity of DSIP, *Annu. Rep. Natl. Inst. Physiol. Sci. Okazaki,* 8, 448, 1987.
124. **Normanton, J. R. and Wolstencroft, J. H.,** Actions of DSIP on thalamic and other central neurons and their possible relation to sleep mechanisms, *Regul. Peptides,* 4, 376, 1982.
125. **Normanton, J. R. and Gent, J. P.,** Comparison of the effect of two 'sleep' peptides, delta sleep-inducing peptide and arginine-vasotocin, on single neurons in the rat and rabbit brain stem, *Neuroscience,* 8, 107, 1983.
126. **Traczyk, W. Z. and Kosinski, S.,** Effect of intracerebroventricular administration of delta-sleep-inducing peptide on hippocampal electrical activity in the rabbit, in *Sleep '84,* Koella, W. P., Rüther, E., and Schulz, H., Eds., Gustav Fischer Verlag, Stuttgart, 1985, 220.
127. **Monnier, M., Hatt, A. M., Cueni, L. B., and Schoenenberger, G. A.,** Humoral transmission of sleep. VI. Purification and assessment of a hypnogenic fraction of "sleep dialysate" (factor delta), *Pfluegers Arch.,* 331, 257, 1972.
128. **Monnier, M., Dudler, L., and Schoenenberger, G. A.,** Humoral transmission of sleep. VIII. Effects of the "sleep factor delta" on cerebral, motor and visceral activities, *Pfluegers Arch.,* 345, 23, 1973.
129. **Graf, M. V., Zadina, J. E., and Schoenenberger, G. A.,** Amphetamine-induced locomotor behavior of mice is influenced by DSIP, *Peptides,* 3, 729, 1982.
130. **Inoué, S.,** Multifactorial humoral regulation of sleep, *Clin. Neuropharmacol.,* 9(Suppl. 4), 470, 1986.
131. **Inoué, S.,** Sleep substances, *Clin. Neurosci.,* 5, 22, 1987.
132. **Inoué, S.,** Interference among sleep substances simultaneously infused in unrestrained rats, in *Sleep '86,* Koella, W. P., Obál, F., Schulz, H., and Visser, P., Eds., Gustav Fischer Verlag, Stuttgart, 1988, 175.
133. **Kimura, M., Honda, K., Okano, Y., Komoda, Y., and Inoué, S.,** Effects of simultaneously infused sleep substances on nocturnal sleep in rats, *Zool. Sci.,* 2, 989, 1985.
134. **Kimura, M., Honda, K., Okano, Y., Komoda, Y., and Inoué, S.,** Interacting sleep-modulatory effects of simultaneously administered sleep substances in rats, *J. Physiol. Soc. Jpn.,* 48, 280, 1986.
135. **Kimura, M., Honda, K., Komoda, Y., and Inoué, S.,** Interacting sleep-modulatory effects of simultaneously administered delta-sleep-inducing peptide (DSIP), muramyl dipeptide (MDP) and uridine in unrestrained rats, *Neurosci. Res.,* 5, 157, 1987.
136. **Burov, Y. V., Yukhananov, R. Y., and Maiskii, A. I.,** Content of delta-sleep-inducing peptide in the brain of rats with different levels of alcohol motivation, *Bull. Exp. Biol. Med. (USSR),* 94, 1240, 1982.
137. **Scherschlicht, R., Schneeberger, J., Steiner, M., and Haefely, W.,** Delta sleep inducing peptide (DSIP) antagonizes morphine insomnia in cats, *Sleep Res.,* 8, 84, 1979.

138. **Graf, M. V., Kastin, A. J., Coy, D. H., and Fischman, A. J.,** Delta sleep inducing peptide reduces CRF-induced corticosterone release, *Neuroendocrinology,* 41, 353, 1985.
139. **Okajima, T. and Hertting, G.,** Delta-sleep-inducing peptide (DSIP) inhibited CRF-induced ACTH secretion from rat anterior pituitary gland in vitro, *Horm. Metabol. Res.,* 18, 497, 1986.
140. **Sun, D. H., Mitsuma, T., Nogimori, T., and Chaya, M.,** Peripheral administration of delta-sleep-inducing peptide inhibits thyrotropin secretion in rats, *J. Aichi Med. Univ. Assoc.,* 14, 36, 1986.
141. **Sahu, A. and Kalra, S. P.,** Delta sleep-inducing peptide (DSIP) stimulates LH release in steroid-primed ovariectomized rats, *Life Sci.,* 40, 1201, 1987.
142. **Nakagaki, K., Ebihara, S., Usui, S., Honda, Y., Takahashi, Y., and Kato, N.,** Effects of intraventricular injection of anti-DSIP serum on sleep in rats, *Jpn. J. Psychopharmacol.,* 6, 259, 1986.
143. **Ashmarian, I. P. and Dovedova, E. L.,** Influence of DSIP on the acetylcholinesterase and monoamineoxydase activity in synaptosoma and mitochondria of rabbit brain in vitro, *Dokl. Akad. Nauk USSR,* 255, 1501, 1980.
144. **Yehuda, S. and Mostofsky, D. I.,** Possible serotonergic mediation of induction of sleep signs by DSIP and by d-amphetamine, *Int. J. Neurosci.,* 16, 221, 1982.
145. **Graf, M. V. and Schoenenberger, G. A.,** DSIP affects adrenergic stimulation of rat pineal N-acetyltransferase in vivo and in vitro, *Peptides,* 7, 1001, 1986.
146. **Yehuda, S., Shredny, B., and Kalechamn, Y.,** Effects of DSIP, 5-HTP and serotonin on the lymphokine system: a preliminary study, *Int. J. Neurosci.,* 33, 185, 1987.
147. **Schneider-Helmert, D., Graf, M., and Schoenenberger, G. A.,** Synthetic delta-sleep-inducing peptide improves sleep in insomniacs, *Lancet,* 1, 1256, 1981.
148. **Schneider-Helmert, D.,** Clinical evaluation of DSIP, in *Sleep: Neurotransmitters and Neuromodulators,* Wauquier, A., Gaillard, J. M., Monti, J. M., and Radulovacki, M., Eds., Raven Press, New York, 1985, 279.
149. **Kaeser, H. E.,** A clinical trial with DSIP, *Eur. Neurol.,* 23, 386, 1984.
150. **Schneider-Helmert, D.,** Effects of DSIP on narcolepsy, *Eur. Neurol.,* 23, 353, 1984.
151. **Larbig, W., Gerber, W. D., Kluck, M., and Schoenenberger, G. A.,** Therapeutic effects of delta-sleep-inducing peptide (DSIP) in patients with chronic, pronounced pain episodes. A clinical pilot study, *Eur. Neurol.,* 23, 372, 1984.
152. **Schneider-Helmert, D.,** DSIP in sleep disturbances, *Eur. Neurol.,* 25(Suppl. 2), 154, 1986.
153. **Schoenenberger, G. A. and Schnieder-Helmert, D.,** Psychophysiological functions of DSIP, *Trends Pharmacol. Sci.,* 4, 307, 1983.
154. **Schneider-Helmert, D.,** Efficacy of DSIP to normalize sleep in middle-aged and elderly chronic insomniacs, *Eur. Neurol.,* 5, 448, 1986.
155. **Hermann, E., Ernst, A., and Schneider-Helmert, D.,** Effects of DSIP onto daytime mood states of insomniacs, *Sleep Res.,* 16, 275, 1987.
156. **Hofman, W. F. and Schneider-Helmert, D.,** The influence of DSIP on the rhythm of spontaneous sleep tendency in insomniacs and normal subjects, *Sleep Res.,* 16, 615, 1987.
157. **Schneider-Helmert, D., Hermann, E., and Schoenenberger, G. A.,** Efficacy of DSIP for withdrawal treatment of low-dose benzodiazepine dependent insomniacs, *Sleep Res.,* 16, 133, 1987.
158. **Schneider-Helmert, D., Schoenenberger, G. A., and Hermann, E.,** Advancing delayed sleep phase by treatment with DSIP, *Sleep Res.,* 16, 222, 1987.
159. **Schneider-Helmert, D.,** DSIP: clinical application of the programming effect, in *Sleep Peptides: Basic and Clinical Approaches,* Inoué, S. and Schneider-Helmert, D., Eds., Japan Scientific Societies Press, Tokyo/Springer-Verlag, Berlin, 1988, 175.
160. **Monti, J. M., Debellis, J., Alterwain, P., Pellejero, T., and Monti, D.,** Study of delta sleep-inducing peptide efficacy in improving sleep on short-term administration of chronic insomniacs, *Int. J. Clin. Pharm. Res.,* 7, 105, 1987.
161. **Schoenenberger, G. A. and Monnier, M.,** Studies on the delta (sleep)-inducing peptide, in *IUPAC Medicinal Chemistry Proceedings,* Cotswold Press, Oxford, 1979, 1010.
162. **Kastin, A. J., Banks, W. A., Zadina, J. E., and Graf, M. V.,** Brain peptides: the dangers of constricted nomenclatures, *Life Sci.,* 32, 295, 1983.

Chapter 4

SLEEP-PROMOTING SUBSTANCE (SPS) AND NUCLEOSIDES

I. FURTHER PURIFICATION OF SPS

As described in Chapters 1 and 2, sleep-promoting substance (SPS) extracted from the water dialyzates of brainstems of 24-h sleep-deprived rats contains a plural number of active components. In addition to uridine, which was first identified as an active component from the SPS complex in 1983 by Komoda et al.,[1,2] there are at least four other partially purified fractions named SPS-A2, SPS-B, and SPS-X[3]. The unidentified fractions have still been under purification procedures and no information is available on their chemical structure. However, our recent bioassays have revealed that SPS-B may contain more than two active subcomponents.

The less purified SPS-B clearly exhibited a bell-shaped dose response curve for slow-wave sleep (SWS), but not for paradoxical sleep (PS) and locomotor activity in our 10-h nocturnal intracerebroventricular (i.c.v.) infusion assay in freely behaving rats[3,4] (see Figure 1). However, PS was not enhanced by the SPS-B samples in a dosage range from 1 to 5 brainstem equivalent units. Locomotor activitiy was dose-dependently suppressed within the dosage range. As to the minimal dose, it was found that a still earlier preparation of crude SPS was effective at a dose as low as 0.1 unit in suppressing nocturnal locomotor activity in rats.[5]

Further purification demonstrated that SPS-B (2 units i.c.v.) could induce a significant increase in both SWS and PS.[3,4,6-10] Interestingly, its effect on PS was considerably long-lasting. As shown in Figure 15 of Chapter 2, changes in PS included a small increase during the first night of SPS-B infusion, a large decrease in the following daytime, and a rebound rise in the second night. Such fluctuations were completely cleared on the third day.[5,6] A similar but less prominent PS-modulatory effect was still observable in further purified samples of SPS-B.[4,9]

Recently, more purified SPS-B (2 units), which was i.c.v. infused for a 10-h nocturnal period in freely behaving rats, elicited a profound sleep-enhancing effect (see Figure 2). The effect appeared shortly after the initiation of the SPS-B infusion and lasted throughout the dark period. The increase in the amount of sleep was caused by the more frequent occurrence of SWS and PS episodes, whereas their duration was less affected. Since natural sleep in rats is episodic and frequently interrupted by wakefulness, especially during the dark period, a prolongation of a single SWS and/or PS episode might not be physiological. In this respect, SPS-B did not alter the natural sleep pattern. Sleep-waking behavior appeared to be quite normal as monitored by a television system. Thus, the excess sleep induced by SPS-B administration could not be distinguished from normal physiological sleep.

II. URIDINE AS AN ACTIVE COMPONENT OF SPS

A. Natural Occurrence and Pharmacology

Uridine is a natural pyrimidine nucleoside widely distributed in the organism as a constituent of RNA. It is characterized by having a barbiturate ring in its structure (see Figure 3). Uridine is distributed not only in bodily tissues but also in body fluids. In our isolation studies of SPS, the amount of uridine in the brainstem was calculated by the values detected in SPS fractions of the brainstems of sleep-deprived rats. It was found that the content of uridine varied from 4 to 10 nmol/brainstem.[1]

Karle et al.[11] measured the serum and plasma levels of uridine by means of reverse-phase,

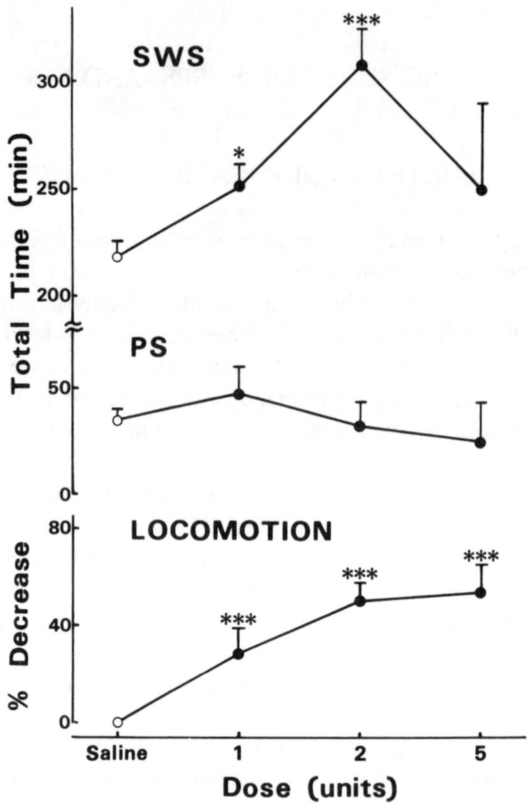

FIGURE 1. Dose response relations in the effects of the less purified, i.c.v.-infused SPS-B on slow-wave sleep (SWS), paradoxical sleep (PS), and locomotor activity in rats. Values (mean ± standard error of mean [SEM]) in the experimental dark period were compared with those in the baseline night. One unit corresponds with one brainstem of 24-h sleep-deprived donor rats. * $p < 0.05$, *** $p < 0.001$. (From Inoué, S., Honda, K., and Komoda, Y., in *Sleep: Neurotransmitters and Neuromodulators*, Wauquier, A., Gaillard, J. M., Monti, J. M., and Radulovacki, M., Eds., Raven Press, New York, 1985, 305. With permission.)

high-pressure liquid chromatography in three different mammalian species. In humans, both plasma and serum levels were in the range 1.9 to 8.4 nmol/ml. In rats, plasma and serum levels ranged from 1.7 to 4.1 nmol/ml and from 3.7 to 9.4 nmol/ml, respectively. In mice, plasma and serum levels measured 1.5 to 4.7 nmol/ml and 8.0 to 11.8 nmol/ml, respectively. Food deprivation for 16 to 24 h had no observable effect on serum or plasma levels of uridine in all these species. Hence it seems likely that uridine levels are regulated and are not a direct reflection of dietary intake of uridine.[11] Dudman et al.[12] developed radioimmunoassay techniques for several nucleosides. They reported that plasma levels of uridine were 21.1 nmol/ml in healthy humans.

Apart from somnogenic activity, which will be dealt with later, exogenously administered uridine is known to exert some pharmacological actions, such as anticonvulsant activity in mice after intraperitoneal (i.p.) injection (5 to 100 mg/kg),[13] depression of spontaneous activity in mice after i.p. or subcutaneous (s.c.) injection (5 to 15 nmol/kg)[14] and after i.c.v. injection (3.8 μmol/mouse),[15] enhancement of the cytotoxic effect of D-glucosamine in rat

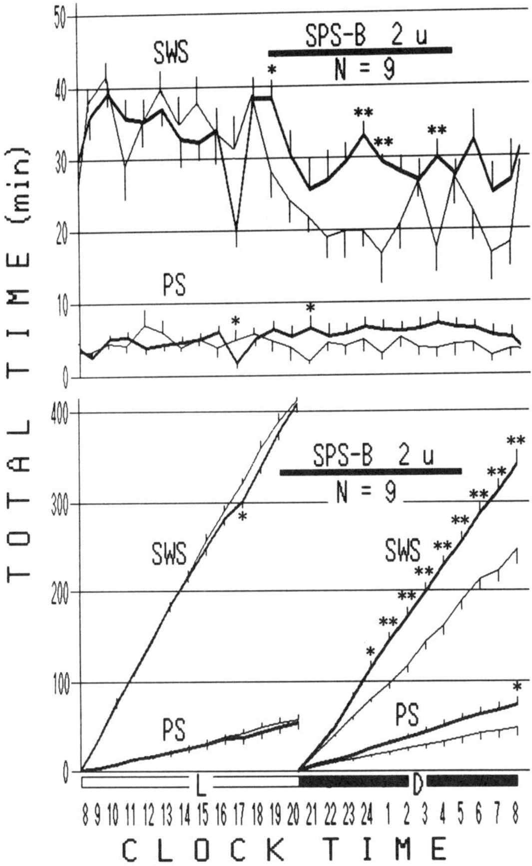

FIGURE 2. Effects of 2 units (equivalent to 2 brainstems of 24-h sleep-deprived donor rats) of SPS-B i.c.v. infused for 10 h at 19.00 to 0.500 h (indicated by a solid bar) on the time-course changes in the amount of sleep in otherwise saline-infused, freely behaving rats. The graph at the top shows the hourly amount of SWS and PS, while the one at the bottom illustrates their cumulative values in the light (L) and dark (D) periods. Thin and thick curves stand for the baseline and the experimental day, respectively. Vertical lines on each hourly value indicate SEM. $*p < 0.05$, $**p < 0.01$. (From Inoué, S., Honda, K., Komoda, Y., Uchizono, K., Ueno, R., and Hayaishi, O., *Proc. Natl. Acad. Sci. U.S.A.*, 81, 6240, 1984. With permission.)

C6 glioma cells *in vitro* (2 m*M*),[16] enhancement of pentobarbital-induced narcosis in mice (0.9 mg/mouse i.c.v),[15] and hyperthermia in humans during continuous intravenous (i.v.) infusion (1 to 2.5 g/m^2/h)[17] and in rabbits after i.v. injection (180 to 700 mg/kg).[17,18]

B. Nocturnal i.c.v. Infusion

Uridine at a dose of 10 pmol induced a rather mild but steady increase in both SWS and PS in freely moving rats, if nocturnally infused into the third cerebral ventricle for a 10-h nocturnal period[3,7,8-10,19-23] (see Figure 4). The effect of 10 pmol uridine was characterized by a long-lasting action (see Figure 14, Chapter 2). During the first recovery night, almost similar increases in SWS and PS were recorded, although the total time of neither SWS nor PS was statistically different from the baseline. Total amounts of sleep time and the other

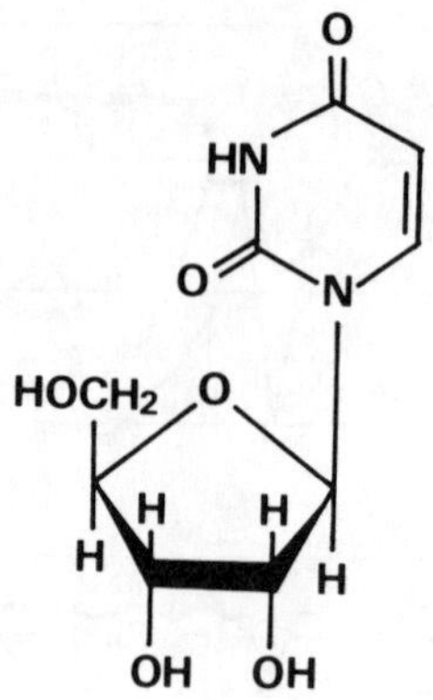

FIGURE 3. Chemical structure of uridine.

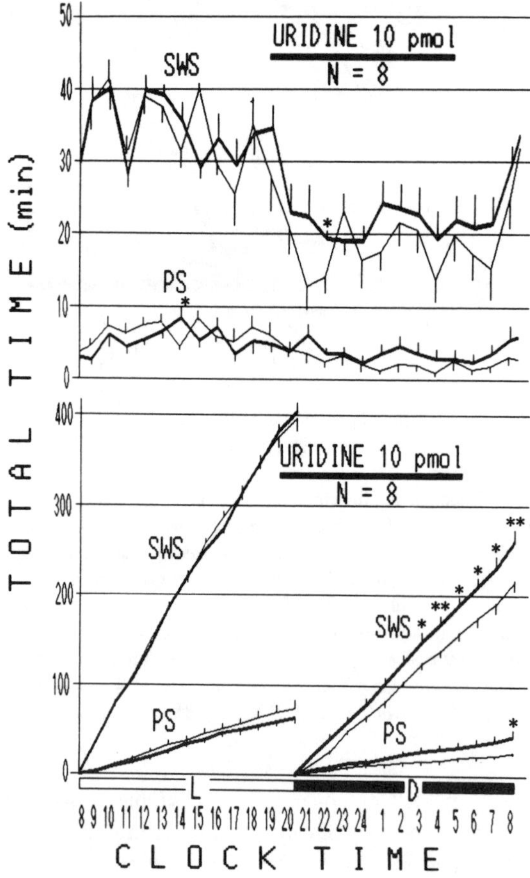

FIGURE 4. Effects of 10 pmol of uridine i.c.v. infused for 10 h at 19.00 to 05.00 h (indicated by a solid bar) on the time-course changes in the amount of sleep in otherwise saline-infused, freely behaving rats. See Figure 2 for other explanations. (From Inoué, S., Honda, K., Komoda, Y., Uchizono, K., Ueno, R., and Hayaishi, O., *Proc. Natl. Acad. Sci. U.S.A.*, 81, 6240, 1984. With permission.)

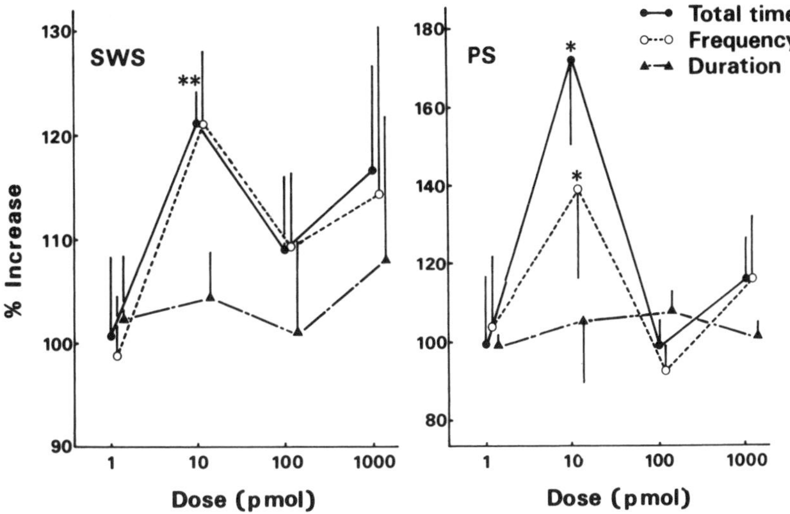

FIGURE 5. Dose response relationship of i.c.v. infusion of uridine in rats. Sleep parameters in the 12-h experimental night were compared with those of the control night. The vertical lines show the range of SEM (1 pmol: n = 6; 10 pmol: n = 9; 100 and 1000 pmol: each n = 5). $*p < 0.05$, $**p < 0.01$. (From Honda, K., Komoda, Y., Nishida, J., Nagasaki, H., Higashi, A., Uchizono, K., and Inoué, S., *Neurosci. Res.*, 1, 243, 1984. With permission.)

parameters during the light period remained stable and unchanged throughout the 4 consecutive days. The increase in the amount of SWS and PS was due mainly to an increase in the frequency of SWS and PS episodes, but not their duration (see Figure 5). This may indicate that uridine-induced sleep is natural and physiological. Sleep-waking behavior was quite normal and no difference could be detected from normal intact rats. No pyrogenic effect of uridine was recorded (see below).

A smaller dose (1 pmol) and larger doses (100 to 1000 pmol) exerted little somnogenic effect. Hence, the dose response curves for sleep parameters exhibit a somewhat bell-shaped form (see Figure 5). No behavioral abnormality was observed during and after the administration of uridine at a dosage range from 1 to 100 nmol. However, 1000 pmol of uridine brought about severe irregularities of circadian locomotor activity rhythms in all recipient rats.

It is noteworthy that the dose of 10 pmol/10 h is the smallest dosage sufficient to induce a significant increase in sleep, so far as our long-term i.c.v. infusion assay in rats has detected up to now. This value means that exogenously supplied uridine was liberated into the cerebrospinal fluid (CSF) at a rate of 17 fmol/min. This figure is 3.6, 200, and 250 times smaller than that of prostaglandin D_2 (PGD_2), muramyl dipeptide (MDP), and delta-sleep-inducing peptide (DSIP), respectively.[3,9,19,20] Since the brain and blood contain considerably high levels of uridine (see above), it is speculated that the uridine content in the brain may be maintained at a certain level and that an extremely slight increase or decrease from this level may trigger sleep and wakefulness, respectively.[1] This speculation has recently been substantiated by Van Belle[24] on the following experimental basis: (1) Van Belle et al.[25] found that the nucleoside-transport inhibitor mioflazine prevented the deamination and washout of nucleosides including uridine and adenonsine in isolated rabbit hearts and (2) Wauquier and associates found that the administration of mioflazine decreased wakefulness and increased SWS in dogs[26-28] and in rats.[27,28] Therefore, it is probable that a change in the uridine content as well as the adenosine content of the brain may eventually result in the state alternation.

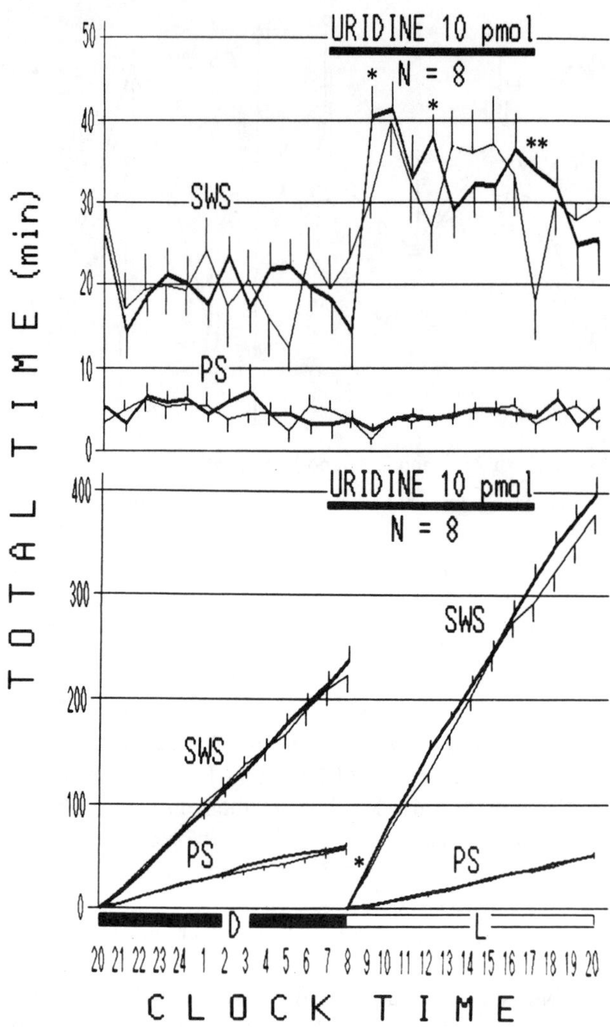

FIGURE 6. Effects of 10 pmol of uridine i.c.v. infused for 10 h at 07.00 to 17.00 h (indicated by a solid bar) on the time-course changes in the amount of sleep in otherwise saline-infused, freely behaving rats. See Figure 2 for other explanations. (From Inoué, S., Honda, K., Komoda, Y., Uchizono, K., Ueno, R., and Hayaishi, O., *Neurosci. Lett.*, 81, 6240, 1984. With permission.)

C. Diurnal i.c.v. Infusion

In contrast to the nocturnal infusion, uridine at a dosage range of 1 to 1000 pmol caused little change in sleep parameters in rats, if the infusion period was shifted by 12 h (10-h diurnal i.c.v. infusion: the infusion period covers the last hour of the dark period and the early and middle phases of the light period)[3,10,21,29] (see Figure 6 and Table 1). During the infusion of 10 pmol of uridine, statistically significant fluctuations from the baseline level appeared on a few points in hourly SWS and PS values, but they were largely due to hour-to-hour variations not directly related to the uridine administration.

As already mentioned in the previous chapter, diurnal i.c.v. infusion of DSIP also resulted in the same ineffectiveness. In addition, Radulovacki and co-workers[30,31] found no somnogenic effects of uridine at higher doses of 1 to 100 nmol during a 6-h diurnal recording period following acute i.c.v. injection in rats.

Table 1
EFFECTS OF 10-H DIURNAL I.C.V. INFUSION OF URIDINE ON PERCENT CHANGES IN SLEEP PARAMETERS (MEAN ± SEM) IN RATS DURING THE LIGHT PERIOD AS COMPARED TO THE BASELINE VALUES

Dose		% SWS			% PS		
(pmol)	N	Total time	Episode number	Episode duration	Total time	Episode number	Episode duration
1	7	106.2 ± 4.7	97.4 ± 4.6	110.3 ± 7.5	109.9 ± 13.9	93.7 ± 10.8	120.3 ± 11.4
10	8	106.4 ± 4.1	121.6 ± 11.6	90.3 ± 6.4	101.3 ± 4.0	112.6 ± 4.9	91.1 ± 3.5
100	5	100.7 ± 2.5	97.8 ± 8.4	105.1 ± 7.4	114.2 ± 16.2	106.3 ± 15.1	107.1 ± 6.7
1000	5	92.5 ± 5.7	116.7 ± 8.6	82.1 ± 10.8	105.4 ± 8.7	107.6 ± 11.6	99.1 ± 5.7

From Inoué, S., Honda, K., and Komoda, Y., in *Sleep '84*, Koella, W. P., Rüther, E., and Schulz, H., Eds., Gustav Fischer Verlag, Stuttgart, 1985, 212. With permission.

The ineffectiveness of the diurnal infusion of exogenously administered sleep promoters is of particular interest from the viewpoint of the circadian mechanism of sleep regulation. The observed fact suggests that a feedback mechanism may exist for the keeping of sleep pressure at a certain time-of-day-dependent normal level, which is preprogrammed in accordance with the phase of the circadian rest activity rhythm.[3,10,20,29,32] Since a large number of neurotransmitter receptors are known to exhibit circadian variations in their activity,[33] the actions of sleep substances may be modulated by the activity of the receptors at molecular levels. Hence, the action of an exogenously supplied sleep substance might be cancelled without inducing supranormal excess sleep.

D. Intraperitoneal Administration

1. Experiments in Rats

As described above, excess sleep may be induced by 10-h i.c.v. infusion of 10 pmol uridine through nocturnal administration, but not through diurnal administration in freely moving rats. This fact suggests the existence of a circadian variation in the effectiveness of exogenously administered sleep substances (see above). Two questions then arise as to (1) whether a single i.p. injection of uridine at a different time of day can clarify such a possibility and (2) whether an i.p. injection of uridine at a dosage equivalent to the optimal dose for i.c.v. infusion can cause similar excess sleep by acting on the brain after passing the blood-brain barrier. Although a crude preparation of SPS was effective by i.p. administration,[1,34,35] it is of interest to investigate the optimal dose of peripherally administered uridine.

Five different doses (0.01 to 100 nmol) of uridine were i.p. injected in rats.[36,37] The clock time of the injection was fixed at either 06.50 h (shortly before the onset of the light period) and 10.50 h (at an early light period) or 18.50 h (shortly before the onset of the dark period). A single i.p. injection of saline solution only at the corresponding clock time on the previous day served as a control treatment.

An injection of 0.1 nmol uridine shortly before the onset of the light period resulted in no significant change in sleep parameters (see Figure 7). However, the same dose injected at 10.50 h induced a significant increase in PS in the first hour, but no significant increase in SWS (see Figure 8). The latency of SWS and PS following the injection was shortened by 4.1 and 9.5 min, respectively, although the differences were statistically not significant. In contrast, as shown in Figure 9, an injection of 0.1 nmol uridine at 18.50 h resulted in a significant increase in SWS in the first hour. PS also increased, but the difference was insignificant. These changes were transient, and the subsequent circadian sleep-waking pattern was almost similar to that of the control. The latency of SWS and PS was shortened by 11.3 and 10.6 min, respectively. The frequency and duration of SWS and PS episodes were little affected. An injection of smaller and larger doses exerted no remarkable effect. Thus, the dose response curves again exhibited a bell-shaped form (see Figure 10).

The circadian variations were thus apparent in the effectiveness of the sleep-promoting potency of the i.p. injection of uridine. This is largely in agreement with the results from a long-term i.c.v. infusion. Accordingly, if timely administered, i.p. injection of uridine did acutely cause a transient excess sleep. It is concluded that systemically administered uridine may pass the blood-brain barrier to evoke an increase in sleep. The optimal dosage of 0.1 nmol was ten times as much as compared to that of 10-h steady i.c.v. infusion. Hence the latter method of administration seems to be far more sensitive to detect the sleep-modulatory effects of the sleep substance in rats.

2. Experiments in Mice

We also observed that a single i.p. injection of uridine at a dose of 1.0 pmol induced a significant increase in both SWS and PS during the following 24-h recording period in mice.[1] A comparable increase in SWS and PS was induced by i.p. injection of 0.05 brainstem unit of SPS-A1, which was isolated from the brainstems of 24-h sleep-deprived rats and identical with uridine.

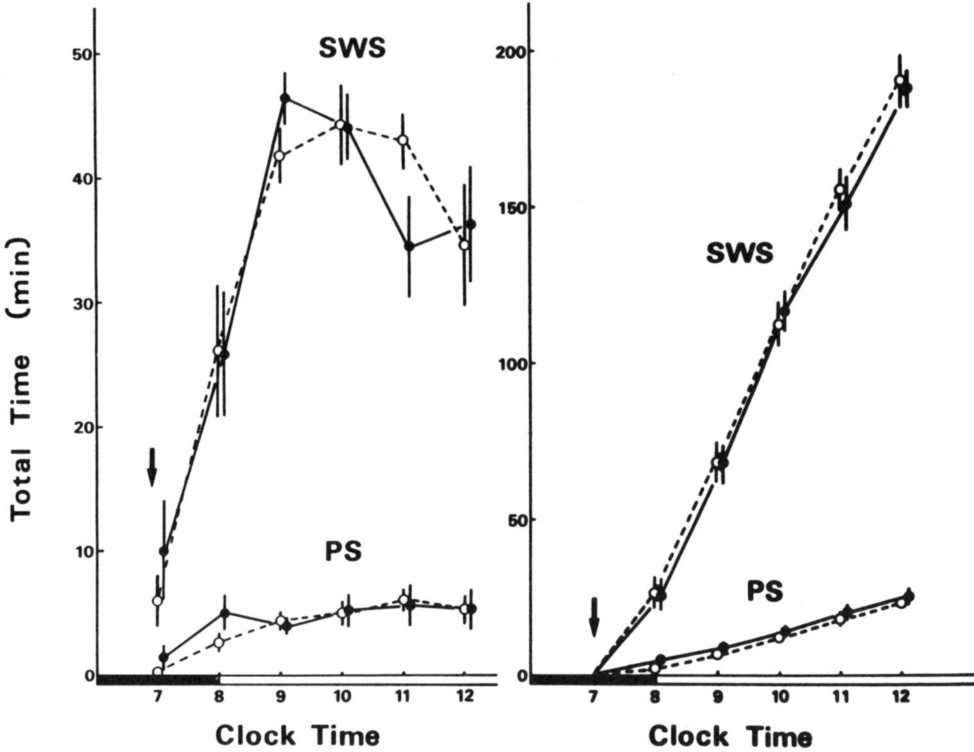

FIGURE 7. Effects of a single i.p. injection of 0.1 nmol uridine at 6.50 h (indicated by an arrow) on the time-course changes in hourly records (left) and cumulative records (right) in SWS and PS in rats (n = 6). Solid and broken lines stand for the control and the experimental day, respectively. The environmental light and dark schedule is presented by an open and closed bar, respectively. (From Honda, K., Okano, Y., Komoda, Y., and Inoué, S., *Neurosci. Lett.*, 62, 137, 1985. With permission.)

III. EXTRA-SLEEP ACTIVITIES OF URIDINE

A. Behavior-Modulatory Activity

As already explained, uridine at a large pharmacological dose may induce depression of spontaneous locomotor activity in mice.[14,15] We also noticed that a continuous nocturnal i.c.v. infusion of uridine at a dosage and rate range of 1 to 1000 pmol/10 h resulted in a reduction in locomotor activity in rats, which occurred in proportion to the increase in sleep and the decrease in wakefulness (unpublished). It was noted, as mentioned before, that the dose of 1000 pmol severely disturbed the circadian rhythmicity of locomotor activity, indicating a toxic reaction in rats.[3,10,19]

Uridine exerted dose-dependently a suppression or an activation of swimming activity in freshwater fish, crucian carp, and topmouth gudgeons.[20,38,39] Behavior-modulating effects were also detected in rhinoceros beetles following the administration of uridine.[20,38,40] The details will be described in Chapter 10.

B. Temperature

Although hyperthermia may be induced by large pharmacological doses of uridine as reported in humans[17] and rabbits[17,18] (see above), neither a pyrogenic nor antipyretic effect of uridine was detected during and after the 10-h nocturnal i.c.v. infusion at a rate of 1 pmol/h in our experiments in unrestrained rats.[3,10,19] Circadian and ultradian variations occurring in brain temperature were little affected by the uridine administration (see Figure 11). Brain temperature was also not affected by 10-h diurnal i.c.v. infusion of 10 pmol of uridine.[3,21]

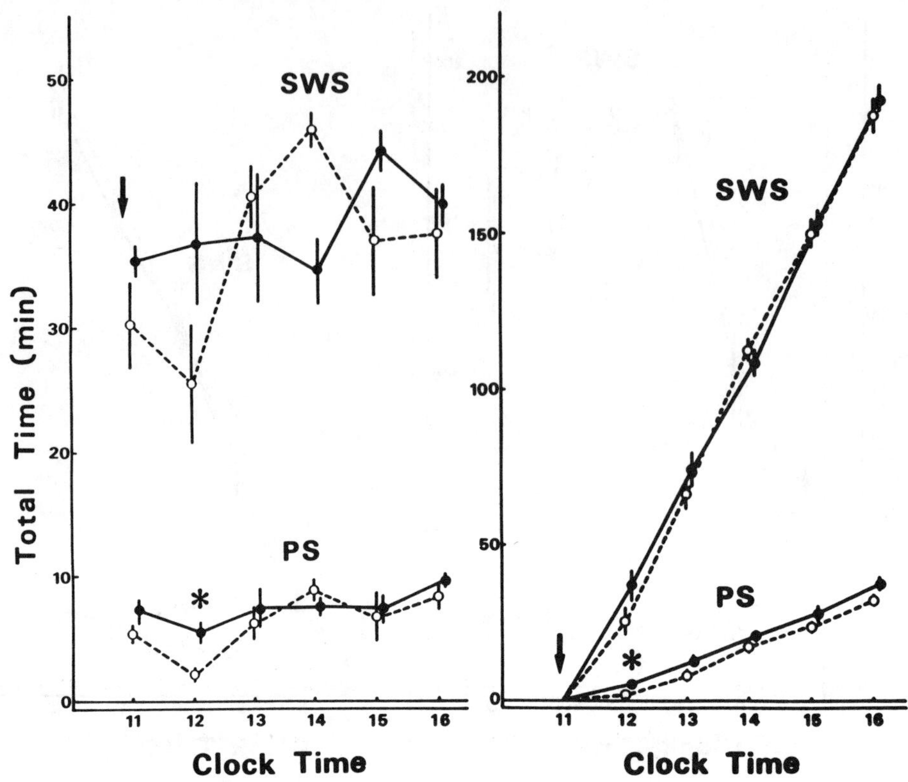

FIGURE 8. Effects of a single i.p. injection of 0.1 nmol uridine at 10.50 h (indicated by an arrow) on the time-course changes in hourly records (left) and cumulative records (right) in SWS and PS in rats (n = 6). See Figure 7 for other explanations. *$p < 0.05$. (From Honda, K., Okano, Y., Komoda, Y., and Inoué, S., *Neurosci. Lett.*, 62, 137, 1985. With permission.)

C. Neuronal Firing Activity

1. Rat Hypothalamus

Inokuchi and Oomura[41,42] examined the effects of electrophoretically applied uridine (derived from SPS and from commercial products) on neuronal activity in the preoptic area (POA) and the posterior hypothalamic area (PHA) in urethane-anesthetized rats. POA and PHA are known to be closely related to the regulation of sleep.[43,44]

Uridine excited and inhibited, respectively, 14 and 13% of tested neurons in the POA (n = 70). The excitatory effect of glutamate (Glu) was blocked or attenuated by the concurrent application of uridine in 32% of the neurons (n = 28), and this modulatory effect was long-lasting. The inhibitory effect of noradrenaline (NA) in uridine-responsive neurons was attenuated, blocked, or reversed by concurrent application of uridine in 28% of the neurons (n = 25). The effect of acetylcholine (Ach), i.e., either enhancement or blockade of excitation, was modified by the concurrent application of uridine in 21% of the neurons (n = 33). Similar modulatory effects of uridine were also observable in the PHA, but the occurrence was slightly less frequent. The difference in modulation rate may depend on differences in neuronal sensitivity to uridine in the examined regions.

On the basis of these observations, the Japanese neurophysiologists concluded that uridine modulates neurotransmission of Glu, NA, and Ach in the POA, and this modulation might be through a metabolic process, since it took a considerably long time to recover the original excitability after the application of uridine.

It should be noted here that Phillis et al.[45] examined the depressant effects of various purines and pyrimidines on corticospinal cells in rats and found very weak or no activity in pyrimidine nucleotides.

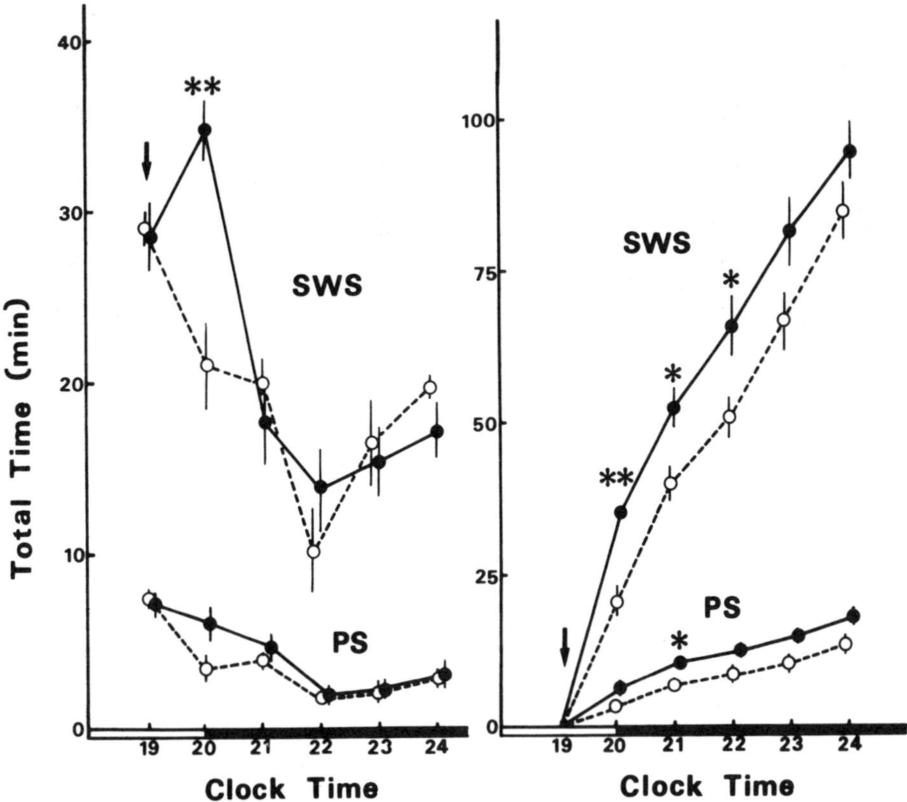

FIGURE 9. Effects of a single i.p. injection of 0.1 nmol uridine at 18.50 h (indicated by an arrow) on the time-course changes in SWS and PS in rats (n = 9). $*p < 0.05$, $**p < 0.01$. See Figure 7 for other explanations. (From Honda, K., Okano, Y., Komoda, Y., and Inoué, S., *Neurosci. Lett.*, 62, 137, 1985. With permission.)

2. *Crayfish Abdominal Ganglion*

Nagasaki et al.[46] and Uchizono et al.[47] developed a sensitive method of estimating the potency of SPS by quantifying the inhibitory activity on unit discharges of isolated abdominal ganglion of crayfish *in vitro*. The crude SPS extracted from the brainstems of 24-h sleep-deprived rats elicited a marked inhibitory effect on the firing activity at a minimal dose of 0.0002 brainstem equivalent units/ml. In contrast, 0.0002 to 0.02 units/ml of brainstem extracts from normal rats exerted entirely no effect. Thus, this technique was conventionally used for the screening of active fractions during the purification procedures of SPS. Later, a similar technique was applied for the crayfish stretch receptor *in vitro* for the same purpose.[48]

Recently, a preliminary experiment by Ozaki et al.[49] has revealed that low doses (5 to 50 n*M*) of synthetic uridine caused either a rapid, short-term suppression or a slow, long-term suppression of neural firing activity in the crayfish stretch receptor *in vitro*. The latter suppression lasted even after washing the receptor with Ringer solution. In contrast, large doses (higher than 1 μ*M*) acted dosage-independently and slowly to enhance the firing activity, which disappeared after washing. Pretreatment with low doses of uridine cancelled the effect of a higher dose of uridine, indicating the presence of a desensitizing mechanism. γ-Aminobutyric acid (GABA) did not affect the action of uridine. It is speculated that uridine may modulate the neuronal activity by affecting the intracellular metabolic activity.

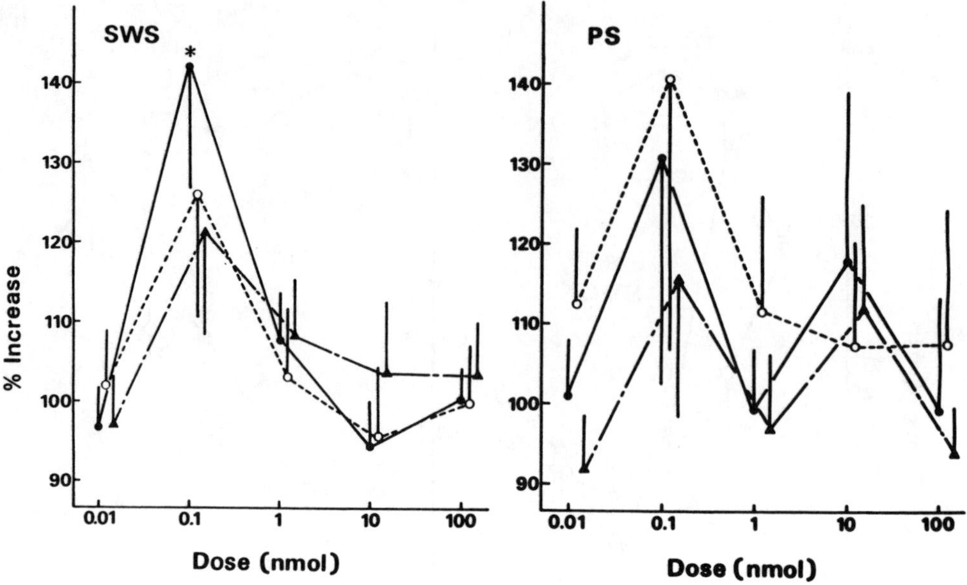

FIGURE 10. Dose response relationship of i.p. injection of uridine in rats at 18.50 h. Total sleep time (closed circles), and frequency (open circles), and duration (closed triangles) of episodes in the first 3 h after the injection were compared with those of the control day. The vertical lines show the range of SEM (each group: n = 9). $*p < 0.05$, $**p < 0.01$. (From Honda, K., Okano, Y., Komoda, Y., and Inoué, S., *Neurosci. Lett.*, 62, 137, 1985. With permission.)

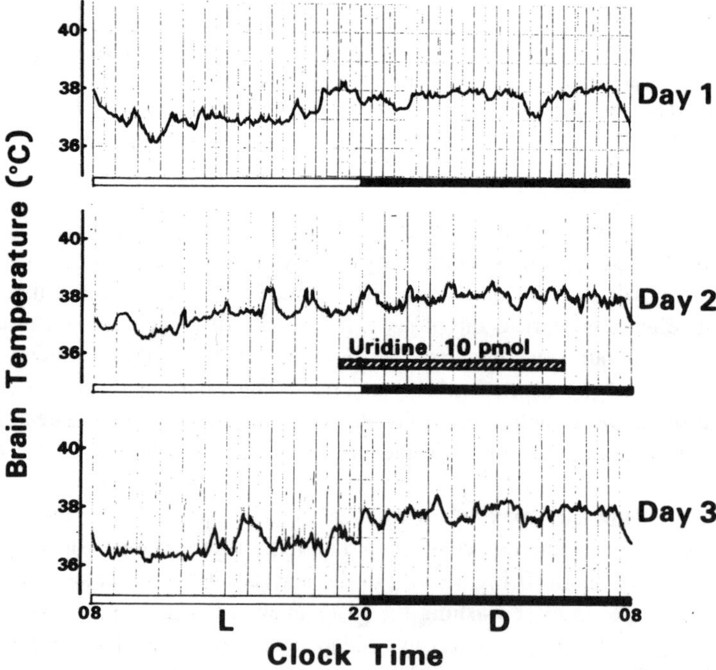

FIGURE 11. Time-course changes in brain temperature in a saline-infused rat before, during, and after 10-h i.c.v. infusion of 10 pmol of uridine (indicated by the horizontal bar on day 2). L and D stand for the environmental light period and the dark period, respectively. (From Honda, K., Komoda, Y., Nishida, S., Nagasaki, H., Higashi, A., Uchizono, K., and Inoué, S., *Neurosci. Res.*, 1, 243, 1984. With permission.)

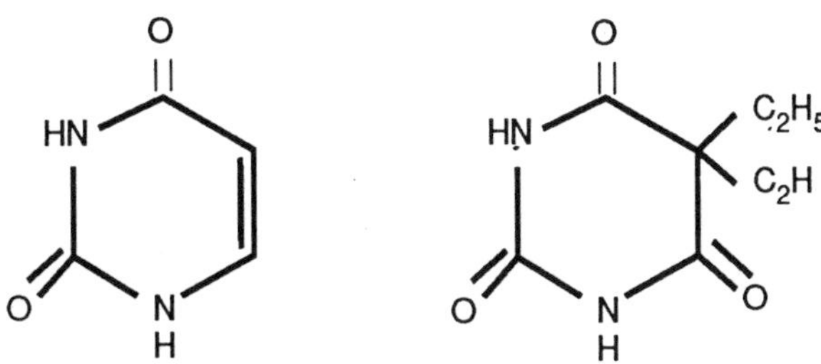

FIGURE 12. Chemical structure of uracil (left) and barbital (right).

IV. RELATED NUCLEOSIDES

A. Pyrimidines and Related Compounds

On the basis that natural pyrimidines and related compounds contain a heterocyclic, six-membered ring which forms part of the structure of many hypnotics (see Figure 12), and that some of these substances exert hypnotic activity,[50] Krooth et al.[14] postulated that uracil and other natural pyrimidines may depress the level of arousal. Indeed, they found that a certain dosage of uridine, 6-azauridine, cytidine, thymidine, thymine, uracil, 6-azauracil, barbital, and phenobarbital could depress the spontaneous activity of mice. However, the minimal dose required for reducing spontaneous activity to 50% of the level observed in animals who received an inactive control substance (D-ribose) was as large as 6.5 mmol/kg for uridine, 14.0 mmol/kg for uracil, 13.8 mmol/kg for cytidine, and so on. An earlier study of Wenzel and Keplinger[51] also mentions that hexobarbital-induced narcosis was markedly potentiated by i.p. (1 mmol/kg) or oral (2 mmol/kg) administration of uracil in mice.

In spite of the structural similarity of uracil to barbital and to uridine, we found no somnogenic activity of uracil at three doses of 1, 10, and 100 pmol after the long-term nocturnal i.c.v. infusion assay in rats[3,22,23,52] (see Figure 13). Consequently, it was suggested that a sugar moiety of uridine may have an importance in eliciting the sleep-promoting effect, or some other uridine metabolites may be active. Recently, we have compared the somnogenic activity of several uridine-related nucleosides by the same bioassay method.[23,53,54] The results are summarized in Table 2.

It has been found that no nucleosides, except for deoxyuridine, exhibited a somnogenic activity comparable to that of 10 pmol of uridine. Deoxyuridine (see Figure 14, left) caused a significant increase in SWS during the dark period. The increase was due largely to the increased occurrence of SWS episodes, but not to their prolongation. PS showed a rise, but the difference was insignificant because of large individual variations. Deoxyuridine, however, required a dose tenfold greater than uridine.

Cytidine, at a dose of 10 pmol (see Figure 15), induced a significant increase in the number of SWS episodes without changing the total SWS time, eventually accompanying a shortened episode duration. This is not in agreement with the results by Radulovacki et al.,[30] who reported that cytidine acutely injected into the lateral cerebral ventricle of rats at a dose of 1 nmol suppressed sleep for a 6-h diurnal recording period. Presumably, the different results may depend on the difference in dosage (10 pmol or 1 nmol), the administration route (via the third cerebral ventricle or the lateral cerebral ventricle), the clock time of administration (10-h nocturnal infusion or acute diurnal injection), and the recording period (12-h dark period or 6-h light period).

Deoxyguanosine (see Figure 14, right) caused a significant prolongation of SWS episodes and a slight reduction of their number. Both compounds increased considerably the amount of PS as compared to the control, but the difference was statistically insignificant. Deoxycytidine (see Figure 15) at a dose of 10 pmol/10 h resulted in a significant increase in PS

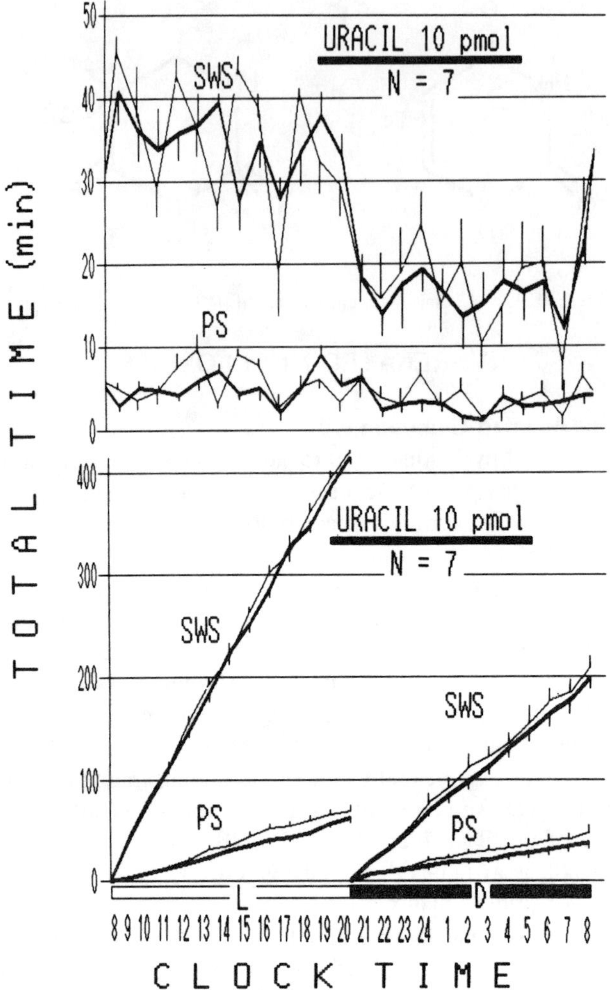

FIGURE 13. Effects of 10 pmol of uracil i.c.v. infused for 10 h at 07.00 to 17.00 h (indicated by a solid bar) on the time-course changes in the amount of sleep in otherwise saline-infused, freely behaving rats. See Figure 2 for other explanations. (From Honda, K., Komoda, Y., and Inoué, S., *Rep. Inst. Med. Dent. Eng.*, 18, 93, 1984. With permission.)

but not in SWS, which was due largely to an increased number of PS episodes, but not a prolongation of their duration.

These results may suggest that several nucleosides play some regulatory role in the induction and maintenance of either SWS or PS or both. Since metabolic pathways of pyrimidine and other nucleosides are closely interrelated, the metabolites of these substances may also be involved in the regulation of sleep. However, judging from the mild somnogenic properties of these substances, it seem likely that the regulatory role of the nucleosides is rather nonspecific.

B. N^3-Benzyluridine

Yamamoto and co-workers[15,55-59] synthesized several uridine-related *N*-substituted derivatives and examined their central depressant activity in mice, N^3-allyluridine and N^3-allyl-

Table 2
EFFECTS OF 10-H NOCTURNAL I.C.V. INFUSION OF NUCLEOSIDES ON PERCENT CHANGES IN SLEEP PARAMETERS (MEAN ± SEM) IN RATS DURING THE DARK PERIOD AS COMPARED TO THE BASELINE VALUES[22,23]

Dose (pmol)	N	%SWS			% PS		
		Total time	Episode number	Episode duration	Total time	Episode number	Episode duration
Cytidine							
10	6	101.4 ± 5.4	121.1 ± 7.1*	83.9 ± 5.3*	182.9 ± 51.5	170.4 ± 55.7	112.6 ± 8.1
Deoxycytidine							
1	3	115.1 ± 11.6	119.3 ± 12.9	95.2 ± 0.9	99.3 ± 17.4	102.3 ± 23.7	95.6 ± 6.2
5	2	97.1 ± 14.7	97.6 ± 10.4	97.7 ± 2.3	77.7 ± 17.7	84.5 ± 28.5	96.3 ± 12.1
10	12	101.6 ± 5.4	105.1 ± 5.8	96.9 ± 4.0	151.5 ± 30.9*	153.6 ± 41.2	103.7 ± 6.2
100	2	98.6 ± 2.5	95.9 ± 10.2	102.8 ± 14.8	90.5 ± 5.1	88.6 ± 2.5	100.0 ± 10.0
Deoxyguanosine							
10	4	114.6 ± 12.5	91.3 ± 7.6	123.9 ± 6.5*	169.1 ± 36.1	169.0 ± 43.8	106.3 ± 8.8
Deoxyuridine							
10	6	95.6 ± 8.6	96.3 ± 7.2	98.9 ± 5.2	118.3 ± 27.6	117.9 ± 33.8	114.4 ± 13.4
100	7	124.6 ± 9.5*	124.0 ± 9.9	100.5 ± 4.3	142.2 ± 27.0	160.4 ± 37.9	91.1 ± 8.4
***N^3*-Benzyluridine**							
10	7	107.2 ± 8.5	115.0 ± 17.9	102.9 ± 12.1	138.3 ± 29.0	143.1 ± 27.5	99.2 ± 9.0
100	6	107.5 ± 12.3	105.9 ± 13.7	102.6 ± 3.1	122.2 ± 13.8	122.9 ± 15.0	100.5 ± 7.2
Uracil							
1	8	95.6 ± 7.3	97.6 ± 6.8	99.6 ± 5.3	124.7 ± 23.9	121.8 ± 24.0	105.7 ± 8.4
10	7	96.4 ± 8.4	95.7 ± 8.7	102.2 ± 8.6	82.7 ± 7.3	93.4 ± 13.3	93.1 ± 8.3
100	7	111.5 ± 10.4	108.4 ± 5.3	102.8 ± 7.8	156.0 ± 35.1	148.4 ± 37.4	107.2 ± 11.4

Table 2 (continued)
EFFECTS OF 10-H NOCTURNAL I.C.V. INFUSION OF NUCLEOSIDES ON PERCENT CHANGES IN SLEEP PARAMETERS (MEAN ± SEM) IN RATS DURING THE DARK PERIOD AS COMPARED TO THE BASELINE VALUES[22,23]

Dose		%SWS			% PS		
(pmol)	N	Total time	Episode number	Episode duration	Total time	Episode number	Episode duration
				Uridine			
1	6	100.6 ± 7.5	98.7 ± 6.1	102.2 ± 6.4	98.8 ± 17.8	103.7 ± 18.0	98.7 ± 3.0
10	8	121.3 ± 3.0**	121.1 ± 6.9	104.2 ± 4.5	172.3 ± 21.7*	138.8 ± 20.3*	105.5 ± 16.0
100	5	108.9 ± 7.0	109.3 ± 7.0	100.8 ± 9.1	98.5 ± 7.3	92.3 ± 7.0	107.8 ± 5.0
1000	5	116.5 ± 10.0	114.3 ± 16.0	108.1 ± 13.8	116.0 ± 10.8	115.9 ± 15.8	100.9 ± 3.9

$^*p < 0.05$; $^{**}p < 0.01$, as compared to the baseline value.

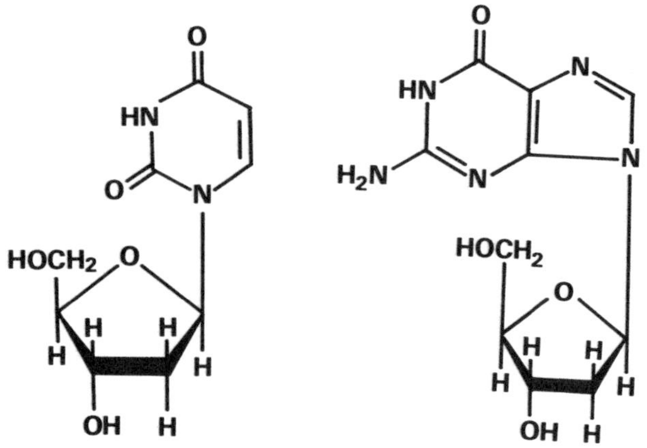

FIGURE 14. Chemical structure of deoxyuridine (left) and deoxyguanosine (right).

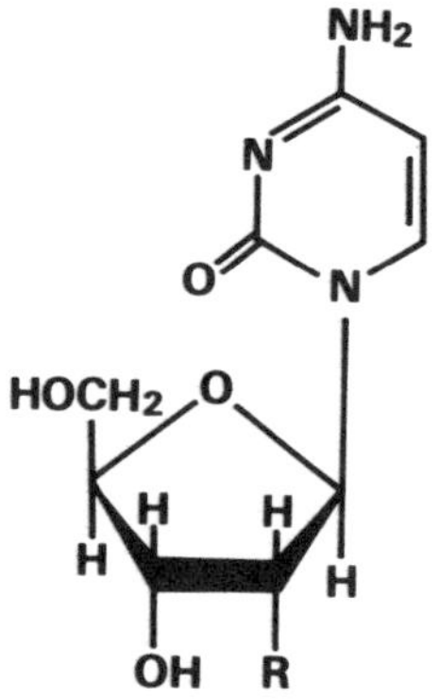

Cytidine : R = OH

Deoxycytidine : R = H

FIGURE 15. Chemical structure of deoxycytidine and cytidine.

thymidine as well as uridine and thymidine were found to potentiate dose-dependently pentobarbital-induced anesthesia, diazepam-induced motor incoordination, and spontaneous activity after acute i.c.v. injection at three doses of 0.6, 0.9, and 1.2 mg/mouse.[15] N^3-benzyluridine (see Figure 16) also exerted a dose-dependent hypnotic activity as revealed by a prolongation of pentobarbital-induced anesthesia,[55] although this compound alone did not induce excess sleep as revealed by our long-term nocturnal i.c.v. infusion technique in rats.[23] The Japanese group headed by Yamamoto[56-58] further synthesized several *N*-allyl-substituted derivatives of pentobarbital, which were either agonists or antagonists for barbiturates.

Currently, no sufficient information is yet available as to the cerebral biochemical and pharmacological mechanisms involved in the antagonism and synergism between uridine- and barbiturate-related substances. However, the close resemblance of chemical structure and hypnogenic activity between these compounds seem to encourage further investigation.

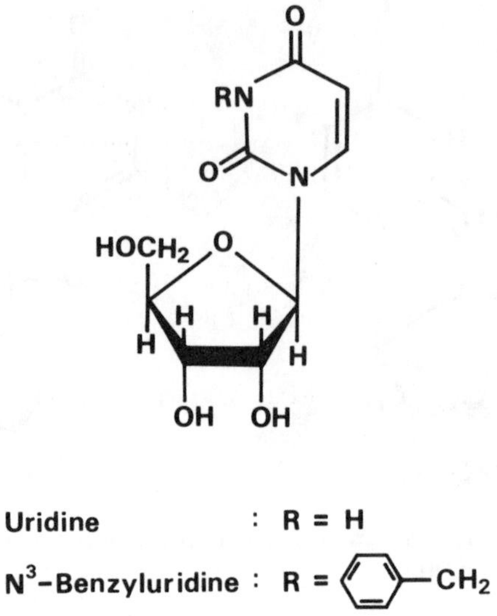

Uridine : R = H

N^3-Benzyluridine : R = C_6H_5–CH_2

FIGURE 16. Chemical structure of N^3-benzyluridine in relation to uridine.

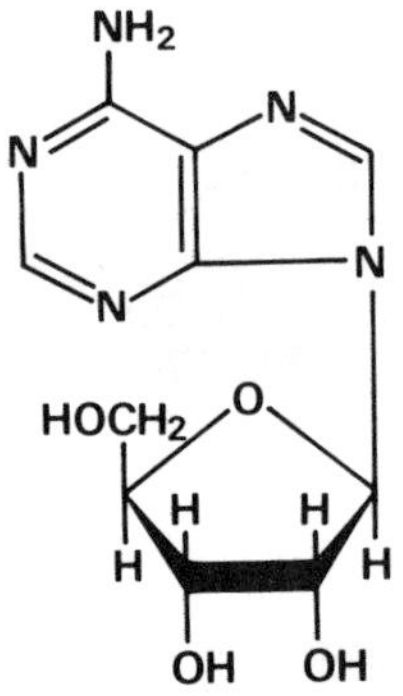

FIGURE 17. Chemical structure of adenosine.

C. Adenosine and Related Substances

Adenosine (see Figure 17) is a naturally occurring purine nucleoside. Its actions as a potent neuromodulator in the central nervous system include sedative, anticonvulsant, hypnogenic, ataxic, and antinociceptive properties.[60] In addition, adenosine exerts an inhibitory action on neuronal firing activity[45] and interacts with other neurotransmitters, neuromodulators, and hormones. For example, its depressant activity on the firing of rat cerebral cortical neurons is potentiated by progesterone.[61] Adenosine at very large doses (1 to 100 mg/kg s.c. or i.p.) causes hypothermia in rats.[62] Thus, adenosine seems to play a considerably general role in regulating the degree of "alertness" in the central nervous system.[60]

Radulovacki and co-workers[30,31,63-67] have extensively studied the hypnogenic activity of adenosine. They found that adenosine at doses of 1 to 100 nmol decreased wakefulness, increased deep SWS, and increased total sleep during a 6-h diurnal period following acute i.c.v. injection in rats in the morning. Sarda et al.[68] confirmed the results: rats injected with adenosine shortly before the dark period onset exhibited a significant increase in SWS (50 mg/kg i.p.) and PS (7 and 50 mg/kg i.p.). Stenberg et al.[69] reported some enhancement of deep SWS and a decrease in wakefulness and PS during the first few hours after i.c.v.

infusion of adenosine (10, 100, and 1000 nmol/1.25 to 12.5 min) in cats. However, the dosage used in these studies was considerably large, and the drug effects seem to have been somewhat pharmacological but not physiological. It should be noted that the dosage required for the enhancement of sleep was 100 to 10,000 times as much as that of uridine, although the latter was steadily infused for a 10-h nocturnal period (see above).

Radulovacki and co-workers[70] have further demonstrated that the action of adenosine is mediated by the stimulation of adenosine A_1 rather than A_2 receptors, because adenosine and its analogs activated A_1 receptors in nanomolar quantities, whereas the activation of A_2 receptors required micromolar concentration of these compounds. Caffeine seems to antagonize the effects of adenosine at the level of both receptors, with differential effects on sleep and behavior.[71] Barraco et al.[72] suggested that there is a possibility of specific benzodiazepine-adenosine interactions in the central nervous system, since they found a marked depression of locomotor activity after concurrent injections of adenosine (1 μmol/kg i.c.v.) and diazepam (0.1 to 5 μmol/kg) in mice.

The elevation of adenosine levels in the brain may cause the enhancement of sleep. Radulovacki et al.[73] found that deoxycoformycin, a potent inhibitor of adenosine deaminase, increased PS and decreased PS latency at a dose of 0.5 mg/kg, and increased deep SWS at a dose of 2 mg/kg. As already mentioned, the nucleoside-transport inhibitor mioflazine decreased wakefulness and increased SWS in dogs[25-27] and in rats.[26,27]

V. INTERACTIONS OF URIDINE WITH OTHER SUBSTANCES

A. Barbiturates

As repeatedly mentioned, uridine has a structural similarity to barbiturates, and hence it may exert some interactions with these unnatural hypnotic compounds. Although the molecular mechanism is still unknown, both agonistic and antagonistic activities are reported by Yamamoto and co-workers.[15,56-58]

B. Other Sleep Factors

Kimura et al.,[74-78] Inoué,[79,80] and Inoué et al.[54,81] in their laboratory have recently demonstrated that uridine may exert complex interactions with other putative sleep substances, such as DSIP and MDP. The details will be described in Chapter 9.

C. Interaction with ACTH

It is known that the incorporation of uridine into brain or brainstem RNA is inhibited by the presence of ACTH and its analog $ACTH_{1-24}$ in rats and mice.[82,83] Since ACTH induces enhancement of arousal (see Chapter 8) and concurrently stress, it might be speculated that reduced sleep caused by ACTH is potentiated by the shortage of uridine and eventually by the blockade of its somnogenic activity.

REFERENCES

1. **Komoda, Y., Ishikawa, M., Nagasaki, H., Iriki, M., Honda, K., Inoué, S., Higashi, A., and Uchizono, K.,** Uridine, a sleep-promoting substance from brainstems of sleep-deprived rats, *Biomed. Res.*, 4(Suppl.), 223, 1983.
2. **Komoda, Y., Ishikawa, M., Nagasaki, H., Iriki, M., Honda, K., Inoué, S., Higashi, A., and Uchizono, K.,** Purification, isolation and identification of sleep-promoting substances (SPS) from brainstems of sleep-deprived rats, *Folia Psychiatr. Neurol. Jpn.*, 39, 210, 1985.

3. **Inoué, S., Honda, K., and Komoda, Y.,** Sleep-promoting substances, in *Sleep: Neurotransmitters and Neuromodulators,* Wauquier, A., Gaillard, J. M., Monti, J. M., and Radulovacki, M., Eds., Raven Press, New York, 1985, 305.
4. **Honda, K. and Inoué, S.,** Effects of sleep-promoting substance on sleep-waking patterns of male rats, *Rep. Inst. Med. Dent. Eng.,* 15, 115, 1981.
5. **Honda, K. and Inoué, S.,** Establishment of a bioassay method for the sleep-promoting substance, *Rep. Inst. Med. Dent. Eng.,* 12, 81, 1978.
6. **Inoué, S., Honda, K., and Komoda, Y.,** A possible mechanism by which the sleep-promoting substance induces slow wave sleep but supresses paradoxical sleep in the rat, in *Sleep 1982,* Koella, W. P., Ed., S. Karger, Basel, 1983, 112.
7. **Inoué, S., Honda, K., Nishida, S., and Komoda, Y.,** Temporal changes in sleep-promoting effects of constantly infused sleep substances in rats, *Sleep Res.,* 12, 81, 1983.
8. **Inoué, S., Nagasaki, H., and Uchizono, K.,** Endogenous sleep-promoting factors, *Trends Neurosci.,* 5, 218, 1982.
9. **Inoué, S., Honda, K., Komoda, Y., Uchizono, K., Ueno, R., and Hayaishi, O.,** Differential sleep-promoting effects of five sleep substances nocturnally infused in unrestrained rats, *Proc. Natl. Acad. Sci. U.S.A.,* 81, 6240, 1984.
10. **Honda, K., Komoda, Y., and Inoué, S.,** Effects of sleep-promoting substances on the rat circadian sleep-waking cycles, in *Endogenous Sleep Substances and Sleep Regulation,* Inoué, S. and Borbély, A. A., Eds., Japan Scientific Societies Press, Tokyo/VNU Science Press BV, Utrecht, 1985, 203.
11. **Karle, J. M., Anderson, L. W., Dietrick, D. D., And Cysyk, R. L.,** Determination of serum and plasma uridine levels in mice, rats, and humans by high pressure liquid chromatography, *Anal. Biochem.,* 109, 41, 1980.
12. **Dudman, N. P., Deveski, W. B., and Tattersall, H. N.,** Radioimmunoassays of plasma thymidine, uridine, deoxyuridine, and cytidine/deoxycytidine, *Anal. Biochem.,* 115, 428, 1981.
13. **Dwivedi, C. and Harbison, R. D.,** Anticonvulsant activities of Δ-8 and Δ-9 tetrahydrocannabinol and uridine, *Toxicol. Appl. Pharmacol.,* 31, 452, 1975.
14. **Krooth, R. S., Hsiao, W. L., and Lam, G. F. M.,** Effects of natural pyrimidines and of certain related compounds on the spontaneous activity in the mouse, *J. Pharmacol. Exp. Ther.,* 207, 504, 1978.
15. **Yamamoto, I., Kimura, T., Tateoka, Y., Watanabe, K., and Ho, I. K.,** Central depressant activities of N^3-allyluridine and N^3-allylthymidine, *Res. Commun. Chem. Pathol. Pharmacol.,* 52, 321, 1986.
16. **Calvaruso, G., Taibi, G., Torregrossa, M. V., Romano, N., and Tesoriere, G.,** Uridine enhances the cytotoxic effect of D-glucosamine in rat C6 glioma cells, *Life Sci.,* 39, 2221, 1986.
17. **Groeningen, C. V., Leyva, A., Peters, G., Kraal, I., and Pinedo, H.,** Development of a uridine (UR) administration schedule for rescue from 5-fluorouracil (5-FU) toxicity, *Proc. Am. Assoc. Cancer Res.,* 25, 169, 1984.
18. **Cradock, J. C., Vishnuvajjala, B. R., Chin, T. F., Hochstein, H. D., and Ackerman, S. K.,** Uridine-induced hyperthermia in the rabbit, *J. Pharm. Pharmacol.,* 38, 226, 1986.
19. **Honda, K., Komoda, Y., Nishida, S., Nagasaki, H., Higashi, A., Uchizono, K., and Inoué, S.,** Uridine as an active component of sleep-promoting substance: its effects on the nocturnal sleep in rats, *Neurosci. Res.,* 1, 243, 1984.
20. **Inoué, S.,** Sleep substances: their roles and evolution, in *Endogenous Sleep Substances and Sleep Regulation,* Inoué, S. and Borbély, A. A., Eds., Japan Scientific Societies Press, Tokyo/VNU Science Press BV, Utrecht, 1985, 3.
21. **Honda, K., Komoda, Y., Nishida, S., Nagasaki, H., Higashi, A., Uchizono, K., and Inoué, S.,** Sleep-promoting effects of uridine and its related substance in rats, *Folia Psychiatr. Neurol. Jpn.,* 39, 212, 1985.
22. **Inoué, S., Honda, K., and Komoda, Y.,** Uridine as an active component of the sleep-promoting substance, in *Sleep '84,* Koella, W. P., Rüther, E., and Schulz, H., Eds., Gustav Fischer Verlag, Stuttgart, 1985, 212.
23. **Honda, K., Naito, K., Komoda, Y., Kimura, T., Yamamoto, I., and Inoué, S.,** Comparison of sleep-promoting activity of uridine-related substances in unrestrained rats, *Sleep Res.,* 16, 62, 1987.
24. **Van Belle, H.,** Adenosine and sleep: can nucleoside transport inhibition be helpful?, in *Book of Abstracts,* Int. Symp. Current Trends in Slow Wave Sleep Res. (Physiological and Pathological Aspects), 1987, 36.
25. **Van Belle, H., Goossens, F., and Wynants, J.,** Formation and release of purine catabolites during hypoperfusion, anoxia, and ischemia, *Am. J. Physiol.,* 252, H886, 1987.
26. **Wauquier, A., Van Belle, H., Vanden Broeck, W. A. E., and Janssen, P. A. J.,** Sleep improvement in dogs after oral administration of mioflazine, a nucleoside transport inhibitor, *Psychopharmacology,* 91, 434, 1987.
27. **Wauquier, A. and Dugovic, C.,** The nucleoside transport inhibitor mioflazine enhances sleep in rats and dogs, in *Book of Abstracts,* Int. Symp. Current Trends in Slow Wave Sleep Res. (Physiological and Pathological Aspects), 1987, 37.

28. **Wauquier, A. and Dugovic, C.,** Slow wave sleep increasing effects of the nucleoside transport inhibitor mioflazine in dogs and rats, *Sleep Res.*, 16, 158, 1987.
29. **Inoué, S., Honda, K., Komoda, Y., Uchizono, K., Ueno, R., and Hayaishi, O.,** Little sleep-promoting effect of three sleep substances diurnally infused in unrestrained rats, *Neurosci. Lett.*, 49, 207, 1984.
30. **Radulovacki, M., Virus, R. M., Rapoza, D., and Crane, R. A.,** A comparison of the dose response effects of pyrimidine ribonucleosides and adenosine on sleep in rats, *Psychopharmacology*, 87, 136, 1985.
31. **Radulovacki, M. and Virus, R. M.,** Purine, 1-methylisoguanosine and pyrimidine compounds and sleep in rats, in *Sleep: Neurotransmitters and Neuromodulators*, Wauquier, A., Gaillard, J. M., Monti, J. M., and Radulovacki, M., Eds., Raven Press, New York, 1985, 221.
32. **Inoué, S., Kimura, M., Honda, K., and Komoda, Y.,** Sleep peptides: general and comparative aspects, in *Sleep Peptides: Basic and Clinical Approaches*, Inoué, S. and Schneider-Helmert, D., Eds., Japan Scientific Societies Press, Tokyo/Springer-Verlag, Berlin, 1988, 1.
33. **Wirz-Justice, A.,** Circadian rhythms in mammalian neurotransmitter receptors, *Prog. Neurobiol.*, 29, 219, 1987.
34. **Nagasaki, H., Iriki, M., Inoué, S., and Uchizono, K.,** The presence of a sleep-promoting material in the brain of sleep-deprived rats, *Proc. Jpn. Acad.*, 50, 241, 1974.
35. **Uchizono, K., Ishikawa, M., Iriki, M., Inoué, S., Komoda, Y., Nagasaki, H., Higashi, A., Honda, K., and McRae-Degueurce, A.,** Purification of sleep-promoting substances (SPS), in *Advances in Pharmacology and Therapeutics II*, Vol. 1, Yoshida, H., Hagiwara, Y., and Ebashi, S., Eds., Pergamon Press, Oxford, 1982, 217.
36. **Honda, K., Okano, Y., Komoda, Y., and Inoué, S.,** Sleep-promoting effects of intraperitoneally administered uridine in unrestrained rats, *Neurosci. Lett.*, 62, 137, 1985.
37. **Honda, K., Okano, Y., Komoda, Y., and Inoué, S.,** Sleep-promoting effects of intraperitoneally administered uridine on unrestrained rats, *Jpn. J. Psychiatr. Neurol.*, 40, 242, 1986.
38. **Inoué, S., Honda, K., Okano, Y., and Komoda, Y.,** Behavior-modulating effect of sleep substances in fish and insects, *Sleep Res.*, 14, 84, 1985.
39. **Inoué, S., Honda, K., and Komoda, Y.,** Effect of sleep substances on circadian swimming activity in fish, *Int. J. Biometeorol.*, 29(Suppl. 1), 118, 1985.
40. **Inoué, S., Honda, K., Okano, Y., and Komoda, Y.,** Behavior-modulating effects of uridine in the rhinoceros beetle, *Zool. Sci.*, 3, 727, 1986.
41. **Inokuchi, A. and Oomura, Y.,** Effects of prostaglandin D_2 and sleep-promoting substance on hypothalamic neuronal activity in the rat, in *Endogenous Sleep Substances and Sleep Regulation*, Inoué, S. and Borbély, A. A., Eds., Japan Scientific Societies Press, Tokyo/VNU Science Press BV, Utrecht, 1985, 215.
42. **Inokuchi, A. and Oomura, Y.,** Effects of sleep-promoting substance on rat hypothalamic neuron activity, *Brain Res. Bull.*, 16, 429, 1986.
43. **Nauta, W. J. H.,** Hypothalamic regulation of sleep in rats. An experimental study, *J. Neurophysiol.*, 9, 285, 1946.
44. **Sterman, M. B. and Shouse, M. N.,** Sleep "centers" in the brain: the preoptic basal forebrain area revisited, in *Brain Mechanisms of Sleep*, McGinty, D. J., Drucker-Colín, R., Morrison, A., and Parmeggiani, P. L., Eds., Raven Press, New York, 1987, 277.
45. **Phillis, J. W., Kostopoulos, G. K., and Limacher, J. J.,** Depression of corticospinal cells by various purines and pyrimidines, *Can. J. Physiol. Pharmacol.*, 52, 1226, 1974.
46. **Nagasaki, H., Iriki, M., and Uchizono, K.,** Inhibitory effect of the brain extract form sleep-deprived rats (BE-SDR) on the spontaneous discharges of crayfish abdominal ganglion, *Brain Res.*, 109, 202, 1976.
47. **Uchizono, K., Nagasaki, H., Iriki, M., Inoué, S., Ishikawa, M., and Komoda, Y.,** Humoral control of sleep, in *Neurohumoral Correlate of Behaviour*, Subrahmanyam, S., Ed., Thompson Press (India), Madras, 1977, 35.
48. **Uchizono, K., Higashi, A., Iriki, M., Nagasaki, H., Ishikawa, M., Komoda, Y., Inoué, S., and Honda, K.,** Sleep-promoting fractions obtained from the brain-stem of sleep-deprived rats, in *Integrative Control Functions of the Brain*, Vol. 1, Ito, M., Tsukahara, N., Kubota, K., and Yagi, K., Eds., Kodansha, Tokyo, 1978, 392.
49. **Ozaki, T., Nishie, H., and Uchizono, K.,** Crayfish single unit firing activity and sleep substance, *Annu. Rep. Natl. Inst. Physiol. Sci., Okazaki*, 8, 449, 1987.
50. **Wells, C. E., Ajmone-Marsan, C., Frei, E., Tuohy, J. H., and Shnider, B. I.,** Electroencephalographic and neurological changes induced in man by administration of 1,2,4-triazine-3,5(2H,4H)-dione (6-azauracil), *Electroencelphalogr. Clin. Neurophysiol.*, 9, 325, 1957.
51. **Wenzel, C. E. and Keplinger, M. L.,** Central depressant properties of uracil and related oxypyrimidines, *J. Am. Pharm. Assoc.*, 44, 56, 1955.
52. **Honda, K., Komoda, Y., and Inoué, S.,** Little sleep-promoting effects of intraventricularly infused uracil in unrestrained rats, *Rep. Inst. Med. Dent. Eng.*, 18, 93, 1984.
53. **Naito, K., Honda, K., Komoda, Y., and Inoué, S.,** Sleep-promoting effects of deoxycytidine in unrestrained rats, *Jpn. J. Psychiatr.*, 42, 128, 1988.

54. **Inoué, S., Honda, K., Kimura, M., Naito, K., Rhee, Y. H., and Komoda, Y.,** Complex interactions of multiple sleep substances, *Jpn. J. Psychiatr. Neurol.*, 43, in press.
55. **Yamamoto, I., Kimura, T., Tateoka, Y., Watanabe, K., and Ho, I. K.,** N^3-benzyluridine exerts hypnotic activity in mice, *Chem. Pharm. Bull.*, 33, 4088, 1985.
56. **Yamamoto, I., Tateoka, Y., Watanabe, K., Nabeshima, T., Fontenot, H. J., and Ho, I. K.,** The prolonging effect of N,N^1-diallylpentobarbital on the drug-induced sleep and motor incoordination, *Res. Commun. Chem. Pathol. Pharmacol.*, 50, 209, 1985.
57. **Tateoka, Y., Kimura, T., Yamazaki, F., Watanabe, K., and Yamamoto, I.,** Central depressant effect of barbituric acid and N,5,5-triallylbarbituric acid in mice, *Yakugaku Zasshi*, 106, 504, 1986.
58. **Yamamoto, I., Tateoka, Y., Watanabe, K., and Ho, I. K.,** N,N^1-diallylpentobarbital: antagonism of barbital in mice and rats, *Life Sci.*, 40, 1439, 1987.
59. **Yamamoto, I., Kimura, T., Tateoka, Y., Watanabe, K., and Ho, I. K.,** Hypnotic action of N-substituted uridine derivatives, *Annu. Rep. Natl. Inst. Physiol. Sci., Okazaki*, 8, 445, 1987.
60. **Williams, M.,** Adenosine — a selective neuromodulator in the mammalian CNS? *Trends Neurosci*, 7, 164, 1984.
61. **Phillis, J. W.,** Potentiation of the depression by adenosine of rat cerebral cortical neurons by progestational agents, *Br. J. Pharmacol.*, 89, 693, 1986.
62. **Wager-Srdar, S. A., Oken, M. M., Morley, J. E., and Levine, A. S.,** Thermoregulatory effects of purines and caffeine, *Life Sci.*, 33, 2431, 1983.
63. **Virus, R. M., Djuricic-Nedelson, M., Radulovacki, M., and Green, R. D.,** The effects of adenosine and 2′-deoxycoformycin on sleep and wakefulness in rats, *Neuropharmacology*, 22, 1401, 1983.
64. **Radulovacki, M., Virus, R. M., Djuricic-Nedelson, M., Baglajewski, E., Meyer, E., and Green, R. D.,** Adenosine and adenosine analogs: effects on sleep in rats, in *Brain Mechanisms of Sleep*, McGinty, D. J., Drucker-Colín, R., Morrison, A., and Parmeggiani, P. L., Eds., Raven Press, New York, 1985, 235.
65. **Radulovacki, M., Virus, R. M., Yanik, G., and Green, R. D.,** Adenosine and sleep in rats, in *Sleep '84*, Koella, W. P., Rüther, E., and Schulz, H., Eds., Gustav Fischer Verlag, Stuttgart, 1985, 17.
66. **Radulovacki, M.,** Role of adenosine in sleep in rats, *Rev. Clin. Basic Pharmacol.*, 5, 327, 1985.
67. **Radulovacki, M., Porter N., Dugich, M., and Green, R. D.,** Adenosine compounds and sleep, in *Book of Abstracts*, Int. Symp. Current Trends in Slow Wave Sleep Res. (Physiological and Pathological Aspects), 1987, 34.
68. **Sarda, N., Gharib, A., Coindet, J., Pacheco, H., and Jouvet, M.,** Possible involvement of the S-adenosyl-L-homocysteine metabolites, adenosine and L-homocysteine, in sleep in rats, *Neurosci. Lett.*, 66, 287, 1986.
69. **Stenberg, D., Porkka-Heiskanen, T., and Salvan, P.,** Intraventricular adenosine and vigilance in cats, in Abstracts, 8th Eur. Congr. Sleep Res., 1986, 347.
70. **Radulovacki, M., Virus, R. M., Djuricic-Nedelson, M., and Green, R. D.,** Adenosine analogs and sleep in rats, *J. Pharmacol. Exp. Ther.*, 228, 268, 1984.
71. **Yanik, G., Glaum, S., and Radulovacki, M.,** The dose-response effects of caffeine on sleep in rats, *Brain Res.*, 403, 177, 1987.
72. **Barraco, R. A., Phillis, J. W., and Delong, R. E.,** Behavioral interaction of adenosine and diazepam in mice, *Brain Res.*, 323, 159, 1984.
73. **Radulovacki, M., Virus, R. M., Djuricic-Nedelson, M., and Green, R. D.,** Hypnotic effects of deoxycoformycin in rats, *Brain Res.*, 271, 392, 1983.
74. **Kimura, M., Honda, K., Okano, Y., Komoda, Y., and Inoué, S.,** Effects of simultaneously infused sleep substances on nocturnal sleep in rats, *Zool. Sci.*, 2, 989, 1985.
75. **Kimura, M., Honda, K., Okano, Y., Komoda, Y., and Inoué, S.,** Interacting sleep-modulatory effects of simultaneously administered sleep substances in rats, *J. Physiol. Soc. Jpn.*, 48, 280, 1986.
76. **Kimura, M., Honda, K., Komoda, Y., and Inoué, S.,** Interacting sleep-modulatory effects of simultaneously administered delta-sleep-inducing peptide (DSIP), muramyl dipeptide (MDP) and uridine in unrestrained rats, *Neurosci. Res.*, 5, 157, 1987.
77. **Kimura, M., Honda, K., Komoda, Y., and Inoué, S.,** Effects of serially administered MDP and uridine on nocturnal sleep in rats, *Zool. Sci.*, 4, 1088, 1987.
78. **Kimura, M. and Inoué, S.,** Simultaneous and serial intracerebroventricular administration of MDP and uridine: differential effects on nocturnal sleep in rats, *Neurosci. Res.*, Suppl. 7, S192, 1988.
79. **Inoué, S.,** Multifactorial humoral regulation of sleep, *Clin. Neuropharmacol.*, 9(Suppl. 4), 470, 1986.
80. **Inoué, S.,** Interference among sleep substances simultaneously infused in unrestrained rats, in *Sleep '86*, Koella, W. P., Obál, F., Schulz, H., and Visser, P., Eds., Gustav Fischer Verlag, Stuttgart, 1988, 175.
81. **Inoué, S., Kimura, M., Honda, K., and Komoda, Y.,** Sleep peptides: general and comparative aspects, in *Sleep Peptides: Basic and Clinical Approaches*, Inoué, S. and Schneider-Helmert, D., Eds., Japan Scientific Societies Press, Tokyo/Springer-Verlag, Berlin, 1988, 1.

82. **Jakoubek, B., Buresová, M., Hájek, I., Etrychová, J., Pavík, A., and Dedicová, A.,** Effect of ACTH on the synthesis of rapidly labelled RNA in the nervous system of mice, *Brain Res.*, 43, 417, 1972.
83. **Gispen, W. H., Reith, M. E. A., Schotman, P., Wiegant, V. M., Zwiers, H., and de Wied, D.,** CNS and ACTH-like peptides: neurochemical response and interaction with opiates, in *Neuropeptide Influences on the Brain and Behavior,* Miller, L. H., Sandman, C. A., and Kastin, A. J., Eds., Raven Press, New York, 1977, 61.

Chapter 5

MURAMYL PEPTIDES AND IMMUNOREACTIVE SUBSTANCES

I. SOMNOGENIC MURAMYL PEPTIDES

A. Outlines

In 1984, Martin et al.[1] and Krueger et al.[2,3] first reported that Factor S, which had been extracted and purified from the cerebrospinal fluid (CSF) of sleep-deprived goats, from bovine and rabbit cerebral tissues, and from human urine (see Chapters 1 and 2), was finally identified as a muramyl tetrapeptide, *N*-acetylglucosaminyl-*N*-1,6-anhydro-*N*-acetylmuramyl-alanyl-glutamyl-diamino-pimelyl-alanine (NAG-1,6-anhydro-NAM-Ala-Glu-Dap-Ala; for the chemical structure, see Figure 6 of Chapter 2). In addition, a muramyl tripeptide lacking the terminal alanine of the above compound (NAG-1,6-anhydro-NAM-Ala-Glu-Dap) was also isolated as an active somnogen from the urinary preparation of Factor S.

Muramyl peptides (MP) are known as components of peptidoglycans, which form the backbone of bacterial cell walls and exert profound pyrogenic and immunostimulatory activities in the mammalian body.[4-6] Consequently, the possibility that MPs may be candidate substances for an endogenous sleep factor has led to the interesting hypotheses, as well as experimental findings, on the following aspects.

1. MPs cannot be biosynthesized in the mammalian tissues and they are of exogenous origin.[5] However, they are detectable in mammalian tissues and body fluids as described in Chapter 2. Moreover, specific binding sites for MPs are demonstrated in murine peritoneal macrophages.[5,7] Recently, Johanssen et al.[8] have found that a chemically unidentified sleep-enhancing substance(s) was produced by macrophages during the digestion of bacteria. This indicates that mammals can process bacterial cell walls to produce biologically active MPs.
2. Specific mammalian amide-synthesizing or hydrolyzing enzymes may exist in the body that have the capacity to control the somnogenic activity of MPs,[9] since structural requirements of MPs for somnogenic activity exist.[2,5,9-14]
3. The somnogenic activity of MPs, which is concerned mainly with the induction of excess slow-wave sleep (SWS), may be mediated through the monokine, interleukin-1 (IL1), since similar excess SWS can be induced sooner by the latter substance than by MPs in rabbits.[5,9-13]
4. The functional roles of exogenous MPs could be analogous to those of vitamins and/or essential amino acids in that they may be incorporated into larger entities in order to exert their normal physiological actions.[4,9-11,13,15]
5. Somnogenic MPs concomitantly accompany pyrogenic and immunostimulatory properties.[9-12] However, the pyrogenicity and somnogenicity of MPs can be dissociated.[3,9-11] Hence, in the normal delivery of MPs, very precise mechanisms may be operative to differentially induce either somnogenicity or pyrogenicity,[3,11] perhaps in concert with other sleep factors and temperature modulators.[16]
6. MP-induced sleep may enhance immunoreaction and then serve a reperculative function in the mammalian body.[13,17-19]

B. Muramyl Tri- and Tetrapeptides as Factor S

The major effect of MPs in rabbits after a 45-min intracerebroventricular (i.c.v.) infusion was to enhance the total time spent in SWS, suppress paradoxical sleep (PS), and elevate the temperature of the brain and peripheries.[2,3,12,13,20,21] Excess SWS was induced by 1 to

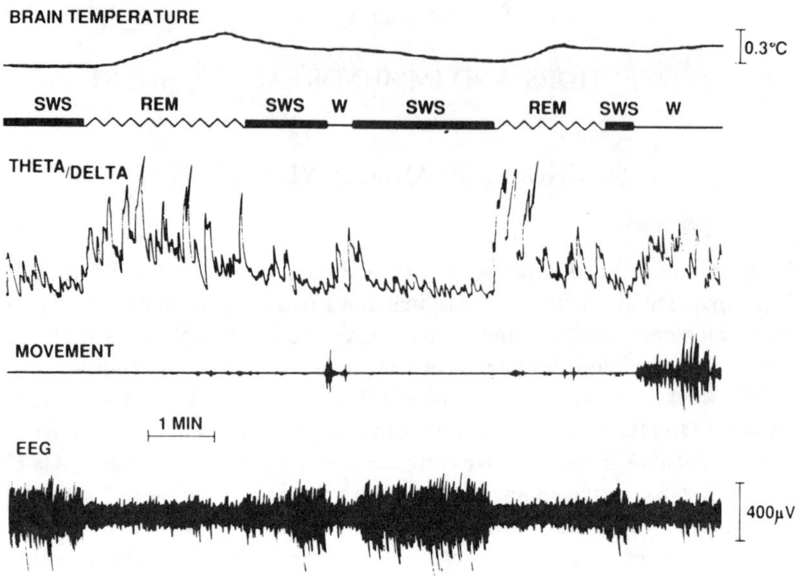

FIGURE 1. Typical polysomnographic record in normal rabbits. REM: rapid-eye-movement sleep (= PS); SWS: slow-wave sleep; W: wakefulness. THETA/DELTA ratio indicates that electroencephalographic (EEG) theta and delta activity is predominant during PS and SWS, respectively. (From Shoham, S., Ahokas, R. A., Blatteis, C. M., and Krueger, J. M., *Brain Res.*, 419, 223, 1987. With permission.)

5 pmol of NAG-1,6-anhydro-NAM-Ala-Glu-Dap-Ala and NAG-1,6-anhydro-NAM-Ala-Glu-Dap during a 2- to 6-h or longer postinfusion period. The increment was usually 40 to 80% over the control level. The rectal and brain temperatures were elevated by about 2°C during a 2- to 6-h postinfusion period. This temperature rise seems to be quite unnatural, since body temperature in normal sleep is characterized rather by a slight decrease during SWS[22,23] (see also Figure 1). In addition, the amplitude of electroencelphalographic (EEG) delta waves became 40 to 50% larger than that occurring during normal physiological sleep. Such supranormal EEG delta waves were largely similar to those observable after sleep deprivation.

In other animal species, urinary Factor S (preparations obtained at the step before the final chemical isolation and identification) reportedly was also somnogenic. In cats,[24] urinary Factor S at doses of 500 to 750 ml human urine equivalent units (less than 1.5 to 2.3 nmol), i.c.v. infused for 25 min at the beginning of the light period, induced a prolonged increase in SWS starting at the third or fourth hour throughout the subsequent 8-h period. PS was suppressed during the first few hours. In rats,[25] urinary Factor S at doses of 20 to 50 ml human urine equivalent units/30 min i.c.v., administered shortly before the dark onset period, induced an increase in SWS during the subsequent 12-h dark period.

Recently, Krueger and associates[14] have extensively investigated the somnogenicity of bacterial peptidoglycans. They compared the somnogenic activity of 14 different compounds that were closely related to the active tri- and tetramuramyl peptides, and found that a disaccharide pentapeptide containing an additional alanine at the C-terminal (NAG-1,6-anhydro-NAM-Ala-Glu-Dap-Ala-Ala) and its analogous, anhydromuramic acid-containing monosaccharide tetrapeptide lacking the glucosamine moiety (1,6-anhydro-NAM-Ala-Glu-Dap-Ala, or anhydromuramyl tetrapeptide [AMTP]) exhibited somnogenic activity similar to the original tetrapeptide identified from Factor S. These peptides were active at a dose as small as 1 pmol. In contrast, replacement of the anhydromuramyl end by a hydrated muramic acid residue resulted in complete inactivity in amounts up to 1000 pmol. From the

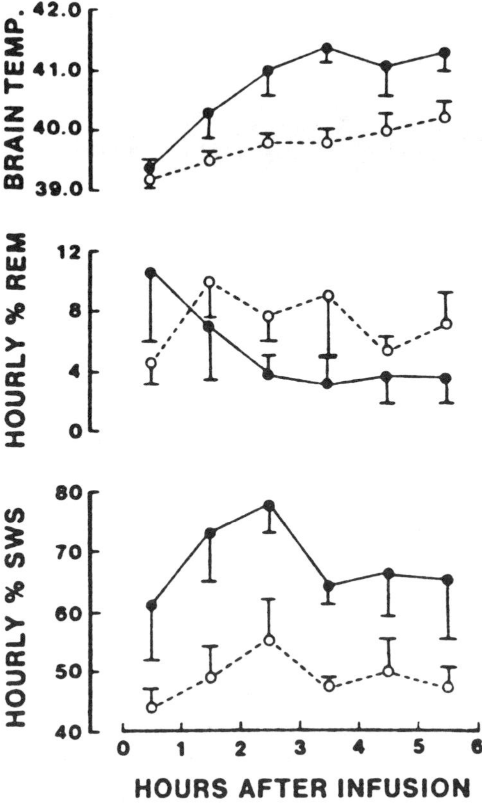

FIGURE 2. Time-course changes in brain temperature and sleep after i.c.v. infusion of anhydromuramyl tetrapeptide (AMTP; 10 pmol for 45 min; solid lines; n = 4) and artificial CSF (0.3 ml for 45 min; broken lines; n = 4). Individual points are means ± standard error of mean (SEM). (From Krueger, J. M., Davenne, D., Walter, J., Shoham, S., Kubillus, S., Rosenthal, R. S., Martin, S. A., and Biemann, K., *Brain Res.*, 403, 258, 1987. With permission.)

results, it is suggested that the anhydromuramic acid end, but not the glucosamine moiety, is essential for maximum somnogenic potency, and that the amidation of carboxyl groups on the peptide-side chain may block MP-mediated somnogenic activity. Johanssen et al.[26] further examined in their rabbit assay the somnogenic and pyrogenic activities of several MP-dimers, in which a dimer containing a (1,6-anhydro)NAM was active at the very high dose of 100 pmol.

Krueger and associates[27] have further examined the effects of AMTP on rabbit SWS, PS, and temperature. The main results are illustrated in Figures 2 to 4. AMTP at 2 doses (1 and 10 pmol i.c.v. for 45 min) significantly increased the precentage of time spent in SWS and brain and rectal temperature, but a smaller dose of AMTP failed to inhibit PS. Amplitudes of EEG delta waves increased to the supranormal levels as observable after sleep deprivation or after the administration of somnogenic MPs and IL1 (see below). AMTP increased the number of SWS episodes and decreased the number of PS episodes. The distribution of episode duration shifted considerably from longer to shorter waking and PS episodes, and from shorter to longer SWS episodes. Thus, AMTP amplified the SWS component of physiological sleep.

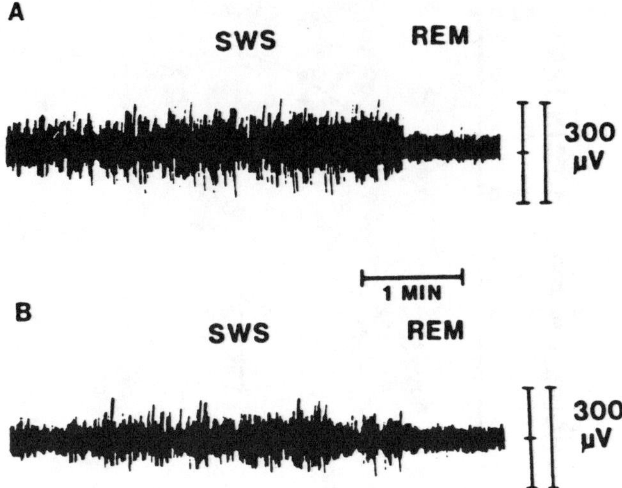

FIGURE 3. Comparison of cortical EEG records from a rabbit taken (A) after the infusion of 1 pmol AMTP and (B) during a no-infusion control period. Supranormal EEG delta waves of increased amplitude are observable after AMTP treatment. (From Krueger, J. M., Davenne, D., Walter, J., Shoham, S., Kubillus, S., Rosenthal, R. S., Martin, S. A., and Biemann, K., *Brain Res.*, 403, 258, 1987. With permission.)

Both brain and rectal temperatures increased significantly after AMTP treatment. However, the transition from one state to another exhibited dynamic temperature changes typical of those occurring between the two states in normal sleep, except for those from SWS to wakefulness, although the absolute values were larger because of fever. Therefore, it seems likely that the thermoregulatory mechanisms responsible for febrile reactions are independent, in part, from those associated with the brain temperature changes that are coupled to the state of vigilance.

As to the site of action of naturally occurring MPs in the brain, little information has been available. However, it should be noted here that García-Arrarás and Pappenheimer[28] found that urinary Factor S effectively induced significant increases in SWS when the sample was microinjected into the restricted regions of the rabbit brain. Effective receptor areas were located from the basal forebrain at the level of the optic chiasma to a region of the brainstem near the mesodiencephalic junction.

C. Muramyl Dipeptide

Muramyl dipeptide (MDP, or *N*-acetylmuramyl-L-alanyl-D-isoglutamine, NAM-L-Ala-D-isoGlu; for the chemical structure, see Figure 5) is the simplest synthetic analog of the somnogenic MPs. MDP is also known for its immunomodulatory and pyrogenic activities. Krueger et al.[9-13,23] demonstrated that MDP and its analog MDP-Lys (NAM-L-Ala-D-isoGlu-L-Lys) dose-dependently increased SWS, the amplitudes of EEG delta waves, and the brain temperature in rabbits (50 to 125 pmol for MDP; 100 to 150 pmol for MDP-Lys) and cats (200 to 500 pmol for MDP and MDP-Lys) in their routine 6-h assay. Suppression of PS was observed in cats.[9] The minimum effective dose in the rabbit assay was 50 pmol.[10] MDP was also effective when administered intravenously (i.v.; 66 to 100 μg/kg), intraperitoneally (i.p.; 160 μg/kg), or orally (5 mg/kg) in rabbits.[29] Pyrogenicity and somnogenicity of MDP can be dissociable. This is evidenced by the fact that pretreatment with acetaminophen, an antipyretic, abolished the pyrogenic effects of peripherally administered MDP (120 μg/kg i.v.) without blocking the sleep-promoting effects in adult rabbits,[8] and that MDP (100 μg/kg i.p.) caused the enhancement of quiet sleep (the precursor of SWS) but no fever in infantile rabbits 7 to 9 d old.[30]

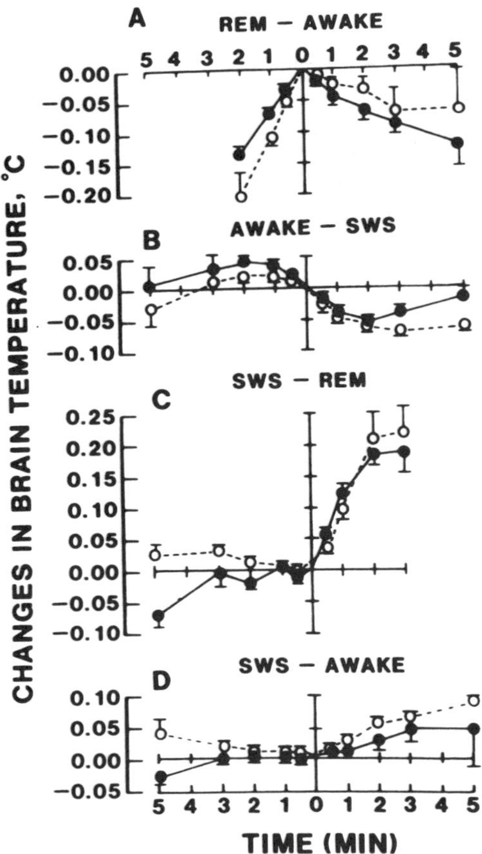

FIGURE 4. Dynamic changes in brain temperature associated with transitions between various states of vigilance. Solid and broken lines indicate results obtained after the infusion of 10 pmol AMTP and those obtained from the same animals without infusion, respectively. Individual points are means ± SEM. (From Krueger, J. M., Davenne, D., Walter, J., Shoham, S., Kubillus, S., Rosenthal, R. S., Martin, S. A., and Biemann, K., *Brain Res.*, 403, 258, 1987. With permission.)

In squirrel monkeys kept under continuous illumination with free-running, circadian rest-activity rhythms, MDP (50 nmol i.v. or i.p.) given either shortly after the beginning of the subjective day or just before the beginning of the subjective night, caused decreases in the percentage of time spent in wakefulness and PS, increases in the percentage of time in SWS, and elevations in body temperature.[31] The SWS-enhancing effect was more marked during the postinjection hours in the subjective day than during those in the subjective night. Although MDP evoked such reactions in the sleep-waking pattern and in body temperature rhythm, it exerted no consistent phase shift in the circadian timing system.

In our routine 10-h nocturnal i.c.v.-infusion assay, MDP at a dose of 2 nmol induced a significant increase in SWS (see Figure 6) and fever (see Figure 7) in rats.[32-34] However, MDP at smaller doses of 0.02 and 0.2 nmol was ineffective.[32] MDP (2 nmol i.c.v. for 10 h) was characterized by a slow SWS-promoting effect. The maximum effect was observed in the middle of the 10-h infusion period, when a marked elevation of brain temperature was observed. MDP at this dose also elicited a slight suppressive effect on PS. Febrile reaction in brain temperature was considerably long-lasting in synchronization with the

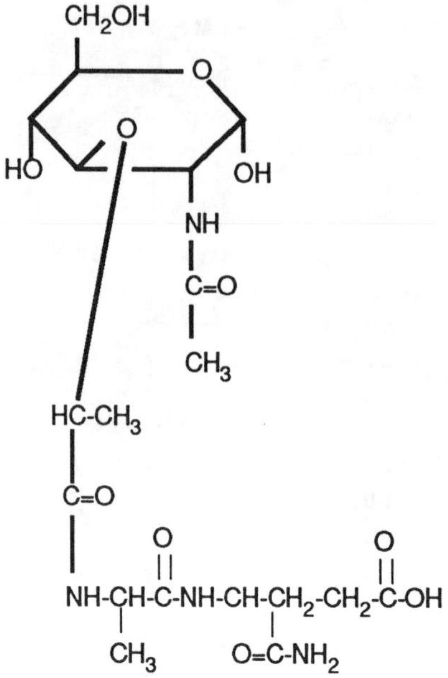

FIGURE 5. Chemical structure of muramyl dipeptide (MDP).

enhanced occurrence of EEG delta waves (see Figure 7). Kadlecová and Masek[35] reported that i.v. injection of MDP at a dose of 25 μg/kg, which provoked little elevation in rectal temperature, caused a significant increase in both SWS and PS in rats during a 5-h nocturnal postinjection period.

Similar to other sleep substances already mentioned (see also Reference 36), MDP did not induce excess sleep when administered in rats during their resting period. According to Fornal et al.,[37] MDP i.p. injected in the morning at doses of 50, 250, and 500 μg/kg in rats failed to affect sleep parameters and rectal temperature during a 6-h diurnal postinjection period, except for a significant decrease in PS and a significant prolongation of PS latency at 250 and 500 μg/kg. Interestingly, Kadlecová and Masek[35] reported a dose dependency of the effects of diurnally administered MDP on PS. According to them, i.v. injection of MDP at 25 μg/kg increased significantly the total time spent in PS during a 5-h diurnal recording period, whereas that at 2000 μg/kg markedly decreased the total PS and the accompanying fever. However, both doses of MDP exerted entirely no effect on SWS.

Shoham et al.[23] recently demonstrated that MDP (0.15 or 15 nmol i.c.v. for 5 min) dose-dependently elevated plasma copper levels in addition to dose-dependent fevers (colonic temperature rises), increases in SWS, and reductions in PS in rabbits during the first 6 h after infusion. In contrast, no changes in plasma copper levels and colonic temperature were observed after i.v. administration of MDP (15 nmol for 30 s). Since plasma copper is a biochemical measure of the host defense response, the syndrome induced by centrally administered MDP seems to indicate an activation of the response at a biochemical level.

As to anatomical levels of MDP action, somnogenic mechanisms of MDP likely involve the midbrain. The recent study of Shoham et al.[38] revealed that pretreatment with MDP (0.125 or 12.5 nmol i.c.v. for 5 min) antagonized the sleep-reducing effect of amphetamine in rabbits and significantly increased the number of SWS episodes, and that pretreatment with MDP (0.125 nmol i.c.v. for 5 min) failed to quickly diminish cortical, low-voltage EEG and reverse the wakefulness-enhancing effect of physostigmine. Thus, an interaction of amphetamine, but not physostigmine, with MDP was evident. Since amphetamine exerts

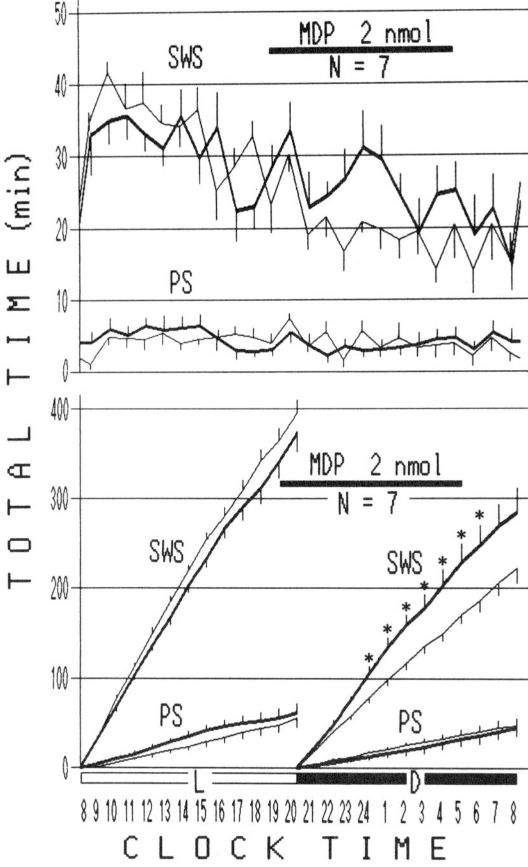

FIGURE 6. Effects of 2 nmol of MDP i.c.v. infused for 10 h at 19.00 through 05.00 h (indicated by a solid bar) on the time-course changes in the amount of sleep in otherwise saline-infused, freely behaving rats. The graph at the top shows the hourly amount of SWS and PS, while the one at the bottom illustrates their cumulative values in the light (L) and the dark (D) periods. The thin and thick curves stand for the baseline and the experimental day, respectively. The vertical lines on each hourly value indicate SEM. $*p < 0.05$. (From Inoué, S., Honda, K., Komoda, Y., Uchizono, K., Ueno, R., and Hayaishi, O., *Proc. Natl. Acad. Sci. U.S.A.*, 81, 6240, 1984. With permission.)

its action at the midbrain level, it is assumed that MDP acts at the midbrain level or below. Indeed, bilateral lesions of preoptic anterior hypothalamic areas did not affect MDP-induced increases in SWS and fever.[39] Furthermore, this assumption is substantiated by the fact that MDP (100 μg/kg i.p.) failed to affect the EEG of newborn rabbits while it did increase their behavioral quiet sleep.[30]

II. SOMNOGENIC ACTIVITY OF OTHER PYROGENS AND IMMUNOSTIMULANTS

A. Interleukin-1 (IL1)

IL1 is a family of unique polypeptides which mediate a variety of host-defense functions and possess several biological properties, including pyrogenic activity. IL1 is liberated by

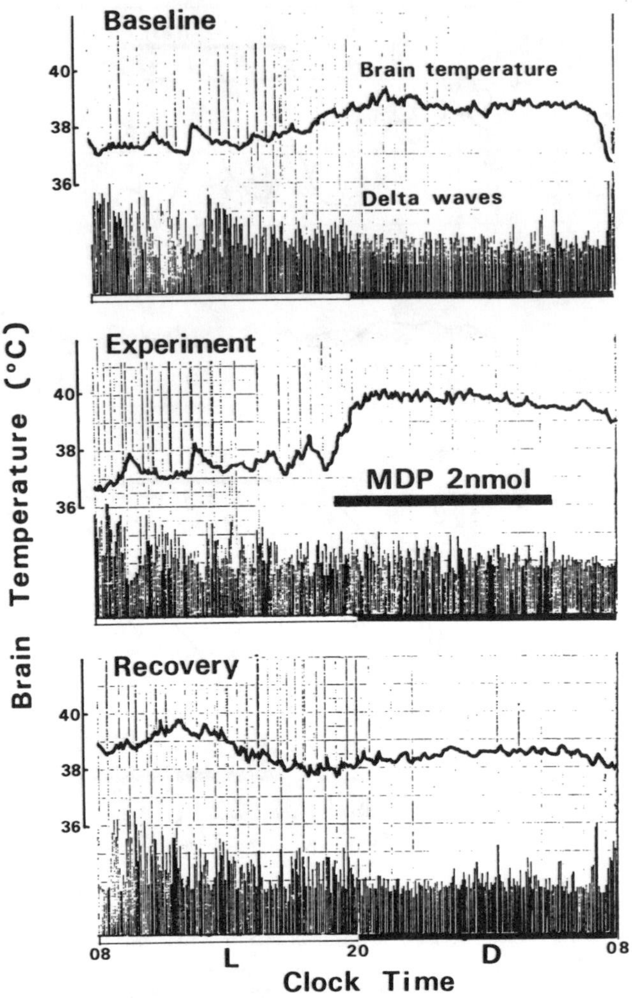

FIGURE 7. Effects of 2 nmol of MDP i.c.v. infused for 10 h at 19.00 through 05.00 h (indicated by a solid bar) on the time course changes in brain temperature and EEG delta wave activity integrated at 5-min intervals in an otherwise saline-infused, freely behaving rat. Fever occurred shortly after the initiation of MDP and lasted for the subsequent dark period (D) and the light period (L) on the next day.

macrophages to activate T-lymphocytes and induce fever by acting on hypothalamic cells. Dinarello and Krueger[40] demonstrated that synthetic MDP and naturally occurring Factor S stimulated human monocytes to produce IL1. In addition, IL1 is known to be produced by astrocytes in the brain.

Krueger et al.[16] first reported that the i.c.v. infusion of purified human macrophage-derived IL1 induced dose-dependent increases in SWS and elevations in rectal temperature in rabbits. However, the i.v. injection of IL1 induced fever and transient increases in SWS, but failed to induce prolonged increases in SWS. Macrophage-derived rabbit IL1 was also somnogenic after i.c.v. administration, but not after i.v. injection in rabbits. This observation was confirmed later by Shoham et al.,[41] who used human recombinant β-IL1 (50 to 5000 ng i.c.v. for 5 min; 1 to 10 μg/kg i.v.). Since anisomycin prevented IL1 from increasing body temperature without affecting its sleep-promoting activity, the somnogenic activity of

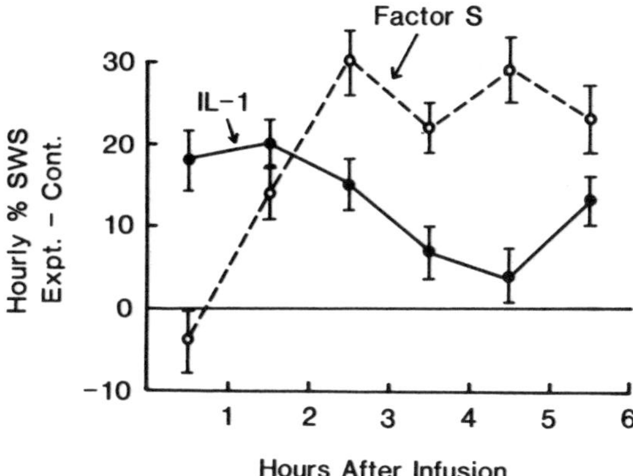

FIGURE 8. Time courses of the effects of urinary Factor S (1 to 5 pmol; n = 11; broken line) and IL1 (5 μl of purified human macrophage-derived IL1; n = 9; solid line) on rabbit SWS. Each value represents the mean ± SEM of individual experimental control differences. (From Krueger, J. M., in *Endogenous Sleep Substances and Sleep Regulation*, Inoué, S. and Borbély, A. A., Eds., Japan Scientific Societies Press, Tokyo/VNU Science Press BV, Utrecht, 1985, 181. With permission.)

IL1 seems not to be secondary to its pyrogenic activity.[16] This possibility was further substantiated by the finding that a nonpyrogenic MDP derivative, murametide, inhibited IL1-induced fever, but not IL1-enhanced SWS and IL1-depressed PS in rabbits.[42] Another indirect evidence is that state-coupled dynamic changes in brain temperature were little affected after IL1 administration in rabbits.[43]

The somnogenic and pyrogenic activities of IL1 have been confirmed by other investigators in other animal species. Using mouse astrocyte-derived IL1 in rats, Tobler et al.[44] reported that IL1 i.c.v. infused for 4 min at the onset of the light period increased SWS significantly, and suppressed PS during a subsequent 4-h recording period. However, the total sleep time did not change. Colonic temperature increased significantly 4 h after IL1 administration. Susic and Totic[45,46] also reported somnogenic and pyrogenic effects of purified human IL1 in cats. Interestingly, a single i.c.v. injection of IL1 brought about significant increases in both SWS and PS, both on the day of administration and on the next day.

In comparison with MPs, IL1 is characterized by its rapid somnogenic activity.[3,16] As illustrated in Figure 8, following IL1 infusions, excess SWS was observed during the first postinfusion hour, whereas excess SWS was not evident until the second hour after infusions of urinary Factor S. There were also differences in the time courses of fever responses. Immediately after IL1 infusions, rabbits were febrile. Following urinary Factor S, fevers developed more slowly. These facts suggest that activities of somnogenic MPs may be mediated by a step involving the production of IL1 or another monokine.[16]

Moldofsky and colleagues[47,48] measured the plasma level of IL1 activity in normal healthy humans, and found that peaks in plasma IL1 appeared shortly after sleep onset and were coincident with stage 3 + 4 sleep. Other immune functions, such as natural killer activity and mitogen responsiveness, also exhibited sleep-related changes. Sleep deprivation for 40 h resulted in a dampened nocturnal mitogen response.[49] The Canadian investigators[50] recently demonstrated in a preliminary experiment that IL1-like activity was detected in the CSF taken from the third cerebral ventricle of cats, and the activity was enhanced during sleep, especially SWS.

B. Endotoxin and Lipid A

Endotoxin, such as lipopolysaccharide (LPS), is another bacterial cell-wall component. Its lipid A moiety, in the form of either monophosphoryl lipid A (MPL) or diphosphoryl lipid A (DPL), is associated with the endotoxic activities of LPS. Both LPS and lipid A can have the capacity to induce the synthesis and release of the cytokine IL1, and many of the biological activities of these substances are regarded to be mediated through IL1. Therefore, these substances also seem to have SWS-enhancing activity. The report of Krueger et al.[51] indicates that this is the case. However, the effects of these substances on sleep are distinctly different from those of MPs and IL1, in that the effects of LPS and lipid A on the course of SWS and PS responses depend on the dose and the route of administration.

The differences are summarized as follows: LPS (5 to 25 ng/kg i.v.) dose-dependently caused increases in SWS, decreases in PS, and fever. After i.v. administration of MPL (8 to 33 μg/kg) or DPL (0.033 to 3.3 μg/kg), excess SWS was observed during the first hour (postinjection). However, after i.c.v. administration of MPL (50 to 100 ng) or DPL (5 to 50 ng), excess SWS did not appear until 3 to 4 h after the infusion. The effects of MPL and DPL on PS was characterized by suppression after i.v. administration, but not after i.c.v. infusion. Thus, it seems likely that unknown but different mechanisms are responsible for SWS, PS, and febrile responses to endotoxin and lipid A, although some of the effects of these substances are mediated by IL1.

C. Interferon

Another cytokine, interferon (IFN) alpha-2, has some homologies with IL1 in amino acid sequence. IFN is increasingly released and synthesized during the immune response to infectious agents. This leukocyte product shares the above-mentioned substances with several biological activities, including antiviral activity, pyrogenicity, and the enhancement of immune functions. The subjective feeling of drowsiness is well known in human patients after treatment with IFN.

Abrams et al.[52] postulated that morning injection of alpha IFN might be responsible for fatigue and sleepiness during daytime and poor sleep during nighttime, since the serum drug levels peaked in the afternoon. Morgano et al.[53] attributed the phenomena to the modification of cortisol levels via an indirect or a direct action of alpha IFN on the adrenal cortex. They injected healthy male volunteers intramuscularly with recombinant human IFN alpha-2a, either in the morning (09.00 h) or in the evening (21.00 h), and measured the serum level of cortisol. The morning administration of alpha IFN induced an increase of serum cortisol levels, peaking at 17.00 h, which significantly modified the physiological circadian rhythm of this hormone. On the contrary, the circadian rhythm of the endogenous cortisol was not significantly affected by the evening administration of alpha IFN. Furthermore, all subjects experienced an improvement in their energy level and better sleep. Thus, a neuro-immuno-endocrine connection may exist in the feeling of fatigue after the morning administration of alpha IFN.

Krueger and co-workers[54] first attempted to measure the effects of IFN on sleep by means of EEG techniques in rabbits. Both i.c.v. infusion (0.5 to 2 $\times$ 10^6 units for 45 min) and i.v. injection (15 $\times$ 10^6 units) of IFN alpha-2 brought about an enhancement of SWS, brain temperature, and EEG delta wave activity, and a slight suppression of PS. Thus, it is evident that IFN has somnogenic and pyrogenic activities.

D. Tumor Necrosis Factor

Another leukocyte product, tumor necrosis factor (TNF), also is a pyrogenic and immunoreactive substance. The production of this substance can be elicited by endotoxin. Shoham et al.[41] examined the effects of recombinant human TNF (rTNF) on rabbit sleep and temperature. It was found that both i.c.v. infusion (0.5 to 5 μg for 5 min) and i.v.

injection (1 to 15 μg/kg) of TNF induced an enhancement of SWS and EEG delta wave activity, and biphasic fevers, and suppressed PS. Thus, the authors concluded that TNF is also an endogenous somnogen.

E. Infectious Agents

Somnogenic substances of low molecular weight seem to be produced by macrophages during the digestion of bacteria.[8,18] After feeding macrophages with viable *Staphylococcus aureus*, the macrophage supernatant was fractionated and tested by the routine rabbit assay: SWS and rectal temperature increased, while PS decreased. In addition, polyriboinosinic: polyribocytidylic acid (poly I:C), a potent synthetic analog to viral, double-stranded RNA, has the capacity to enhance SWS and fever and suppress PS in rabbits.[55] The minimum effective doses were 0.3 μg/kg i.v. and 1 ng i.c.v. Therefore, both bacterial and viral infectious agents possibly contribute to sleep regulatory processes after an interaction with mammalian cells.

F. Histocompatibility Antigen

Narcolepsy is characterized by elevated daytime sleepiness and sleep paralysis. This sleep disorder is regarded as an autoimmune disease. A certain histocompatibility antigen, HLA-DR2, is found in almost all narcoleptics. Interestingly, several putative sleep substances are involved that alter the expression of this antigen:[18] prostagladins inhibit it, whereas MPs and endotoxin enhance it. Vasoactive intestinal polypeptide (VIP) is suspected to be involved. Moreover, the presence of HLA-DR2 can potentiate cell sensitivity to IL1. Although humoral mechanisms are unknown as to the induction of narcolepsy, these facts also suggest a close association between immune activity and sleep.

III. INTERACTIONS OF MPs WITH OTHER SLEEP FACTORS

Since the extensive studies conducted by Krueger and associates as overviewed above, it has been convincingly established that the immunoactive substances, including MPs, IL1, endotoxin, IFN, and TNF are actually involved in the induction of SWS. This fact further suggests that SWS is linked to the immune response. The question then arises as to what interrelationship exists among these and other somnogenic substances in the regulation of normal physiological sleep and of sleep as a restorative defense response to immune processes. The answer is still left open at present. However, Krueger's school[18,41,51,54] has developed the speculation that a common biochemical-physiological mechanism may exist to enhance SWS as a result of the interactions among several somnogens. In the background, there are the following evidences:

1. MPs, TNF, and lipid A can induce the production of IFN and IL1.
2. MPs, IFN, IL1, TNF, and endotoxin can alter arachidonic acid metabolism and stimulate the production of prostaglandin D_2 (PGD_2).
3. PGD_2 can stimulate the release of VIP.
4. MPs can compete with serotonin for binding sites and alter brainstem serotonin turnover.
5. Endotoxin can induce the synthesis and release of IL1 and can alter hypothalamic PGD_2.
6. IL1 can alter IFN metabolism.
7. Poly I:C can induce the production of IFN.
8. MPs and endotoxin can alter the synthesis of HLA-DR2.
9. HLA-DR2 can potentiate cell sensitivity to IL1.

Other possible interactions may exist between these immunoactive somnogens and nucleosides, as has been evidenced by our combined infusion studies of MDP and uridine in rats.[56-58] This will be described in-depth in Chapter 9.

IV. RELATED ASPECTS

A. Germ-Free Animals and the Ontogency of Sleep

If sleep, especially SWS, is triggered by bacteria-derived MPs, germ-free animals or humans may be insomniacs or hyposomniacs, or at least short sleepers. Such a possibility has not yet been examined. However, two reports published in 1987 offer some related information on this matter. One is concerned with the sleep of rabbit neonates,[30] and the other deals with human short sleepers.[59]

Rabbits are born germ-free and nonimmunoreactive. Neonates acquire immunoreactivity during the early course of postnatal life. The blood-brain barrier is not completely established. The sleep-waking states of neonatal rabbits are characterized by the frequent occurrence of active sleep, the precursor of PS, and the far less frequent occurrence of wakefulness and quiet sleep, the precursor of SWS. The two sleep states differentiate into the adult type during the first postnatal month. SWS appears in accordance with the maturation of cortical EEG slow waves. Thus, the ontogeny of immunoreactivity and SWS is largely synchronized.

According to Davenne and Krueger,[30] MDP (100 μg/kg i.p.) failed to accelerate the maturation of EEG slow waves, but increased quiet sleep and decreased active sleep in infantile rabbits (7 to 9 d of age) without causing febrile reactions. Since no follow-up study was done, no information is available on the long-term effects of MDP administration at an early juvenile stage on the subsequent maturation of sleep-waking mechanisms. However, it seems likely that bacterial products or MPs derived exogenously from maternal milk and food do not contribute to the maturation of the cerebral cortex and the development of adult-type sleep-waking patterns, but it can affect archaic areas of the brain to enhance quiet sleep/SWS.

B. Bacterial Flora and Human Sleep

Rhee and Kim[59] have recently found a positive correlation between sleep time and intestinal bacterial flora in humans. In psychiatric patients, insomniacs (n = 49) sleeping 2.2 h/night on average showed 3.0×10^3 of mean colony count, whereas noninsomniacs (n = 20) sleeping 7.5 h/night showed a mean colony count of 4.7×10^4. In patients in the general ward, short sleepers (n = 46) having a mean sleep time of 3.8 h/night showed 1.0×10^4 of mean colony count, whereas long sleepers (n = 34) having a mean sleep time of 10.4 h/night showed 4.5×10^4 of mean colony count. In healthy aged subjects (n = 9), the mean sleep time and mean colony count were 4.8 h/night and 7.5×10^3, respectively. In healthy young adults (n = 20), the mean sleep time and mean colony count were 7.2 h/night and 5.0×10^4, respectively. Aministrations of antibiotics in healthy age-matched adults (n = 20) significantly decreased the bacterial population (1.5×10^3) and concomitantly sleep time (5.2 h/night). It was also found that 92% of the short sleepers in the general ward had been treated with a prolonged therapy of antibiotics or antimicrobial drugs, while only 4% of the long sleepers had been similary treated. Thus, it was suggested that quantitative changes in bacterial flora may modify the duration of sleep both in insomniac patients and in healthy humans, and that bacterial peptideglycans, i.e., MPs, may be involved in the modification.

C. Unknown Somnogenic Material

Moldofsky et al.[47] published a preliminary report that a sleep-promoting material was demonstrated in the purified peritoneal dialyzates of patients with end-stage renal disease. This material induced excessive SWS and fever in rabbits, similar to those elicited by somnogenic MPs. No information is currently available as to the chemical structure of the material and its relation to MPs.

V. IMMUNE THEORY OF SLOW-WAVE SLEEP

There are numerous interactions between the central nervous system and the immune system as the first line of defense against a potentially dangerous environment.[60] Not only MPs and the MP-related above substances, but VIP and proopiomelanocortin-derived peptides (see Chapter 8) also modulate the immune function and states of vigilance. The hypothalamus and raphe system, which are know to be linked with sleep regulation, are involved in immune function and temperature regulation. Aging accompanies changes in both the immune function and the EEG delta sleep. Prolonged sleep deprivation leads to alterations in immune function, as well as increased CSF levels of Factor S. Infectious diseases are accompanied by a subjective feeling of sleepiness. Taking all of these facts into consideration, Krueger et al.[17] have proposed that sleep serves as an immune function. This means that sleep may be an adjunct to the immune system that aids in the recovery from environmental challenges encountered during waking activity. Such a relationship is consistent with the folk knowledge that bed rest aids in the recovery from and the prevention of pathological states.

Moldofsky et al.[47] support this view on the basis of their findings that IL1 and other immune activities were elevated during sleep.

REFERENCES

1. **Martin, S. A., Karnovsky, M. L., Krueger, J. M., Pappenheimer, J. R., and Biemann, K.,** Peptideglycans as promoters of slow-wave sleep. I. Structure of the sleep-promoting factor isolated from human urine, *J. Biol. Chem.,* 259, 12652, 1984.
2. **Krueger, J. M., Karnovsky, M. L., Martin, S. A., Pappenheimer, J. R., Walter, J., and Biemann, K.,** Peptideglycans as promoters of slow-wave sleep. II. Somnogenic and pyrogenic activities of some naturally occurring muramyl peptides; correlations with mass spectrometric structure determination, *J. Biol. Chem.,* 259, 12659, 1984.
3. **Krueger, J. M.,** Muramyl peptides and interleukin-1 as promoters of slow wave sleep, in *Endogenous Sleep Substances and Sleep Regulation,* Inoué, S. and Borbély, A. A., Eds., Japan Scientific Societies Press, Tokyo/VNU Science Press BV, Utrecht, 1985, 181.
4. **Adam, A. and Lederer, E.,** Muramyl peptides: immunomodulators, sleep factors, and vitamins, *Med. Res. Rev.,* 4, 111, 1984.
5. **Karnovsky, M. L.,** Muramyl peptides in mammalian tissues and their effects at the cellular level, *Fed. Proc., Fed. Am. Soc. Exp. Biol.,* 45, 2556, 1986.
6. **Werner, G. H., Floc'h, F., Migliore-Samour, D., and Jollès, P.,** Immunomodulating peptides, *Experientia,* 42, 521, 1986.
7. **Silverman, D. H. S., Krueger, J. M., and Karnovsky, M. L.,** Specific binding sites for muramyl peptides on murine macrophages, *J. Immunol.,* 136, 2195, 1986.
8. **Johanssen, L., Wecke, J., and Krueger, J. M.,** Macrophages produce a sleep-enhancing substance(s) during the digestion of bacteria, *Book of Abstracts,* Int. Symp. Current Trends in Slow Wave Sleep Res., (Physiological and Pathological Aspects), 42, 1987.
9. **Krueger, J. M., Pappenheimer, J. R., and Karnovsky, M. L.,** Sleep-promoting effects of muramyl peptides, *Proc. Natl. Acad. Sci. U.S.A.,* 79, 6102, 1982.
10. **Krueger, J. M., Walter, J., Karnovsky, M. L., Chedid, L., Choay, J. P., Lefrancier, and Lederer, E.,** Muramyl peptides. Variation of somnogenic activity with structure, *J. Exp. Med.,* 159, 68, 1984.
11. **Krueger, J. M.,** Endogenous sleep factors, in *Sleep: Neurotransmitters and Neuromodulators,* Wauquier, A., Gaillard, J. M., Monti, J. M., and Radulovacki, M., Eds., Raven Press, New York, 1985, 319.
12. **Krueger, J. M.,** Somnogenic activity of muramyl peptides, *Trends Neurosci.,* 8, 218, 1985.
13. **Krueger, J. M.,** Muramyl peptide enhancement of slow-wave sleep, *Meth. Find. Exp. Clin. Pharmacol.,* 8, 105, 1986.
14. **Krueger, J. M., Rosenthal, R. S., Martin, S. A., Walter, J., Davenne, D., Shoham, S., Kubillus, S., and Biemann, K.,** Bacterial peptidoglycans as modulators of sleep. I. Anhydro forms of muramyl peptides enhance somnogenic potency, *Brain Res.,* 403, 249, 1987.

15. **Chedid, L., Bahr, G. M., Riveau, G., and Kruger, J. M.,** Specific absorption with monoclonal antibodies to muramyl dipeptide of the pyrogenic and somnogenic activities of rabbit monokine, *Proc. Natl. Acad. Sci. U.S.A.,* 81, 5888, 1984.
16. **Krueger, J. M., Walter, J., Dinarello, C. A., Wolff, S. M., and Chedid, L.,** Sleep-promoting effects of endogenous pyrogen (interleukin-1), *Am. J. Physiol.,* 246, R994, 1984.
17. **Krueger, J. M., Walter, J., and Levin, C.,** Factor S and related somnogens: an immune theory for slow-wave sleep, in *Brain Mechanisms of Sleep,* McGinty, D. J., Drucker-Colín, R., Morrison, A., and Parmeggiani, P. L., Eds., Raven Press, New York, 1987, 253.
18. **Krueger, J. M., Toth, L. A., Cady, A. B., Johanssen, L., and Obál, F., Jr.,** Immunomodulation and sleep, in *Sleep Peptides: Basic and Clinical Approaches,* Inoué, S. and Schneider-Helmert, D., Eds., Japan Scientific Societies Press, Tokyo/Springer-Verlag, Berlin, 1988, 95.
19. **Krueger, J. M. and Karnovsky, M. L.,** Sleep and immune response, *Ann. N. Y. Acad. Sci.,* 496, 510, 1987.
20. **Krueger, J. M.,** Somnogenic muramyl peptides, in *Biological Properties of Peptidoglycan,* Seidl, P. H. and Schleifer, K. H., Eds., Walter de Gruyter, Berlin, 1986, 329.
21. **Krueger, J. M., Karaszewski, J. H., Davenne, D., and Shoham, S.,** Somnogenic muramyl peptides, *Fed. Proc., Fed. Am. Soc. Exp. Biol.,* 45, 2552, 1986.
22. **Obál, F., Jr.,** Thermoregulation and sleep, *Exp. Brain Res.,* Suppl. 8, 157, 1984.
23. **Shoham, S., Ahokas, R. A., Blatteis, C. M., and Krueger, J. M.,** Effects of muramyl dipeptide on sleep, body temperature and plasma copper after intracerebral ventricular administration. *Brain Res.,* 419, 223, 1987.
24. **García-Arrarás, J.,** Effects of sleep-promoting factor from human urine on sleep cycles of cats, *Am. J. Physiol.,* 241, E269, 1981.
25. **Krueger, J. M., Bacsik, J., and García-Arrarás, J.,** Sleep-promoting material from human urine and its relation to factor S from brain, *Am. J. Physiol.,* 238, E116, 1980.
26. **Johannsen, L., Cady, A. B., and Krueger, J. M.,** Muramyl peptide structure — somnogenic activity relationships, *Sleep Res.,* 16, 96, 1987.
27. **Krueger, J. M., Davenne, D., Walter, J., Shoham, S., Kubillus, S., Rosenthal, R. S., Martin, S. A., and Biemann, K.,** Bacterial peptidoglycans as modulators of sleep. II. Effects of muramyl peptides on the structure of rabbit sleep, *Brain Res.,* 403, 258, 1987.
28. **García-Arrarás, J. and Pappenheimer, J. R.,** Site of action of sleep-inducing muramyl peptide isolated from human urine: microinjection studies in rabbit brains, *J. Neurophysiol.,* 49, 528, 1983.
29. **Krueger, J. M.,** Sleep-promoting factor S: characterization and analogs, in *Sleep 1982,* Koella, W. P., Ed., S. Karger, Basel, 1983, 107.
30. **Davenne, D. and Krueger, J. M.,** Enhancement of quiet sleep in rabbit neonates by muramyl dipeptide, *Am. J. Physiol.,* 253, R646, 1987.
31. **Wexler, D. B. and Moore-Ede, M. C.,** Effects of a muramyl dipeptide on the temperature and sleep-wake cycles of the squirrel monkey, *Am. J. Physiol.,* 247, R672, 1984.
32. **Inoué, S., Honda, K., Nishida, S., and Komoda, Y.,** Temporal changes in sleep-promoting effects of constantly infused sleep substances in rats, *Sleep Res.,* 12, 81, 1983.
33. **Inoué, S., Honda, K., Komoda, Y., Uchizono, K., Ueno, R., and Hayaishi, O.,** Differential sleep-promoting effects of five sleep substances nocturnally infused in unrestrained rats, *Proc. Natl. Acad. Sci. U.S.A.,* 81, 6240, 1984.
34. **Inoué, S.,** Sleep substances: their roles and evolution, in *Endogenous Sleep Substances and Sleep Regulation,* Inoué, S. and Borbély, A. A., Eds., Japan Scientific Societies Press, Tokyo/VNU Science Press BV, Utrecht, 1985, 3.
35. **Kadlecová, O. and Masek, K.,** Muramyl dipeptide and sleep in rat, *Meth. Find. Exp. Clin. Pharmacol.,* 8, 111, 1986.
36. **Inoué, S., Honda, K., Komoda, Y., Uchizono, K., Ueno, R., and Hayaishi, O.,** Little sleep-promoting effect of three sleep substances diurnally infused in unrestrained rats, *Neurosci. Lett.,* 49, 207, 1984.
37. **Fornal, C., Markus, R., and Radulovacki, M.,** Muramyl dipeptide does not induce slow-wave sleep or fever in rats, *Peptides,* 5, 91, 1984.
38. **Shoham, S., Davenne, D., and Krueger, J. M.,** Muramyl dipeptide, amphetamine, and physostigmine: effects on sleep of rabbits, *Physiol. Behav.,* 41, 179, 1987.
39. **Shoham, S., Blatteis, C. M., Ahokas, R. A., and Krueger, J. M.,** Effects of bilateral preoptic anterior hypothalamic damage on muramyl dipeptide-induced sleep, fever and hypercupremia, *Sleep Res.,* 16, 145, 1987.
40. **Dinarello, C. A. and Krueger, J. M.,** Induction of interleukin-1 by synthetic and naturally-occurring muramyl peptides, *Fed. Proc., Fed. Am. Soc. Exp. Biol.,* 45, 2545, 1986.
41. **Shoham, S., Davenne, D., Cady, A. B., Dinarello, C. A., and Krueger, J. M.,** Recombinant tumor necrosis factor and interleukin-1 enhance slow-wave sleep in rabbits, *Am. J. Physiol.,* 253, R142, 1987.

42. **Cady, A. B., Johanssen, L., Riveau, G., Chedid, L., Dinarello, C. A., and Krueger, J. M.,** Interleukin-1-induced febrile but not slow wave sleep responses are inhibited by the muramyl peptide murametide, *Sleep Res.*, 16, 79, 1987.
43. **Walter, J., Davenne, D., Shoham, S., Dinarello, C. A., and Krueger, J. M.,** Brain temperature changes coupled to sleep states persist during interleukin-1-enhanced sleep, *Am. J. Physiol.*, 250, R96, 1986.
44. **Tobler, I., Borbély, A. A., Schwyzer, M., and Fontana, A.,** Interleukin-1 derived from astrocytes enhances slow wave activity in sleep, *Eur. J. Pharmacol.*, 104, 191, 1984.
45. **Susic, V. and Totic, S.,** Short and long-term effects of interleukin-1 on sleep and temperature in the cat, *Sleep Res.*, 16, 150, 1987.
46. **Totic, S. and Susic, V.,** Effects of interleukin-1 on the body temperature and sleep in the cat, *Sleep Res.*, 16, 151, 1987.
47. **Moldofsky, H., Gorczynski, R. M., Lue, F. A., and Keystone, E.,** Interleukins (IL-1 and IL-2) and immune functions during sleep, *Sleep Res.*, 13, 42, 1984.
48. **Moldofsky, H., Lue, F. A., Eisen, J., Keystone, E., and Gorczynski, R. M.,** The relationship of interleukin-1 and immune functions to sleep in humans, *Psychosom. Med.*, 48, 309, 1986.
49. **Moldofsky, H., Lue, F. A., Saskin, P., Davidson, J. R., and Gorczynski, R. M.,** The effect of sleep deprivation on immune functions in humans. I. Mitogen and natural killer cell activities, *Sleep Res.*, 16, 531, 1987.
50. **Lue, F. A., Bail, M., Gorczynski, R., and Moldofsky, H.,** Sleep and interleukin-1-like activity in cat cerebrospinal fluid, *Sleep Res.*, 16, 51, 1987.
51. **Krueger, J. M., Kubillus, S., Shoham, S., and Davenne, D.,** Enhancement of slow wave sleep by endotoxin and lipid A, *Am. J. Physiol.*, 251, R591, 1986.
52. **Abrams, P. G., McClamrock, E., and Foon, K. A.,** Evening administration of alpha interferon, *N. Engl. J. Med.*, 312, 443, 1985.
53. **Morgano, A., Puppo, F., Criscuolo, D., Lotti, G., and Indiveri, F.,** Evening administration of alpha interferon: relationship with the circadian rhythm of cortisol, *Med. Sci. Res.*, 15, 615, 1987.
54. **Krueger, J. M., Dinarello, C. A., Shoham, S., Davenne, D., Walter, J., and Kubillus, S.,** Interferon alpha-2 enhances slow-wave sleep in rabbits, *Int. J. Immunopharmacol.*, 9, 23, 1987.
55. **Endsley, J., Ahokas, R. A., Majde, J. A., Blatteis, C. M., and Krueger, J. M.,** Enhancement of slow-wave sleep by polyriboinosinic: polyribocytidic acid, *Fed. Proc., Fed. Am. Soc. Exp. Biol.*, 46, 1128, 1987.
56. **Inoué, S., Kimura, M., Honda, K., and Komoda, Y.,** Sleep peptides: general and comparative aspects, in *Sleep Peptides: Basic and Clinical Approaches,* Inoué, S. and Schneider-Helmert, D., Eds., Japan Scientific Societies Press, Tokyo/Springer-Verlag, Berlin, 1988.
57. **Kimura, M., Honda, K., Komoda, Y., and Inoué, S.,** Interacting sleep-modulatory effects of simultaneously administered delta-sleep-inducing peptide (DSIP), muramyl dipeptide (MDP) and uridine in unrestrained rats, *Neurosci. Res.*, 5, 157, 1987.
58. **Kimura, M., Honda, K., Komoda, Y., and Inoué, S.,** Effects of serially administered MDP and uridine on nocturnal sleep in rats, *Zool. Sci.*, 4, 1088, 1987.
59. **Rhee, Y.-H. and Kim, H.-I.,** The correlation between sleeping-time and numerical change of intestinal normal flora in psychiatric insomnia patients, *Bull. Nat. Sci. Chungbuk Natl. Univ.*, 1, 159, 1987.
60. **Veldhuid, H. D., Croiset, G., Ballieux, R. E., Heijnen, C. J., and De Wied, D.,** Emotional stress and modulation of immunological reactivity, in *Hypothalamic Dysfunction in Neuropsychiatric Disorders,* Goodwin, F. K. and Costa, E., Eds., Raven Press, New York, 1987, 237.

Chapter 6

PROSTAGLANDINS

I. PROSTAGLANDINS AS NEUROMODULATORS

A. Natural Occurrence and Biochemistry

Prostaglandins (PG) occur in most mammalian tissues and are released on tissue stimulation or activation. The concentrations of individual PGs vary considerably in the tissues and species. Certain PGs, such as PGD_2, PGE_2, and $PGF_{2\alpha}$ are natural constituents of the brain in various mammals.[1-3] PGD_2 is known to be the major substance among PGs in the rat brain.[4] The amount of PGD_2 measured by radioimmunoassay was 3.42 ng/brain, while that of PGE_2 and PGF_2 were 1.32 and 0.96 ng/brain, respectively.[5] PGD_2 was distributed at high concentrations in the hypothalamus (2.1 ng/g), the pituitary (7.7 to 24.7 ng/g), and the pineal gland (40.4 ng/g) of rats.[5] In the mouse brain, it is reported that basal concentrations were below 2.11 ng/g for PGD_2, below 0.38 ng/g; for PGE_2, and below 2.64 ng/g for $PGF_{2\alpha}$.[6] In the brain of spontaneously convulsing gerbils, basal levels of PGD_2 and $PGF_{2\alpha}$ were highest in the hypothalamus (9.67 and 5.19 ng/g, respectively) and the cerebral cortex (8.01 and 5.27 ng/g, respectively).[7]

PGs are produced from arachidonic acid located in the cell wall by a cyclooxygenase at the first step of biosynthesis. Subsequent steps are shown in Figure 1. Both PGD_2 synthetase and PGD_2 dehydrogenase are found in the rat and swine brain, and the former enzyme exhibited high concentrations, especially in the hypothalamus and thalamus.[8-10] Interestingly, in night-active animals like rats, PGD_2 synthetase exhibits a clear circadian variation in its activity, while PGD_2 dehydrogenase does not[11] (see Figure 2). Moreover, the cellular localization of PGD synthetase immunohistochemically detected in developing and mature rat brains differed considerably and age-dependently migrated from the neurons in neonates to oligodendrocytes in adult rats.[12] Hence, it seems likely that the major site of PGD_2 synthesis changes from neurons to oligodendrocytes, and that PGD_2 plays different roles in these cells at each developmental and mature stage.

Specific binding proteins or receptors for PGD_2, PGE_2, and/or $PGF_{2\alpha}$ are found to be distributed in restricted brain regions including the cerebral cortex, the amygdala, and the hypothalamus in rats,[13,14] Japanese monkeys,[15] and humans.[16] PGD_2 seems to be rapidly taken up intact by the brain and to rapidly disappear. Brain uptake of PGD_2 in mice was dose-dependent over the dose range from 0.1 to 1.0 mg/kg after intravenous (i.v.) administration. The half-life of PGD_2 in the brain and blood was 1.1 and 0.9 min, respectively. Thus, PGD_2 can pass the blood-brain barrier. However, the permeability of the blood-brain barrier to PGD_2 is rather low, since about 0.08% of 1.0 mg of PGD_2 could be detected in the brain 30 s after its i.v. injection.[17]

B. Central Nervous System and PGs

As the central actions of PGs, pharmacological and/or physiological effects on behavior, body temperature, food intake, analgesia, pituitary hormone secretion, respiration, and spinal cord reflex have been documented.[1,2,18] Sedative properties are commonly demonstrated after the administration of the E series of PGs, but not $PGF_{2\alpha}$. As early as 1964, Horton[19] first observed that intracerebroventricular (i.c.v.) injection of PGE_1 (3 to 20 μg/kg), PGE_2 (12 μg/kg), and PGE_3 (12 μg/kg) induced sedation, stupor, and signs of catatonia. Slight sedation was also produced after i.v. injection of PGE_1 (20 μg/kg). They also observed in chicks respiratory depression, profound sedation, and the loss of normal posture and the righting reflex after i.v. injection of the three kinds of PGs (10 to 400 μg/kg). Gilmore and Shaikh[20]

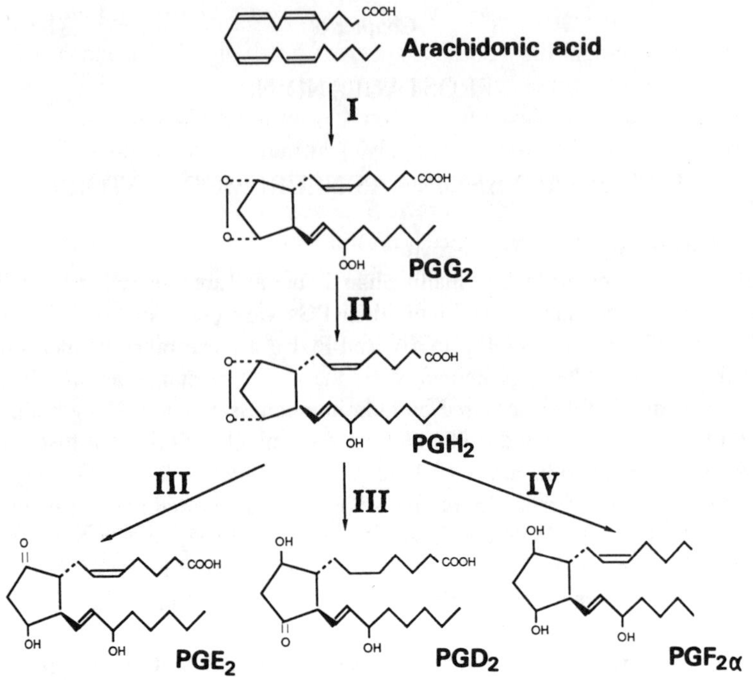

FIGURE 1. Steps in biosynthesis of prostaglandins (PGs) from arachidonic acid. I: cyclooxygenase; II: hydroperoxydase; III: isomerase (synthetase); IV: reductase.

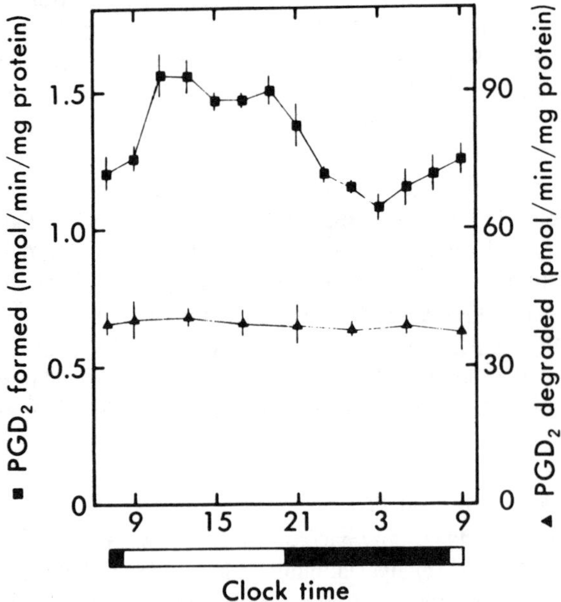

FIGURE 2. Circadian rhythm of PGD_2 synthetase (square) and PGD_2 dehydrogenase (triangle) in the rat brain. (From Ueno, R., Hayaishi, O., Osama, H., Honda, K., Inoué, S., Ishikawa, Y., and Nakayama, T., in *Endogenous Sleep Substances and Sleep Regulation*, Inoué, S. and Borbély, A. A., Eds., Japan Scientific Societies Press, Tokyo/VNU Science Press BV, Utrecht, 1985, 193. With permission.)

reported that a single subcutaneous (s.c.) injection of 0.5 mg of PGE_1 or PGE_2 in rats rapidly exerted a sedative or tranquilizing effect and concomitantly induced an increased respiratory rate, flushing of the extremities, and mild diarrhea.

Using electroencephalographic (EEG) monitoring techniques, Haubrich et al.[21] observed that intraperitoneal (i.p.) injection of PGE_1 (1 mg/kg) induced a paradoxical sleep (PS)-like state in rats. Desiraju[22] demonstrated somnolence and marked decreases in spontaneous neuronal discharges in the prefrontal cortex, following postural disturbance, catatonia, and stupor after i.c.v. injection of PGE_1 (50 μg/kg) in rhesus monkeys. Nistico and Marley[23,24] also observed that sedation, behavioral and EEG sleep, and fever were induced after i.c.v. or intrahypothalamic injection of PGE_1 (2 to 10 μg), PGE_2 (2.8 to 42 nmol), and PGA_1 (8.4 to 56 nmol). In contrast, neither sedation nor sleep, but hypothermia occurred after $PGF_{2\alpha}$ administration (14 to 56 nmol). Masek et al.[25] found that PGE_1 (50 μg/kg i.c.v.). caused a significant elevation of wakefulness and a reduction of PS in rats, whereas Dzoljic[26] showed that the same treatment with PGE_1 (50 to 100 μg/kg i.c.v.) also resulted in an elevation of wakefulness, but a significant reduction of slow-wave sleep (SWS) in rats. However, in cats, no such change was induced by PGE_1 (50 μg/kg i.c.v.) except for increases in ponto-geniculo-occipital (PGO) spikes, even during behavioral wakefulness and SWS.

On the contrary, the central actions of PGD_2 had been unclear, although this PG was assumed to act as a neuromodulator,[8] before Ueno et al.[27,28] first attempted to microinject PGD_2 into the preoptic area (POA) of rats. They found that PGD_2 (6 nmol/kg) elicited hypothermia as revealed in brain and colonic temperature, while the same dose of PGE_2 and $PGF_{2\alpha}$ induced hyperthermia.[27] In addition, profound EEG sleep occurred after microinjection of PGD_2 (0.3 to 5 nmol/rat) into the POA with a concomitant induction of hypothermia and bradycardia.[28] PGE_2 and $PGF_{2\alpha}$ exerted a far weaker effect on sleep that PGD_2.

It should be noted that these results were obtained in rats on a small platform under restrained conditions. Interestingly, we found no somnogenic activity during a long-term diurnal i.c.v. infusion of PGD_2 in unrestrained rats[11,29] and, recently, Matsumura et al.[30] have found that weak hyperthermia, but neither hypothermia nor enhancement of sleep, occurred after an acute microinjection of PGD_2 (2.5 nmol/rat) into the POA in loosely restrained rats, which were housed in a small box without the allowance of body turning. Further details will be described below.

C. PGD_2 and Hypothalamic Neuronal Firing Activity

Neuronal activities are modified by the local administration of PGD_2 to the POA and the posterior hypothalamic area (PHA), both of which are known to be closely related to the regulation of sleep.[31,32] Inokuchi and Oomura[33,34] examined the effects of electrophoretically applied PGD_2 on neuronal activity and on the interactions with some neurotransmitters in the lateral POA and the PHA of urethane-anesthetized rats.

PGD_2 excited and inhibited 20 and 26%, respectively of the tested neurons in the POA (n = 105).[33] Bidirectional responses were observed in 6% of them. The direct effects often showed desensitization after repeated applications. PGD_2-excitatory neurons were significantly excited in response to acetylcholine (Ach). The excitatory effect of Ach on POA neurons was attenuated, blocked, or reversed by concurrent applications of PGD_2 in 38% of the neurons. The excitatory and inhibitory effects of noradrenaline (NA) was modulated by concurrent applications of PGD_2 in 58% of the neurons. Changes of the NA responses were also observed after a PGD_2 infusion into the third ventricle at doses of 50 to 150 nmol. In addition, neurotransmission in the POA following the stimulation of the ventral noradrenergic bundle was modified by the application of PGD_2 in 43% of the neurons. Similar modulatory effects of PGD_2 were also observed in the PHA (n = 32), but little interactions with Ach and NA occurred.

On the basis of these observations, the Japanese neurophysiologists have concluded that

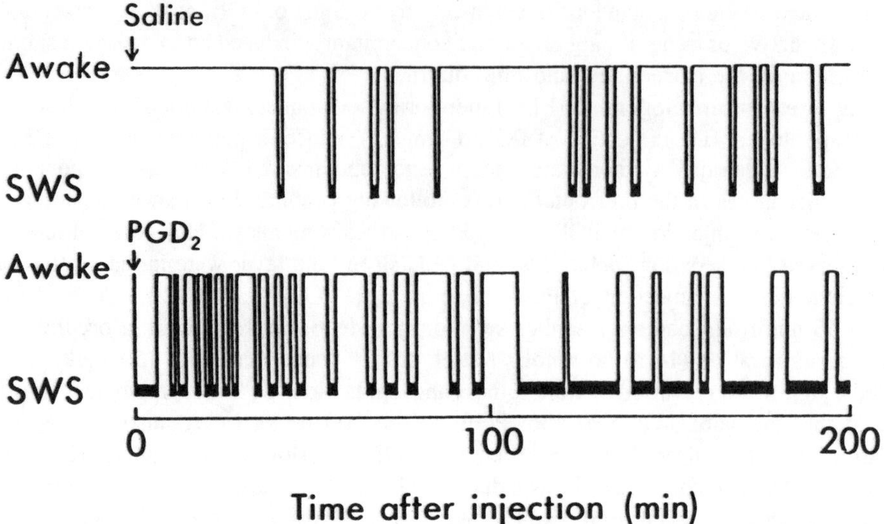

FIGURE 3. Hypnogram illustrating the time course of sleep-waking state alternations after the preoptic microinjection of saline (top) and PGD_2 (2.5 nmol, bottom) in a rat. (From Ueno, R., Hayaishi, O., Osama, H., Honda, K., Inoué, S., Ishikawa, Y., and Nakayama, T., in *Endogenous Sleep Substances and Sleep Regulation*, Inoué, S. and Borbély, A. A., Eds., Japan Scientific Societies Press, Tokyo/VNU Science Press BV, Utrecht, 1985, 193. With permission.)

PGD_2 is a neuromodulator, and probably modulates neurotransmission in the POA, and that, through this mechanism, PGD_2 plays an important role in sleep-waking and temperature regulation.

II. SLEEP-ENHANCING ACTIVITY OF PGD_2

A. Experiments in Restrained Rats

As mentioned above, Ueno et al.[28] observed an enhancement of SWS after microinjection of PGD_2 into the POA of rats. In this study, rats were restrained on a small platform without the allowance of free movement. The microinjection of saline or PGD_2, and subsequent recordings of EEG and colonic temperature, were done during daytime. Thus, in spite of their resting phase, the animals were forced to be alert during the experimental session for up to 6 h. Consequently, the rats slept little and no sign of PS was observed (see Figure 3). Nevertheless, PGD_2 at the dose range of 0.3 to 5 nmol promptly and dose-dependently induced SWS, which episodically occurred during a 200-min postinjection period much more frequently than in saline-injected rats. In contrast, microinjection of PGD_2 into the PHA exerted little effect on sleep, and microinjection of PGE_2 and $PGF_{2\alpha}$ into the POA slightly increased the amount of sleep (see Figure 4). It is also noted that microinjection of PGD_2 into the thalamus and cerebral cortex resulted in no change in the amount of sleep.[35]

These results indicate that the POA is a site of action of PGD_2 for inducing sleep. Since anatomical and neurophysiological evidence indicates that the POA plays an important role in sleep regulation,[31,32] and since high concentrations of PGD_2 receptors are demonstrated in this region (see above), the site of action of PGD_2 for sleep induction coincided with the sleep center.[11]

Although the acute microinjection of PGD_2 into the rat POA induced a profound increase in SWS, animals restrained under such condition might not be able to exhibit physiological sleep. Indeed, as mentioned above, the rats showed prolonged wakefulness and no occurrence of PS, which cannot be observed in this species during daytime under normal conditions.

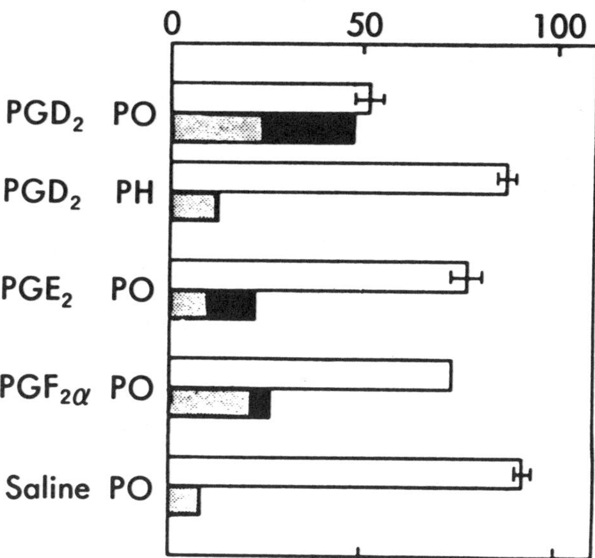

FIGURE 4. The ratios (percent of 3-h recording time) spent in wakefulness (open), light SWS (hatched) and deep SWS (closed) after the preoptic (PO) and posterior hypothalamic (PH) microinjection of saline and PGs (2.5 nmol) in rats (mean ± standard deviation, each n = 4 except for n = 2 in the $PGF_{2\alpha}$ group). (From Ueno, R., Hayaishi, O., Osama, H., Honda, K., Inoué, S., Ishikawa, Y., and Nakayama, T., in *Endogenous Sleep Substances and Sleep Regulation*, Inoué, S. and Borbély, A. A., Eds., Japan Scientific Societies Press, Tokyo/VNU Science Press BV, Utrecht, 1985, 193. With permission.)

It was further demonstrated in later studies that unrestrained[29] and semirestrained[30] rats did not show increases in diurnal sleep in response to PGD_2 administration. Therefore, another more convincing experimental approach to the somnogenic property of PGD_2 was conducted in our laboratory.

B. Experiments in Unrestrained Rats

1. Nocturnal Infusion

Using our routine 10-h nocturnal i.c.v. infusion technique, we demonstrated that small amounts of PGD_2 (0.036 to 3.6 nmol/rat) did enhance physiological sleep in freely behaving rats.[11,35-37] The somnogenic effect depended on the dose of PGD_2; a small dose of 3.6 pmol/rat was ineffective. The minimum dose required for a significant increase in SWS and PS was 0.036 and 0.36 nmol respectively. The results are summarized in Table 1.

As illustrated in Figure 5, 10-h nocturnal i.c.v. infusion of PGD_2 was characterized by a rapid increase in SWS and a gradual increase in PS. However, more striking and significant changes took place in the second half of the infusion period. Hence it is suggested that PGD_2 may primarily induce sleep and secondarily activate sleep-maintaining mechanisms.[37,38] The increases in the amount of sleep were due to the more frequent occurrence of SWS and PS episodes, but not due to their prolongation.[37] Thus, the normal episodic structure of sleep cycles remained unaltered during the PGD_2 infusion period.

We also found that nocturnal infusion of $PGF_{2\alpha}$ (0.36 nmol i.c.v.) increased slightly but significantly the amount of SWS by 15% without changing the amount of PS.[11,35] On the other hand, a nocturnal infusion of PGE_2 (0.36 nmol i.c.v.) failed to exhibit somnogenic activity in rats (see Table 1).

Table 1
EFFECT OF PGs ON SLEEP IN RATS

Dose (nmol/10 h)	n	% SWS	% PS
Saline			
0	20	100	100
PGD_2			
0.0036	3	96.5 ± 2.3	89.1 ± 7.9
0.036	4	122.4 ± 3.6**	110.3 ± 11.9
0.36	8	133.3 ± 3.9**	156.2 ± 16.0*
3.6	4	125.5 ± 4.8**	178.3 ± 12.1**
$PGF_{2\alpha}$			
0.36	6	115.2 ± 6.0*	100.4 ± 16.3
PGE_2			
0.36	8	96.3 ± 5.1	104.3 ± 11.7

*$p < 0.05$; ** $p < 0.001$, as compared to saline control.

From Ueno, R., Honda, K., Inoué, S., and Hayaishi, O., *Proc. Natl. Acad. Sci. U.S.A.*, 80. 1735, 1983. With permission.

2. *Diurnal Infusion*

Similar to many other putative sleep substances already mentioned in previous chapters, PGD_2, given during the daytime, is ineffective in inducing excess sleep, at least in rats under unrestrained conditions of our sleep assay system.[11,29,37,38] In freely behaving rats, an optimal dose of PGD_2 (0.36 nmol), which can induce profound increases in both SWS and PS at our routine nocturnal i.c.v. infusion, failed to elicit somnogenic properties when it was i.c.v. infused during a 10-h light period (see Figure 6). Although several hourly values of SWS exceeded the baseline level, largely due to hour-to-hour fluctuations, no increase was induced in the total amount of diurnal sleep, both SWS and PS, by the diurnal PGD_2 infusion. Subsequent nocturnal sleep also was not affected. A smaller and even a larger dose, 0.036 and 3.6 nmol, respectively, similarly failed to exhibit enhancement of sleep. These results show a sharp contrast not only to those from the nocturnal infusion of the same doses of PGD_2, but also to those from the acute microinjection of PGD_2 into the POA in restrained rats.

The circadian differences may lead to the following assumption: under the condition that physiologically required sleep is satisfied, the somnogenic activity of an excessive supply of PGD_2 might be cancelled by the endogenous regulatory mechanisms, and no supranormal excess sleep is induced by the exogenously supplemented PGD_2. Such aspects appear to be common in the sleep modulation of sleep substances in general.

In this connection, biosynthesis of PGD_2 is known to show a clear circadian variation in the rat brain (see above). The activity of PGD_2 synthetase in the rat brain is higher during the light period than during the dark period, whereas the activity of PGD_2 dehydrogenase did not change over the light-dark cycle[11] (see also Figure 2). Thus, sufficient endogenous PGD_2 is physiologically available during the light period.

C. Experiments in Rabbits

The somnogenic activity of PGD_2 was confirmed by Krueger,[39] who reported that i.c.v. infusion of PGD_2 (250 pmol) increased significantly SWS and rectal temperature in his routine rabbit assay (see Figure 7). A dose of 25 pmol had neither somnogenic nor pyrogenic effects. The time course of PGD_2 effects was similar to that observed following interleukin-1 infusions, but not similar to that following muramyl peptide administration (see Chapter 5). Interestingly, PGD_2 elicited hyperthermia in rabbits. This is consistent with the obser-

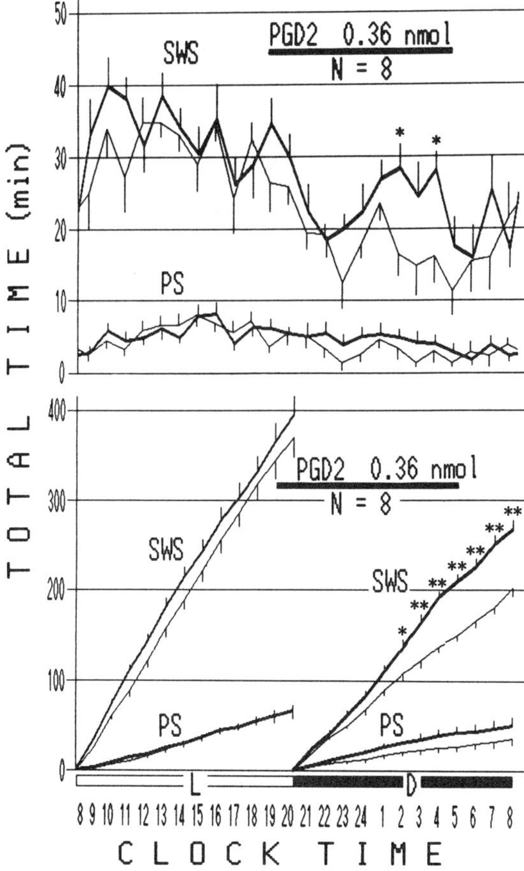

FIGURE 5. Effects of 0.36 nmol of PGD_2 i.c.v. infused for 10 h at 19.00 through 05.00 h (indicated by a solid bar) on the time-course changes in the amount of sleep in otherwise saline-infused, freely behaving rats. The top graph shows the hourly amount of SWS and PS, while the bottom graph illustrates their cumulative values in the light (L) and the dark (D) periods. The thin and thick curves stand for the baseline and the experimental day, respectively. The vertical lines on each hourly value indicate the standard error of mean (SEM). $*p < 0.05$, $**p < 0.01$. (From Inoué, S., Honda, K., Komoda, Y., Uchizono, K., Ueno, R., and Hayaishi, O., *Proc. Natl. Acad. Sci. U.S.A.*, 81, 6240, 1984. With permission.)

vation of Matsumura et al.[30] in semirestrained rats, but not with that of Ueno et al.[27,28] in restrained rats (see above). Such a difference may indicate that PGD_2 acts differentially on the regulatory systems concerning body temperature and sleep.[39]

D. Experiments in Monkeys

Ueno et al.[40-42] have recently reported that PGD_2 is involved in the regulation of sleep in rhesus monkeys. On a monkey chair, animals were i.c.v. infused with PGD_2 for a 6-h diurnal period through a chronically implanted cannula in the lateral cerebral ventricle. Control monkeys were infused with artificial cerebrospinal fluid (CSF) in a similar manner. Although sensitivities to PGD_2 were slightly different among individual monkeys, i.c.v. infusion of this substance at 5.4 to 810 nmol for 6 h dose-dependently increased the amount

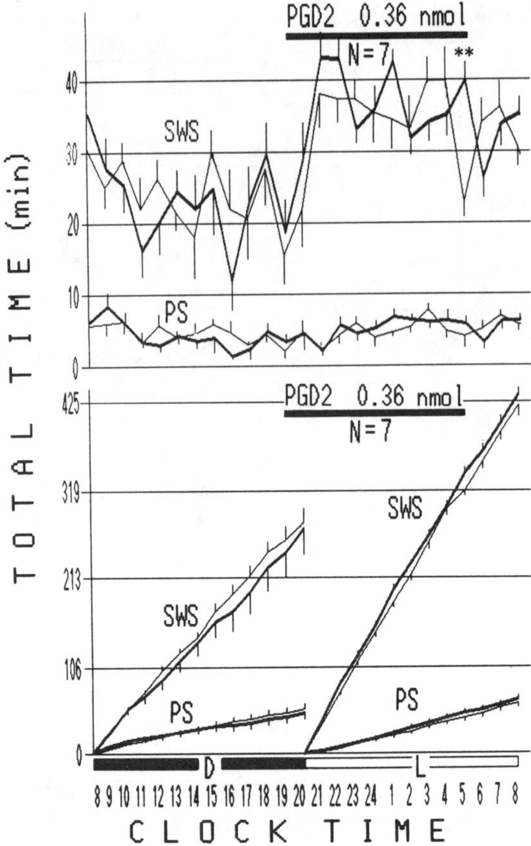

FIGURE 6. Effects of 0.36 nmol of PGD_2 i.c.v. infused for 10 h at 07.00 through 17.00 h (indicated by a solid bar) on the time-course changes in the amount of sleep in otherwise saline-infused, freely behaving rats. For other explanations, see Figure 5. (From Inoué, S., Honda, K., Komoda, Y., Uchizono, K., Ueno, R., and Hayaishi, O., *Neurosci. Lett.*, 49, 207, 1984. With permission.)

of sleep. The maximal amount of sleep, i.e., 216 min/6 h, induced by the infusion of the largest dose, was considerably longer than that occurring during the infusion of the control solution (45 to 60 min/6 h). The PGD_2-induced sleep appeared physiological, manifesting physiological proportions of light and deep SWS and PS. Electromyographic (EMG) and electrooculographic monitorings revealed that rapid eye movement (REM) was synchronized with atonia during PS. The PGD_2 infusion exerted no influence on sleep during the subsequent night and the circadian sleep-waking patterns. Although the large dose of PGD_2 (540 nmol) increased the heart rate and rectal temperature, the lower doses did not affect these parameters. The somnogenic activity of PGD_2 was specific for the molecular structure of PGD, since 9β-PGD_2, a stereoisomer of PGD_2, PGD_2-methylester, and PGD_1 had little effect on sleep. In contrast, PGE_2 or $PGF_{2\alpha}$ decreased SWS and caused fever.

Ueno and co-workers[41,42] have demonstrated further that a 6-h diurnal infusion of PGD_2 into the third cerebral ventricle markedly reduced the effective dosage required for somnogenicity: an amount 1000-fold less as compared with the infusion into the lateral cerebral ventricle sufficiently induced excess sleep in monkeys. This indicates that the POA located close to the third ventricle also plays an important role in sleep regulation in monkeys. EEG

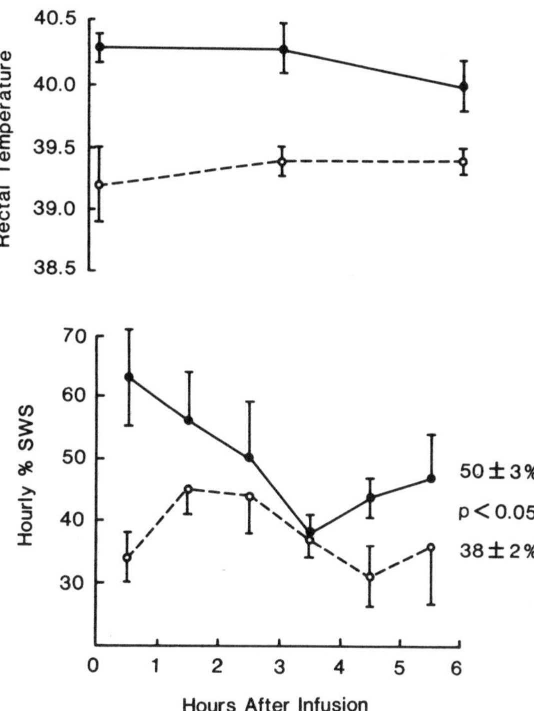

FIGURE 7. Effects of 0.25 nmol of PGD_2 (solid lines with closed circles) i.c.v. infused for 45 min in the morning on body temperature and SWS in rabbits. Broken lines with open circles are control values. Hourly means accompany SEM. (From Krueger, J. M., in *Endogenous Sleep Substances and Sleep Regulation,* Inoué, S. and Borbély, A. A., Eds., Japan Scientific Societies Press, Tokyo/VNU Science Press BV, Utrecht, 1985, 181. With permission.)

power spectral analysis revealed that PGD_2-induced SWS was largely similar to that observable in untreated animals.[42] Thus, PGD_2 and possibly other PGs, such as PGE_2 and PGF_2, seem to serve as cerebral sleep regulators, not only in rodents, but also in primates.

III. AWAKING ACTIVITY OF PGE_2

A. Experiments of Semirestrained Rats

We have already referred to some controversial reports on the central effects of PGE_2 on sleep-waking states. The induction of sedation after PGE_2 administration does not imply the occurrence of physiological sleep. Recently, Matsumura et al.[30] found that microinjection of PGE_2 at doses of 0.25 to 2500 pmol into the POA dose-dependently reduced the amount of SWS and PS in rats during a 60-min postinjection period at daytime. Concurrently, fever was induced and persisted for more than 90 min following the microinjection of PGE_2 at doses larger than 25 pmol, as revealed by marked elevations in rectal temperature. However, sleep reduction occurred without concomitant hyperthermia at the dose of 0.25 pmol. The reductions of sleep were due mainly to the shortening of SWS episodes and the less frequent occurrence of PS episodes. These observations were done in semirestrained rats housed in a small box without the allowance of body turning, mainly for the purpose of measuring rectal temperature. Furthermore, the observation period was restricted to a few hours of daytime. Consequently, a further experimental approach has been one in our laboratory in unrestrained rats with a long-term observation period.

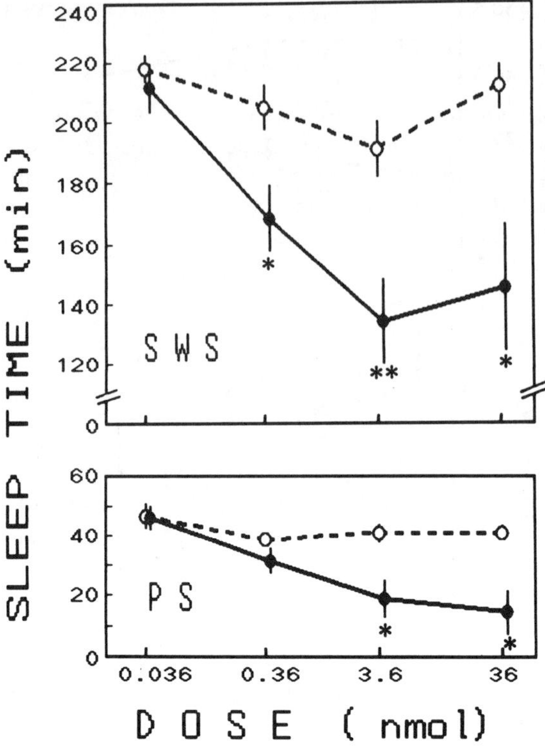

FIGURE 8. Dose-response relationship of the sleep-suppressing effects of PGE_2. The total time of SWS and PS during 6-h i.c.v. infusion of PGE_2 (solid line) was compared to that during a corresponding 6-h period (at 11.00 through 17.00 h) under continued saline infusion (broken line) on the previous day. Vertical lines indicate SEM. $*p < 0.05$, $**p < 0.01$. Number of rats: n = 6 each for 0.036 nmol and 36 noml; n = 8 for 0.36 nmol; n = 7 for 3.6 nmol.

B. Experiments in Unrestrained Rats

Matsumura et al.[43,44] have confirmed the sleep-suppressive properties of PGE_2 in freely moving rats. It has been found that a 6-h diurnal i.c.v. infusion (into the third cerebral ventricle) of PGE_2 at doses of 0.36 to 36 nmol reduced significantly in a dose-dependent manner the total amount of sleep in freely behaving rats during the infusion period (see Figure 8). A small dose of 36 pmol/rat was ineffective. The minimum dose required for a significant reduction of SWS and PS was 0.36 and 3.6 nmol, respectively. The reduction of SWS and PS was followed by drastic rebounds during the subsequent dark period, but the amounts of SWS and PS returned thereafter to the baseline level. These results indicate that PGE_2 has an awaking action centrally evoked without any irreversible damage to the regulatory mechanism involved in sleep and wakefulness, and that a site of action is located in the POA.

C. Experiments in Monkeys

Onoe et al.[42] have recently reported that PGE_2 and $PGF_{2\alpha}$ (5.4 to 540 nmol diurnally infused for 6-h into the lateral cerebral ventricle) slightly but dose-dependently suppressed SWS with sedation and elevated rectal temperature and heart rate in rhesus monkeys on a monkey chair.

IV. SUPPRESSION OF SLEEP BY PROSTAGLANDIN SYNTHESIS INHIBITORS

A. Prostaglandin Synthesis Inhibitors

The presence of somnogenic and antisomnogenic PGs in the brain leads to the question as to whether or not the inhibition of PG synthesis *in vivo* provokes any change in the amount of sleep. PGs are biosynthsized from arachidonic acid by cyclooxygenase (see Figure 1). The activity of this enzyme is inhibited by some nonsteroidal, anti-inflammatory drugs, including diclofenac sodium (DF) and indomethacin (IM).[6] Krueger et al.[45] reported that the i.c.v. administration of IM resulted in a reduction of normal sleep in rabbits with some accompanying toxic reactions. Hence, we undertook extensive studies on the effects of PG synthesis inhibitors on sleep in unrestrained rats and their sleep-waking states were continuously monitored.[46-49]

Since a large dose of DF and IM induces a wide spectra of toxic effects, including hypothermia[50] and ulcerogenicity[51,52] in rats, we considered the following points in order to minimize the possibility that the toxic actions of PG synthesis inhibitors by themselves affect sleep.

1. Both DF and IM should be adopted as mutual controls. DF seems preferable, since it is milder in its toxicity than IM.[53]
2. The drugs should be administered either peripherally through i.p., oral, or i.v. routes, or centrally through i.c.v. with different subtoxic dosages.
3. Short- and long-term effects of the drugs should be compared by a single acute administration and a continued infusion, respectively.
4. Sleep-waking states before, during, and after the drug administration should be uninterruptedly observed in freely behaving rats.

B. Peripheral Administration of DF and IM

1. Oral Administration

A single oral administration of DF and IM at an early phase of the light period transiently decreased diurnal sleep in unrestrained rats. Although a marked reduction in sleep was provoked by manual handlings within the first hour, the treatment with the drugs still lowered the amount of sleep as compared with the treatment with vehicles. DF administered at a dose of 5 mg/kg (see Figure 9) and IM administered at 10 mg/kg reduced SWS and PS significantly. The suppression of sleep lasted for a few hours of daytime and no further change could be detected in the amount of hourly sleep thereafter. Higher and lower doses of DF (1 to 2.5 and 10 to 20 mg/kg, respectively) and IM (5 to 20 mg/kg, respectively) also modified sleep, but the differences from the baseline were statistically insignificant, except for 10 mg/kg of DF, which suppressed PS significantly.

2. I.P. Injection

Manual handlings at the time of i.p. injection before noon also resulted in a transient alert state and an eventual reduction in sleep within the first hour. However, an even more dramatic reduction was induced by the administration of IM (5 mg/kg). SWS and PS decreased significantly to 62% and 38%, respectively, of the baseline. The suppression of sleep lasted only a few hours of daytime and no more difference could be detected in hourly sleep amounts thereafter. The administration of IM at 10 mg/kg exerted a slight but insignificant sleep-suppressing effect.

3. I.V. Infusion

Continuous diurnal 10-h i.v. infusion of DF (0.4 mg) significantly affected sleep in freely behaving rats (see Figure 10). The effect was considerably long-lasting over 2 d. On the

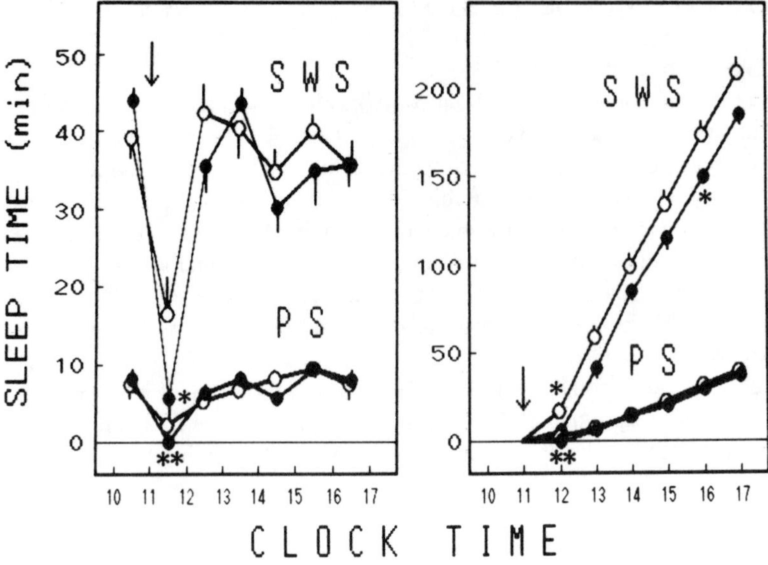

FIGURE 9. Effects of oral administration of 5 mg/kg DF at 11.00 h (indicated by an arrow) on the time course of the hourly amount of sleep (left) and the cumulative amount of sleep (right) in unrestrained rats (n = 5). Open and closed circles stand for the baseline and the experimental day, respectively. The baseline values were obtained on the previous day of the experiment by orally administering the vehicle only. The vertical lines on each hourly value indicate SEM. $*p < 0.05$, $**p < 0.01$.

experimental day, diurnal sleep was suppressed significantly: the decrement of SWS and PS was 70.2 (16.1% of the baseline) and 17.2 min (20.8%), respectively. In contrast, nocturnal sleep showed a rebound rise: the increment of SWS and PS was 78.2 (39.0%) and 18.0 min (77.3%), respectively. Thus, the loss of sleep during the daytime was compensated for during the dark period. On the following recovery day, SWS recovered to the baseline level, but PS decreased significantly during the light period, whereas SWS and PS tended to increase during the dark period. DF, at doses of 0.04 and 4 mg, exerted less and insignificant sleep modulation.

The same treatment with IM (0.4 mg for a 10-h i.v. infusion) apparently reduced diurnal sleep (see Figure 11). Both SWS and PS were suppressed significantly: the decrement during the 12-h light period was 41.7 min (9.4% of the baseline) and 16.8 min (20.1%), respectively. However, the sleep-suppressive effect of IM lasted only during the infusion period and no more modification was detectable thereafter. A 50-fold smaller dosage exerted entirely no effect on sleep.

C. Central Administration of DF

Diurnal 10-h i.c.v. infusion of 0.04 mg of DF significantly affected sleep (see Figure 12). The effect was long-lasting and almost similar to that of the 10-h i.v. infusion of 0.4 mg of DF. Diurnal SWS and PS on the experimental day decreased by 51.3 (13.4% of the baseline) and 12.0 min (17.7%), respectively, whereas the following nocturnal SWS and PS increased by 84.3 (49.9%) and 24.0 min (87.9%), respectively. A significant decrease was also observed in diurnal PS on the recovery day. Higher and lower doses of DF exerted a less marked and insignificant sleep modulation (see Figure 13).

D. Possible Mechanisms

The PG synthesis inhibitors, DF and IM, which were either peripherally or centrally and either acutely or chronically administered in freely behaving rats during their resting phase,

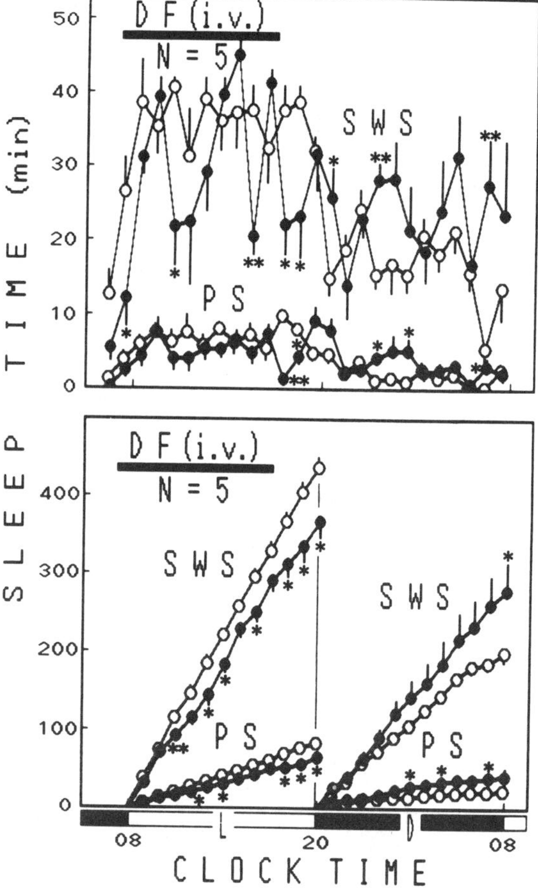

FIGURE 10. Effects of 0.4 mg of DF i.v. infused for 10 h at 07.00 through 17.00 h (indicated by a solid bar) on the time-course changes in the amount of sleep in otherwise saline-infused, freely behaving rats. The top graph shows the hourly amount of SWS and PS, while the bottom graph illustrates their cumulative values in the light (L) and the dark (D) periods. The open and closed circles stand for the baseline and the experimental day, respectively. The vertical lines on each hourly value indicate SEM. $*p < 0.05$, $**p < 0.01$.

provoked a concomitant suppression of sleep, either transiently or prolongedly. The effective doses were lower than those which may induce toxic reactions like those for hypothermia (IM: 10 to 20 mg/kg i.p.)[50] and ulcerogenicity (IM: 25 mg/kg oral).[51] The effective doses for the long-term administration were far below the toxic levels. Consequently, the sleep suppression might not be a secondary effect due to the toxic reactions induced by these drugs. Rather, it seems likely that DF and IM could inhibit the brain from synthesizing PGs, and that the eventual cerebral depletion of PGs might be responsible for the reduction in diurnal sleep in rats.

In rabbits, Krueger et al.[45] noticed a reduction of sleep following the administration of IM, although some toxic reactions occurred simultaneously. In healthy human volunteers, Horne and Shackell[54] observed that aspirin, another PG synthesis inhibitor, counterbalanced the increase in sleep after bath-heating. Since there was no difference between aspirin and placebo conditions in the body temperature rise during heating, the counterbalancing effect

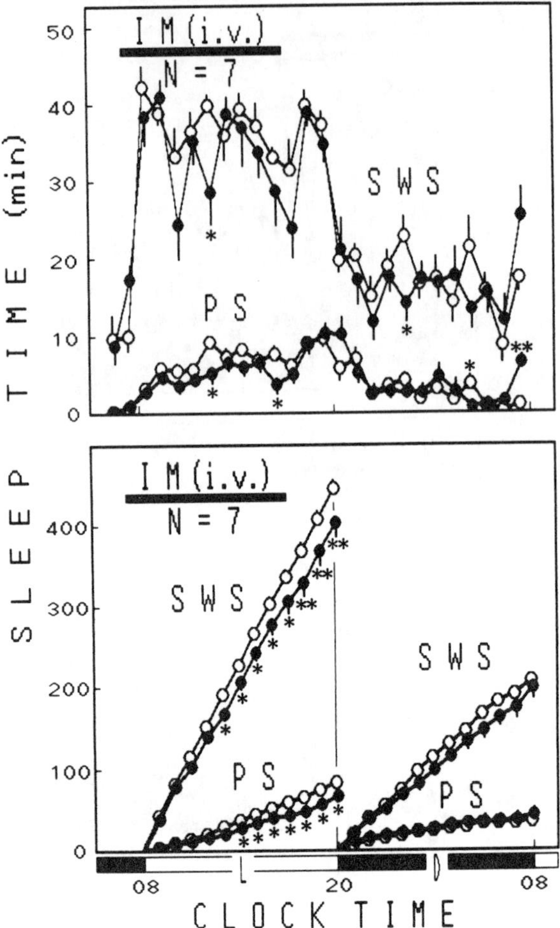

FIGURE 11. Effects of 0.4 mg of IM i.v. infused for 10 h at 07.00 through 17.00 h (indicated by a solid bar) on the time-course changes in the amount of sleep in otherwise saline-infused, freely behaving rats. For other explanations, see Figure 10.

of aspirin could not be attributed to the antipyretic action of aspirin. Thus, it was suggested that brain warming through body heating increased brain PG levels for the enhancement of sleep, but this action was blocked by aspirin.

As already mentioned, there are several evidences concerning the sleep-enhancing action of PGs, especially of PGD_2, and the activity of PGD_2 synthetase in the rat brain is higher during the light period than during the dark.[11] The supplementation of exogenously supplied PGD_2 during the diurnal period causes no change in diurnal sleep of rats[29] in dramatic contrast to a marked sleep-promoting effect of a nocturnal infusion on nocturnal sleep.[36,37] This fact may also indicate that the synthesis and release of PGD_2 in the rat brain is high enough during the light period to induce sleep at the maximum level. Hence it is likely that the suppression of diurnal sleep by IM and DF was due largely to the inhibition of the PGD_2 supply *in vivo*.

The maximum total sleep time deprived by DF and IM was 9 to 18 min/h in our short-term injection studies and 58 to 88 min/12 h in our long-term infusion studies. Since our rats sleep normally 40 to 50 min/h around noon and 450 to 530 mn during the 12-h light period, the amount of sleep loss was equal to 15 to 30% of normal hourly sleep for our acute treatments and 12 to 17% of normal diurnal sleep for our chronic ones. This means

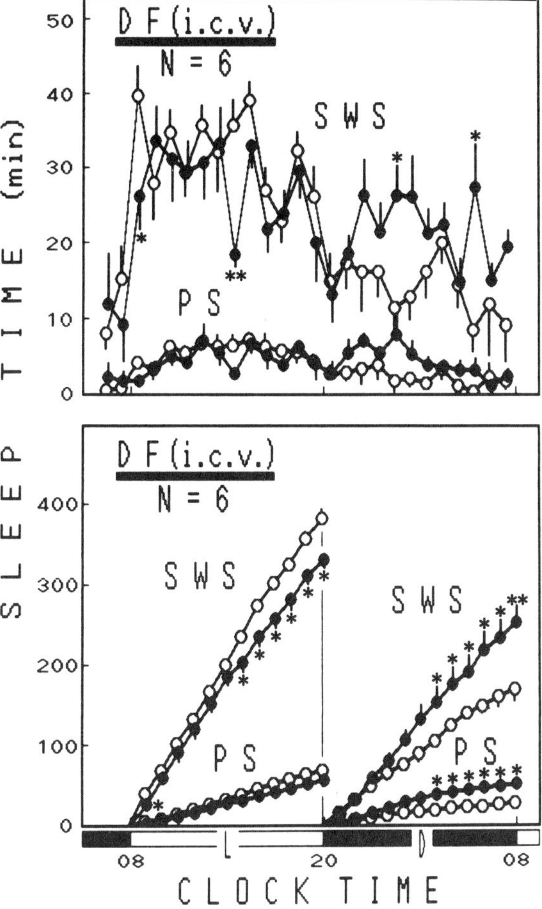

FIGURE 12. Effects of 0.04 mg of DF i.c.v. infused for 10 h at 07.00 through 17.00 h (indicated by a solid bar) on the time-course changes in the amount of sleep in otherwise saline-infused, freely behaving rats. For other explanations, see Figure 10.

that under the condition of PG depletion the rats showed incomplete insomnia, still being able to sleep to a considerable extent. Furthermore, a bell-shaped dose-response relationship was reproducibly obtained in our experiments. A limited range of the effective dosage is commonly observed in some sleep substances, such as delta-sleep-inducing peptide (DSIP)[55] and uridine.[56] Since a large number of endogenous sleep substances seem to be involved in the regulation of physiological sleep,[57,58] the sleep observed in the state of PG depletion was probably induced and maintained by the other sleep substances.

In comparison with IM, DF exhibited a stronger and more long-lasting effect. The suppression of diurnal sleep by DF was followed by a rebound sleep during the dark period. This may indicate that a homeostatic sleep-regulatory mechanism was operative to compensate for the sleep loss induced by DF. It seems likely that a feedback action of the PG synthesis-release system is involved in the regulatory process, and that the sleep-regulatory mechanisms are not impaired by the treatment with the drugs.

V. RELATED MATTERS

A. Depression and PGD_2

Ueno[59] quantified the amounts of PGs by radioimmunoassay in saliva from healthy humans and patients with the major depressive disorder, the minor depressive disorder, and neurotic

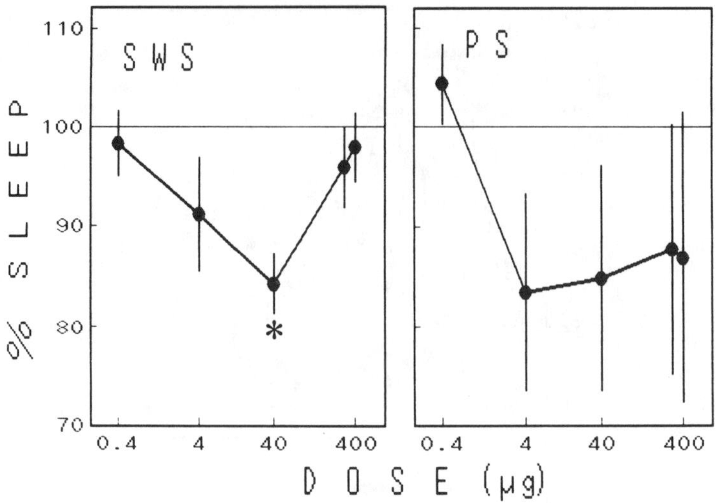

FIGURE 13. Dose-response relationship of the sleep-suppressing effects of DF. The total time of SWS and PS during 10-h i.c.v. infusion of DF was compared to that during a corresponding 10-h period (at 07.00 through 17.00 h) under continued saline infusion on the previous day (100%). The vertical lines indicate SEM. $*p < 0.05$. Number of rats: 4 for 0.4 μg, 7 for 4 μg, 6 for 40 μg, 6 for 120 μg, and 2 for 400 μg.

disorders of various kinds. The salivary concentrations of PGD_2, PGE_2, and $PGF_{2\alpha}$ were significantly higher in patients with the major depressive disorder than in normal adults and patients with other disorders. The salivary level of these PGs was dependent on the severity of the disease, since high levels occurred during the depressive phase, whereas normal levels appeared with the recovery of the disease.

B. Narcolespy and PGs

PGs are known to inhibit the expression of a major immunocompatibility antigen, HLA-DR2,[60] which is found in most narcoleptic patients. On the other hand, it has recently been found that the administration of IM dramatically suppressed the occurrence of narcolepsy in dogs.[60a] These facts indicate that PGs may be involved in the manifestation of narcolepsy, although causal relationships between these phenomena remain unknown.

C. Interactions with Other Sleep Substances

Scott et al.[61] demonstrated that the perfusion of leukocytic pyrogen prepared from guinea pig blood potentially stimulated the guinea pig hypothalamic tissue slices to produce PGE_2, $PGF_{2\alpha}$, and other prostanoids in a dose-dependent manner, although the tissue concentrations of the PGs were considerably low. On the other hand, it has been recently found that lipopolysaccharide and muramyl dipeptide, an analog of bacterial peptidoglycans (see Chapter 5), stimulated astrocytes of the first generation in a cell culture to produce PGD_2, whereas DSIP and uridine did not.[60a] This fact may substantiate Krueger's immune theory in which muramyl peptides are enrolled to alter arachidonic acid metabolism and stimulate the production of PGD_2 (see Chapter 5). Finally, it is suggested that PGD_2 plays a stimulatory role in regulating vasoactive intestinal polypeptide release from the hypothalamus into hypophyseal portal blood and causes prolactin secretion from the pituitary in rats.[62]

VI. RECIPROCAL ROLES OF PGD_2 AND PGE_2 IN SLEEP REGULATION

On the basis of the above and other experimental evidences, Hayaishi et al.[11,36,44] have developed the PG-dependent humoral theory of sleep regulation. The outlines are summarized as follows:

1. PGD_2 and PGE_2 are chemically defined natural constituents in the brain. The enzymes responsible for cerebral metabolism of these PGs are established.
2. PGD_2 synthesis in the brain exhibits a circadian rhythm depending on sleep-waking cycles.
3. A site of action of both PGD_2 and PGE_2 is located in the POA, in which the binding proteins specific for both PGs are concentrated.
4. PGD_2 can modulate the firing activities of neurons and interact with neurotransmitters in the POA.
5. PGD_2 can increase the amount of sleep in a dose-dependent manner, whereas PGE_2 can dose-dependently increase wakefulness.
6. PGD_2 infusion at a rate as low as 60 fmol/min (= 0.036 nmol/10 h) into the third cerebral ventricle of the rat can cause excess SWS, and the concentration of endogenous PGD_2 (a few ng/g tissue) in the brain appears sufficient to induce sleep.
7. Sleep induced by PGD_2 is indistinguishable from physiological sleep as judged by EEG, EMG, locomotor activities, and the behavior of recipient animals.
8. PG synthesis inhibitors decrease diurnal sleep in rats.
9. PGD_2 is not pyrogenic, but slightly decreases body temperature at least in restrained rats, as observed in physiological sleep, whereas PGE_2 is pyrogenic.
10. Reciprocal activities of PGD_2 and PGE_2 may regulate the state of vigilance: PGD_2 enhances sleep, whereas PGE_2 enhances wakefulness as a centrally acting neuromediator.

REFERENCES

1. **Coceani, F.,** Prostaglandins and the central nervous system, *Arch. Intern. Med.,* 133, 119, 1974.
2. **Marley, E., Poole, S., and Stephenson, J. D.,** Prostaglandins and the central nervous system, *Postgrad. Med. J.,* 53, 649, 1977.
3. **Wolfe, L. S. and Coceani, F.,** The role of prostaglandins in the central nervous system, *Annu. Rev. Physiol.,* 41, 669, 1979.
4. **Abdel-Halim, M. S., Hamberg, M., Sjöquist, B., and Anggard, E.,** Identification of prostaglandin D_2 as a major prostaglandin in homogenates of rat brain, *Prostaglandins,* 14, 633, 1977.
5. **Narumiya, S., Ogorochi, T., Nakao, K., and Hayaishi, O.,** Prostaglandin D_2 in rat brain, spinal cord, and pituitary: basal levels and regional distribution, *Life Sci.,* 31, 2093, 1982.
6. **Förstermann, U., Heldt, R., Knappen, F., and Hertting, G.,** Potential anticonvulsive properties of endogenous prostaglandins formed in mouse brain, *Brain Res.,* 240, 303, 1982.
7. **Seregi, A., Förstermann, U., Heldt, R., and Hertting, G.,** The formation and regional distribution of prostaglandin D_2 and $F_{2\alpha}$ in the brain of spontaneously convulsing gerbils, *Brain Res.,* 337, 171, 1985.
8. **Shimizu, T., Mizuno, N., Amano, T., and Hayaishi, O.,** Prostaglandin D_2, a neuromodulator, *Proc. Natl. Acad. Sci. U.S.A.,* 76, 6231, 1979.
9. **Shimizu, T., Yamamoto, S., and Hayaishi, O.,** Purification and properties of prostaglandin D synthetase from rat brain, *J. Biol. Chem.,* 254, 5222, 1979.
10. **Watanabe, K., Shimizu, T., Iguchi, S., Wakatsuka, H., Hayashi, M., and Hayaishi, O.,** An NADP-like prostaglandin D dehydrogenase in swine brain, *J. Biol. Chem.,* 255, 1779, 1980.
11. **Ueno, R., Hayaishi, O., Osama, H., Honda, K., Inoué, S., Ishikawa, Y., and Nakayama, T.,** Prostaglandin D_2 regulates physiological sleep, in *Endogenous Sleep Substances and Sleep Regulation,* Inoué, S. and Borbély, A. A., Eds., Japan Scientific Societies Press, Tokyo/VNU Science Press BV, Utrecht, 1985, 193.

12. **Urade, Y., Kaneko, T., Fujimoto, N., Watanabe, Y., Mizuno, N., and Hayaishi, O.,** Purification, characterization, and immunohistochemistry of rat prostaglandin S synthetase, in *Advances in Prostaglandin, Thromboxane, and Leukotriene Research,* Vol. 15, Hayaishi, O. and Yamamoto, S., Eds., Raven Press, New York, 1985, 549.
13. **Yamashita, A., Watanabe, Y., and Hayaishi, O.,** Autoradiographic localization of a binding protein(s) specific for prostaglandin D_2 in rat brain, *Proc. Natl. Acad. Sci. U.S.A.,* 80, 6114, 1983.
14. **Dray, F., Wisner, A., Bommelaer, M. C., Heaulme, M., Viossat, I., Gerozissis, K., and Renard, C. A.,** Hypothalamic prostaglandin E_2 receptors: biochemical characteristics and episodic fluctuations during rat estrus cycle, in *Advances in Prostaglandin, Thromboxane, and Leukotriene Research,* Vol. 15, Hayaishi, O. and Yamamoto, S., Eds., Raven Press, New York, 1985, 555.
15. **Watanabe, Y., Kaneko, T., and Hayaishi, O.,** Localization of prostaglandin bindings in the central nervous system, in *Advances in Prostaglandin, Thromboxane, and Leukotriene Research,* Vol. 15, Hayaishi, O. and Yamamoto, S., Eds., Raven Press, New York, 1985, 553.
16. **Watanabe, Y., Tokumoto, H., Yamashita, A., Narumiya, S., Mizuno, N., and Hayaishi, O.,** Specific bindings of prostaglandin D_2, E_2 and $F_{2\alpha}$ in postmortem human brain, *Brain Res.,* 342, 110, 1985.
17. **Suzuki, F., Hayashi, H., and Hayaishi, O.,** Transport of prostaglandin D_2 into brain, *Brain Res.,* 385, 321, 1986.
18. **Ueno, R.,** Prostaglandin and autonomic nervous systems, *Jpn. J. Neuropsychopharmacol.,* 8, 195, 1986.
19. **Horton, E. W.,** Actions of prostaglandins E_1, E_2 and E_3 on the central nervous system, *Br. J. Pharmacol.,* 22, 189, 1964.
20. **Gilmore, D. P. and Shaikl, A. A.,** The effect of prostaglandin E_2 in inducing sedation in the rat, *Prostaglandins,* 2, 143, 1972.
21. **Haubrich, D. R., Perez-Cruet, J., and Reid, W. D.,** Prostaglandin E_1 causes sedation and increases 5-hydroxytryptamine turnover in rat brain, *Br. J. Pharmacol.,* 48, 80, 1973.
22. **Desiraju, T.,** Effect of intraventricularly administered prostaglandin E1 on the electrical activity of cerebral cortex and behavior in the unanesthetized monkey, *Prostaglandins,* 3, 859, 1973.
23. **Nistico, G. and Marley, E.,** Central effects of prostaglandin E_1 in adult fowls, *Neuropharmacology,* 12, 1009, 1973.
24. **Nistico, G. and Marley, E.,** Central effects of prostaglandins E_2, A_1 and $F_{2\alpha}$ in adult fowls, *Neuropharmacology,* 15, 737, 1976.
25. **Masek, K., Kadlecová, O., and Pöschlová, N.,** Effect of intracisternal administration of prostaglandin E_1 on waking and sleep in the rat, *Neuropharmacology,* 15, 491, 1976.
26. **Dzoljic, M. R.,** Prostaglandins and sleep. Awaking effect of prostaglandins and sleep pattern of essential fatty acids deficient (EFAD) rats, *Prostaglandins,* 15, 317, 1978.
27. **Ueno, R., Narumiya, S., Ogorochi, T., Nakayama, T., Ishikawa, Y., and Hayaishi, O.,** Role of prostaglandin D_2 in the hypothermia of rats caused by bacterial lipopolysaccharide, *Proc. Natl. Acad. Sci. U.S.A.,* 79, 6093, 1982.
28. **Ueno, R., Ishikawa, Y., Nakayama, T., and Hayaishi, O.,** Prostaglandin D_2 induces sleep when microinjected into the preoptic area of conscious rats, *Biochem. Biophys. Res. Commun.,* 109, 576, 1982.
29. **Inoué, S., Honda, K., Komoda, Y., Uchizono, K., Ueno, R., and Hayaishi, O.,** Little sleep-promoting effect of three sleep substances diurnally infused in unrestrained rats, *Neurosci. Lett.,* 49, 207, 1984.
30. **Matsumura, H., Goh, Y., Ueno, R., Sakai, T., and Hayaishi, O.,** Awaking effect of PGE_2 microinjected into the preoptic area of rats, *Brain Res.,* 444, 265, 1988.
31. **Nauta, W. J. H.,** Hypothalamic regulation of sleep in rats. An experimental study, *J. Neurophysiol.,* 9, 285, 1946.
32. **Sterman, M. B. and Shouse, M. N.,** Sleep "centers" in the brain: the preoptic basal forebrain area revisited, in *Brain Mechanisms of Sleep,* McGinty, D. J., Drucker-Colín, R., Morrison, A., and Parmeggiani, P. L., Eds., Raven Press, New York, 1987, 277.
33. **Inokuchi, A. and Oomura, Y.,** Effects of prostaglandin D_2 and sleep-promoting substance on hypothalamic neuronal activity in the rat, in *Endogenous Sleep Substances and Sleep Regulation,* Inoué, S. and Borbély, A. A., Eds., Japan Scientific Societies Press, Tokyo/VNU Science Press BV, Utrecht, 1985, 215.
34. **Inokuchi, A. and Oomura, Y.,** Effects of prostaglandin D_2 on the activity of hypothalamic neurons in the rat, *Jpn. J. Physiol.,* 36, 497, 1986.
35. **Ueno, R., Hayaishi, O., Inoué, S., and Nakayama, T.,** Prostaglandin D_2: a cerebral sleep-regulating substance in rats, in *Advances in Prostaglandin, Thromboxane, and Leukotriene Research,* Vol. 15, Hayaishi, O. and Yamamoto, S., Eds., Raven Press, New York, 1985, 581.
36. **Ueno, R., Honda, K., Inoué, S., and Hayaishi, O.,** Prostaglandin D_2, a cerebral sleep-inducing substance in rats, *Proc. Natl. Acad. Sci. U.S.A.,* 80, 1735, 1983.
37. **Inoué, S., Honda, K., Komoda, Y., Uchizono, K., Ueno, R., and Hayaishi, O.,** Differential sleep-promoting effects of five sleep substances nocturnally infused in unrestrained rats, *Proc. Natl. Acad. Sci. U.S.A.,* 81, 6240, 1984.

38. **Inoué, S.**, Sleep substances: their roles and evolution, in *Endogenous Sleep Substances and Sleep Regulation*, Inoué, S. and Borbély, A. A., Eds., Japan Scientific Societies Press, Tokyo/VNU Science Press BV, Utrecht, 1985, 3.
39. **Krueger, J. M.**, Muramyl peptides and interleukin-1 as promoters of slow wave sleep, in *Endogenous Sleep Substances and Sleep Regulation*, Inoué, S. and Borbély, A. A., Eds., Japan Scientific Societies Press, Tokyo/VNU Science Press BV, Utrecht, 1985, 181.
40. **Ueno, R., Onoe, H., Matsumura, H., Hayaishi, O., Fujita, I., Nishino, H., and Oomura, Y.**, Regulation of sleep by prostaglandins in conscious Rhesus monkeys, *Sleep Res.*, 16, 38, 1987.
41. **Ueno, R.**, Prostaglandin D_2, a cerebral sleep-inducing substance in monkeys, Abstracts of Research Report Meeting, Hayaishi Bioinformation Transfer Project, Research Development Corporation of Japan, 1987, 3.
42. **Onoe, H., Ueno, R., Fujita, I., Nishino, H., Oomura, Y., and Hayaishi, O.**, Prostaglandins D_2, a cerebral sleep-inducing substance in monkey, *Proc. Natl. Acad. Sci. U.S.A.*, 85, 4082, 1988.
43. **Matsumura, H., Goh, Y., Honda, K., Ueno, R., Sakai, T., Inoué, S., and Hayaishi, O.**, Central awaking action of PGE_2 in rats, *Final Program*, Taipei Conference on Prostaglandin and Leukotriene Research, Taiperi, 1988, 68.
44. **Matsumura, H., Honda, K., Goh, Y., Ueno, R., Sakai, T., Inoué, S., and Hayaishi, O.**, Awaking effect of prostaglandin E_2 in freely moving rats, *Brain Res.*, in press.
45. **Krueger, J. M., Pappenheimer, J. R., and Karnovsky, M. L.**, Sleep-promoting effects of muramyl peptides, *Proc. Natl. Acad. Sci. U.S.A.*, 79, 6102, 1982.
46. **Ueno, R., Osama, H., Naito, K., Hayaishi, O., Honda, K., and Inoué, S.**, Effects of prostaglandin synthesis inhibitors on sleep in rats, *Seikagaku*, 57, 925, 1985.
47. **Inoué, S., Naito, K., and Honda, K.**, Prostaglandins and sleep, *Jpn. J. Neuropsychopharmacol.*, 8, 217, 1986.
48. **Naito, K., Ueno, R., Hayaishi, O., Honda, K., and Inoué, S.**, Suppression of sleep by prostaglandin synthesis inhibitors in unrestrained rats, *Sleep Res.*, 16, 64, 1987.
49. **Naito, K., Osama, H., Ueno, R., Hayaishi, O., Honda, K., and Inoué, S.**, Suppression of sleep by prostaglandin synthesis inhibitors in unrestrained rats, *Brain Res.*, 453, 329, 1988.
50. **Scales, W. E. and Kluger, M. J.**, Effect of antipyretic drugs on circadian rhythm in body temperature, *Am J. Physiol.*, 253, R306, 1987.
51. **Ligumsky, M., Sestieri, M., Karmeli, F., and Rachmilewitz, D.**, Protection by mild irritants against indomethacin-induced gastric mucosal damage in the rat: role of prostaglandin synthesis, *Isr. J. Med. Sci.*, 11, 807, 1986.
52. **Rainsford, K. D.**, Structural damages in eicosanoid metabolites in the gastric musoca of rats and pigs induced by anti-inflammatory drugs of varying ulcerogenicity, *Int. J. Tissue React.*, 8, 1, 1986.
53. **Scholer, D. W., Ku, E. C., Boettcher, I., and Schweizer, A.**, Pharmacology of diclofenac sodium, *Am. J. Med.*, 80, 34, 1986.
54. **Horne, J. A. and Shackell, B. S.**, Slow wave sleep elevations after body heating: proximity to sleep and effects of aspirin, *Sleep*, 10, 383, 1987.
55. **Schoenenberger, G. A.**, Characterization, properties and multivariate functions of delta-sleep-inducing peptide (DSIP), *Eur. Neurol.*, 23, 321, 1984.
56. **Honda, K., Komoda, Y., Nishida, S., Nagasaki, H., Higashi, A., Uchizono, K., and Inoué, S.**, Uridiine as an active component of sleep-promoting substance: its effects on the nocturnal sleep in rats, *Neurosci. Res.*, 1, 243, 1984.
57. **Inoué, S.**, Sleep substances: their roles and evolution, *Seikagaku*, 55, 445, 1983.
58. **Inoué, S.**, Multifactorial humoral regulation of sleep, *Clin. Neuropharmacol.*, 9 (Suppl. 4), 470, 1986.
59. **Ueno, R.**, Increased level of salivary prostaglandins in patients with major depression, Abstracts of Research Report Meeting, Hayaishi Bioinformation Transfer Project, Research Development Corporation of Japan, 1986, 3.
60. **Uranue, E. R. and Allen, P. M.**, The basis for the immunoregulatory role of macrophages and other accessory cells, *Science*, 236, 551, 1987.

60a. **Hayaishi, O.**, personal communication.

61. **Scott, I. M., Fertel, R. H., and Boulant, J. A.**, Leukocytic pyrogin effects on prostaglandins in hypothalamic tissue slices, *Am. J. Physiol.*, 253, R71, 1987.
62. **Shimatsu, A., Kato, Y., Matsushita, M., Ohta, H., Kabayama, Y., Yanaihara, M., and Imura, H.**, Prostaglandins D_2 stimulates vasoactive intestinal polypeptide release into rat hypophysial portal blood, *Peptides*, 5, 395, 1984.

Chapter 7

PARADOXICAL SLEEP (PS) FACTORS

I. PS FACTORS IN CEREBROSPINAL FLUID

A. Bromine Compounds

1. Natural Occurrence

In 1973, Yanagisawa and Yoshikawa[1] reported that a bromine compound physiologically occurring in human cerebrospinal fluid (CSF) and blood was finally identified as 1-methylheptyl-γ-bromoacetoacetate (γ-Br or MHBAA; for the chemical structure, see Figure 2 in Chapter 2). This substance is regarded as a derivative of short-chain fatty acids, which reportedly increase paradoxical sleep (PS) in cats.[2,3] According to Yanagisawa and coworkers,[4,5] γ-Br was biosynthesized from ^{14}C-butyrate in rat tissues *in vivo* and distributed specifically in the brain, pituitary gland, and retina in rats. Further microdetermination studies of the Japanese biochemists[6] revealed that human plasma levels of γ-Br ranged from 236 to 621 pmol/ml. Okudaira et al.[7] demonstrated that similar levels of γ-Br were present in the plasma and CSF of cats. They also detected circadian fluctuations in plasma and CSF levels of γ-Br with dependence on sleep-waking cycles in both species.

2. PS-Enhancing Effect of γ-Br

Torii and colleagues[8] demonstrated that intravenous (i.v.) injection of L-γ-Br, the L form of synthetic γ-Br, at a dose of 0.1 mg/kg significantly promoted PS but not slow-wave sleep (SWS) during a 1-h postinjection period in encéphale isolé cats. The same dose of D-γ-Br had no effect on SWS and PS. Larger doses (1 to 5 mg/kg) of L-γ-Br and D-γ-Br elicited no enhancement of sleep.

Okudaira et al.[7] clearly demonstrated a close correlation between the state of vigilance and the plasma and/or CSF levels of γ-Br in cats and humans. The findings are the following. (1) The higher anxiety score accompanied the lower plasma level of γ-Br in humans. (2) The lower percentage of rapid-eye-movement (REM) sleep (= PS) during nocturnal bed rest in humans accompanied the higher plasma level of γ-Br. (3) The plasma level of γ-Br increased in the morning shortly after total sleep deprivation in humans and more markedly after REM sleep deprivation in cats and humans. REM sleep deprivation in cats also brought about an increase of γ-Br in CSF. (4) The plasma levels of γ-Br tended to increase before REM sleep onset and appeared to be expended during REM episodes (see Figure 1).

Recent studies of Yamane et al.[9] and Okundaira et al.[10,11] indicate that γ-Br looks like a competitive inhibitor of acetylcholinesterases, since microinjection of 683 nmol of synthetic γ-Br into the rostral pontine reticular formation (caudal to the ventral tegmental nucleus and rostral to the genu of the facial nerve and the abducens nucleus) induced a significant decrease in the latency to sleep onset, an increase in REM sleep during the first 6 h, and an increase in the total sleep time during 24 postinjection hours. The activation of the cholinergic mechanism by extrinsic γ-Br, which inhibited the acetylcholinesterase activity, might account for the reactions.

3. Analogs of γ-Br

Torii et al.[12] found that oral administration of calcium *N*-2-ethylhexyl-β-oxybutyramide (M-2), an analog of γ-Br, dose-dependently increased PS and reduced wakefulness during a 24-h recording period in cats. A dose of 100 mg/kg was most effective.

Another analog, butoctamide hydrogen succinate (BAHS; see Figure 2), the decalcificated monomer of M-2, also exhibited a capacity to specifically increase REM sleep in young

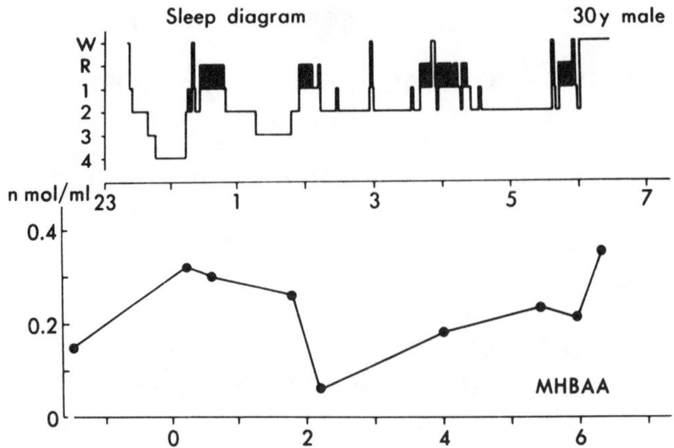

FIGURE 1. The correlation of plasma levels of γ-Br (MHBAA) with the stages of sleep in a human subject. γ-Br seems to increase before REM sleep onset and expend during REM sleep. W, wakefulness; R, REM sleep. The abscissae represent the time of day. (From Okudaira, N., Koike, H., Inubushi, S., Abe, A., Yamane, M., Yanagisawa, I., and Torii, S., in *Endogenous Sleep Substances and Sleep Regulation*, Inoué, S. and Borbély, A. A., Eds., Japan Scientific Societies Press, Tokyo/VNU Science Press BV, Utrecht, 1985, 237. With permission.)

```
                      O            C2H5
                      ||           |
BAHS :  CH3-CH-CH2-C-NH-CH2-CH-CH2-CH2-CH2 -CH3
            |
        CH2-CO-O
        |
        CH2-COOH
```

FIGURE 2. Chemical structure of butoctamide hydrosuccinate (BAHS).

healthy humans. Okudaira et al.[13] found that BAHS at a dose of 600 mg per night given 1 h prior to retiring resulted in significant increases in REM sleep at the expense of stage 1 + 2 sleep without changing the total sleep time. The maximum percentage of BAHS-induced REM sleep was 34%. A carryover effect of BAHS was observed during the withdrawal period. Recently, Okudaira et al.[14] have reported that BAHS elicited a weak EEG delta-wave enhancing activity during the last one third of nocturnal sleep in healthy young humans. Hayashi et al.[15] examined the effects of BAHS in aged subjects and essentially confirmed the findings in young adults.

In Down's syndrome children, who are characterized by a reduced occurrence of REM sleep, an increased occurrence of undifferentiated sleep, and a prolonged latency to the first REM episode, Gruber et al.[16] found a significant increase in REM sleep as well as a decrease in undifferentiated sleep and latency to the first REM after oral administrations of BAHS (600 mg per night). In addition, a preliminary report has been published as to the modulatory effects of BAHS on nocturnal sleep in mental retardates.[17]

Interestingly, a dimer of γ-Br, di-*l*-methylheptyl-2,5-dioxocyclohexane-1,4-dicarboxylate (DOC), which is known to be biosynthesized in incubated cat CSF[1], enhanced both SWS and PS in rats.[18] In our preliminary experiment, it has been demonstrated that 2.3 nmol of synthetic D-DOC, the D form of DOC, which was i.c.v. infused for a 10-h nocturnal period, exerted a dramatic elevation of nocturnal SWS (46.0% of baseline value) and nocturnal PS (55.4%).[18] The change was due largely to increases in the episode frequency of SWS and

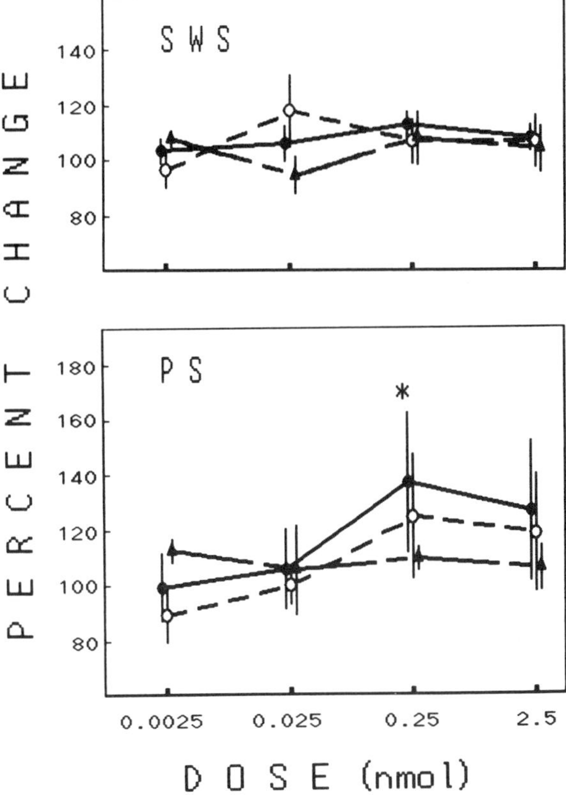

FIGURE 3. Dose-response relationship for the somnogenic effects of L-DOC on SWS and PS in rats. Values are means ± standard error of mean (SEM) for total time during a 12-h dark period (closed circle), number of episodes (triangle), and duration of episodes (open circle), as compared with the baseline values during the previous dark period under saline infusion. Number of rats: five for 0.0025 nmol; eight for 0.025 nmol; five for 0.25 nmol; six for 2.5 nmol. $*p < 0.05$.

PS. Power spectral analysis of an electroencephalogram (EEG) revealed that delta activity was enhanced during the infusion period. PS was severely suppressed during the light period of the following 2 consecutive days and returned to the baseline level on the third recovery day. Smaller doses (0.023 to 0.23 nmol for a 10-h infusion) were ineffective for enhancing SWS and PS. However, the effect seems to be too long-lasting and considerably unphysiological, since severe disturbances of locomotor activity and other behaviors were concomitantly observed.

In contrast, we have recently found that L-DOC i.c.v. infused for 10 h during the same nocturnal period enhanced PS but not SWS at a dose of 250 pmol.[18a] Smaller (2.5 to 25 pmol) and larger (0.25 to 2.5 nmol) doses were ineffective on any parameters of both SWS and PS. As illustrated in Figure 3, the dose-response curve for PS enhancement exhibits a bell-shaped form.

B. Unidentified PS Factors

In 1969, Oswald[19] suggested that protein synthesis in the brain is closely related to the occurrence of PS. In line with his hypothesis, Drucker-Colín and associates[20-22] extensively studied on the "REM sleep proteins" found in the midbrain of REM sleep-deprived cats,

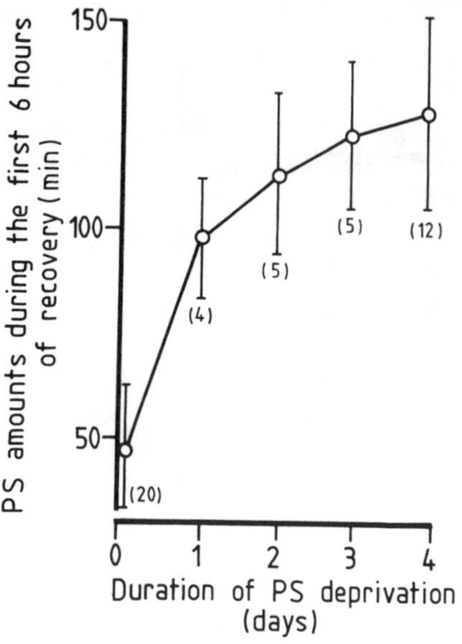

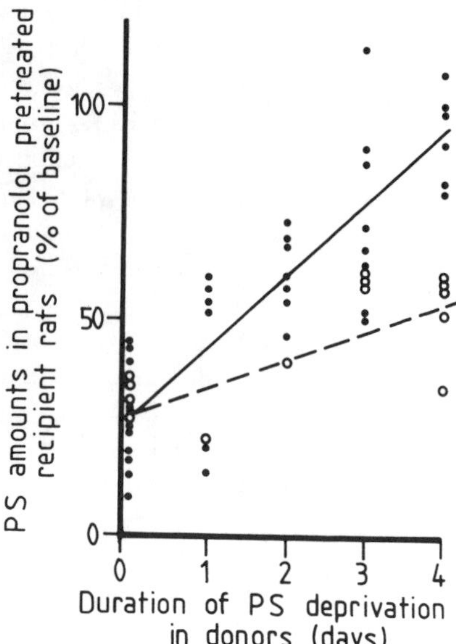

FIGURE 4. Effects of a PS-inducing factor on the amount of PS in PS-deprived donor and recipient rats. Left: The occurrence of PS during the first 6 h of the recovery period increased in proportion to the length of PS deprivation in donor rats. The typical PS rebounds observed suggest a progressive accumulation of some PS factor. Values are means ± standard deviation for a total of 14 rats. The number of tests are shown in parentheses. Right: PS restoration in propranolol-pretreated recipient rats following i.c.v. transfer of 20 μl of fresh (closed) and of frozen (open) CSF from PS-deprived donors. The frozen CSF had been kept for 1 to 15 d at −20°C before transfer. The amounts of PS were calculated for the first 6-h postadministration period and expressed as the percentage of the baseline recordings. (From Adrien, J. and Dugovic, in *Endogenous Sleep Substances and Sleep Regulation,* Inoué, S. and Borbély, A. A., Eds., Japan Scientific Societies Press, Tokyo/VNU Science Press BV, Utrecht, 1985, 227. With permission.)

as overviewed in Chapter 1. Further evidences (as follows) obtained in his and other laboratories indicate that specific proteins appear to be involved in PS regulation. (1) Protein synthesis inhibitors prevent the occurrence of PS without affecting SWS in several mammalian species.[23-28] Especially, the occurrence of PS and the associated single and multiple unit firing activity of neurons within the cat midbrain reticular formation reduced after either perfusion or oral administration of protein synthesis inhibitors.[29,30] (2) The antireticular formation antibody produced a specific decrease in PS without altering SWS in cats.[31] (3) An inhibitor of peptidases, bacitracin, produced an increase in PS in cats.[32] (4) The CSF of PS-deprived cats[33,34] and rats[35,36] by the water tank method could restore PS in pharmacologically PS-deprived recipient rats pretreated with parachlorophenylalanine (PCPA) and propranolol or α-methyldopa, respectively (see Figure 4).

No attempt has been made to chemically isolate and identify the specific REM sleep proteins. In this connection, Adrien and Dugovic[36] reported that delta-sleep-inducing peptide (DSIP) at doses of 25 (i.v.) or 100 μg/kg (s.c.; subcutaneously) could not restore PS in propranonol-induced, PS-deprived rats. Hence, DSIP did not substitute the CSF of the PS-deprived rats. Drucker-Colín et al.[37] paid attention to piperidine and growth hormone (GH) at first, but have recently emphasized that vasoactive intestinal polypeptide (VIP) and cholecystokinin (CCK) are responsible for the regulation of PS (see Chapter 8).

II. PS-PROMOTING ACTIVITY OF KNOWN SUBSTANCES

A. Growth Hormone

It is well established that an elevated amount of GH is associated with stage 3 + 4 sleep of the first sleep cycle in humans.[38] This sleep-onset GH peak may also appear shortly after sleep deprivation in dogs.[39] It is interesting that DSIP is involved in mediating the SWS-related GH release. Takahashi et al.[40] observed that DSIP (4 to 40 μg/kg i.c.v.) induced a sharp increase in blood levels of GH in a sleep-deprived dog. According to Iyer and McCann,[41] an injection of DSIP (0.1 to 10 μg) into the third cerebral ventricle of ovariectomized rats dose-dependently induced an elevation of the plasma level of GH. The release of GH is presumably mediated by the hypothalamus via a dopaminergic mechanism, since it could be blocked by pretreatment of the animals with pimozide, a dopamine receptor blocker. Cultured pituitary cells from ovariectomized rats exhibited a dose-related (bell-shaped) increase in GH release in static incubations with DSIP (1 p*M* to 10 n*M*). The close correlation of the GH peak with deep SWS leads to the assumption that GH release stimulates the biosynthesis of proteins in the brain, which in turn trigger the occurrence of PS, either directly or indirectly, via certain PS factors.

Indeed, Stern et al.[42] demonstrated that intraperitoneal (i.p.) injection of bovine GH at doses of 50 to 1000 μg induced selective increases in PS in cats during the first 3-h postinjection period. Interestingly, when the effect of GH on sleep was blocked by PS deprivation for the first 3 h, the PS-enhancing effect of GH appeared in the subsequent recovery sleep.

Drucker-Colín et al.[24] also reported the enhancement of PS during the first 3-h period following rat GH administration (0.1 to 1 mg/kg i.p.) in rats. It was further noted that PS reduction induced by anisomycin, a protein synthesis inhibitor, was blocked by the concurrent administration of GH.

Mendelson et al.[43] administered human GH (2 or 5 units s.c.) to healthy young volunteers at night. The lower dose had no effect, whereas the higher dose significantly decreased stage 3 + 4 sleep from 86 to 70 min and increased the amount of REM sleep from 98 to 110 min. No other sleep parameters were affected by the GH administration. However, the American investigators attributed these results to pharmacological or pathological reactions to exogenously supplied GH rather than to physiological ones.[44,45]

B. Somatostatin

Danguir,[46,47] on the other hand, suspected that the above enhancement of PS might be induced not by GH itself, but by somatostatin (SRIF), a hypothalamic neuropeptide, since GH administration reportedly induces the stimulation of SRIF release via a feedback control mechanism for GH secretion. He demonstrated evidences of this assumption by the following experiments. Chronic i.c.v. infusion of SRIF (20 μg/d) for 2 consecutive days elicited a selective increase of the daily amount of PS in freely moving rats without affecting the SWS quantity (see Figure 5). In order to eliminate the possibility that SRIF might trigger the release of GH, which actually induced the PS enhancement, a selective SRIF-depleting drug, cysteamine, was used to block the action of SRIF. This drug dose-dependently suppressed the occurrence of PS during the i.c.v. infusion period. Similar results were obtained after chronic i.c.v. infusion of a specific SRIF antiserum. From the results, Danguir proposed that SRIF may be involved in the regulation of PS, presumably at the hypothalamic level. The further discussion by the French researcher will be dealt with in the next chapter.

However, the earlier observations are not consistent with the proposal of Danguir. Havlicek et al.[48,49] reported that a marked suppression of sleep, both SWS and PS, and an increase in exploratory and motor behavior were induced after i.c.v. or supracortical infusions of SRIF (10 μg) over 10 to 15 min during a 2-h postinfusion period in normal and hypophysectomized rats. Similar but less prominent effects were observed after i.p. infusions of SRIF at 100 μg. Takahashi et al.[50] observed a phase advancement of the PS-acrophase but no effect on the amount of PS during chronic infusion of SRIF by an osmotic minipump (1 or 10 nmol/h i.c.v.; 10 nmol/h s.c.) for 7.5 d in rats. They also noted that an i.c.v. infusion of SRIF at 10 nmol/d brought about a slight but significant decrease in the amount of total sleep time and SWS.

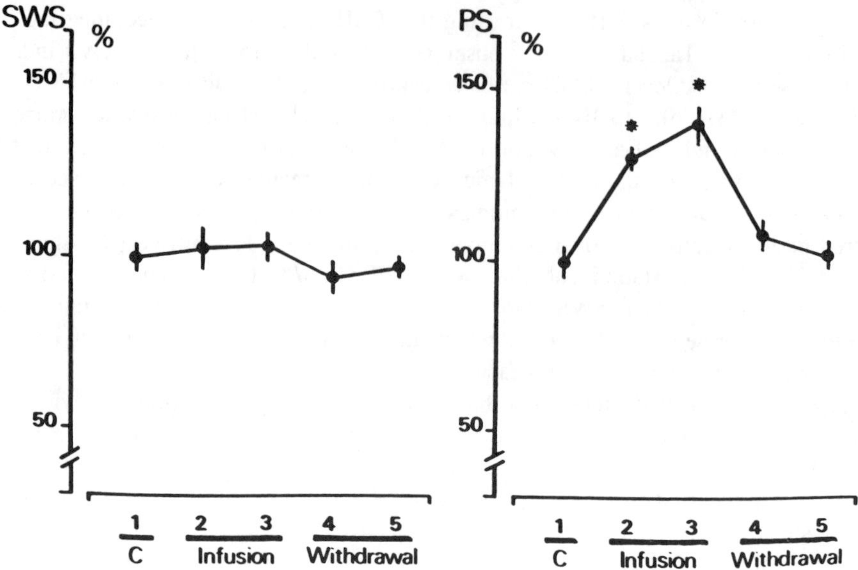

FIGURE 5. Effects of somatostatin (20 μg/d i.c.v. for 2 d) on daily amounts of SWS (left) and PS (right) in rats. Values on days 2 to 5 are compared with those of the control day (day 1; C) and expressed as percentages (mean ± SEM). $*p < 0.01$. (From Danguir, J., in *Sleep Peptides: Basic and Clinical Approaches,* Inoué, S. and Schneider-Helmert, D., Eds., Japan Scientific Societies Press, Tokyo/Springer-Verlag, Berlin, 1988, 53. With permission.)

C. Corticotropin-Like Intermediate Lobe Peptide

It is well known that proopiomelanocortin (POMC) is the prohormone of adrenocorticotropic hormone (ACTH), which is in turn cleaved into α-melanophore-stimulating hormone α-MSH; = $ACTH_{1-13}$) and corticotropin-like intermediate lobe peptide (CLIP; = $ACTH_{18-39}$). Chastrette and Cespuglio[51-54] have demonstrated a possibility that the POMC-derived hormones differentially play an interesting role in the regulation of sleep-waking states. According to them, CLIP selectively induced PS in rats after an i.c.v. injection at the end of the light period. The dose-response curve showed a bell-shaped form. CLIP at 10 ng induced a significant increase in the amount of nocturnal PS (see Figure 6). The increment was 59% above the baseline. PS enhancement was observable even during the following night (44% above the baseline). Doses of 1, 20, and 100 ng were ineffective, while a 5-ng dose induced a slight but insignificant increase in SWS and PS. The sleep-modulatory effects of the other POMC-derived hormones will be dealt with in the next chapter.

D. Prolactin and Pituitary Extracts

Jouvet et al.[55] have recently suggested that the pituitary gland isolated from the hypothalamic control by disconnection can release some PS-enhancing factors. The French investigators demonstrated that a significant increase in PS occurred in chronic pontine cats with an isolated pituitary. Since the isolated pituitary secretes an elevated level of prolactin (PRL) from the anterior lobe of the pituitary gland and the POMC-derived peptides from the intermediate lobe freed from the hypothalamic dopamine inhibition, it is of particular interest to examine the somnogenic effects of these substances. Repeated s.c. injections, but not continuous s.c. infusion by an osmotic minipump, of ovine PRL (1 to 15 units/d) dramatically increased PS up to 170 min/d in hypothalamo-hypophysectomized chronic pontine cats. Otherwise, such cats exhibited the complete absence of PS. Interestingly, the administration of PRL, regardless of injection and infusion, significantly prolonged the survival of hy-

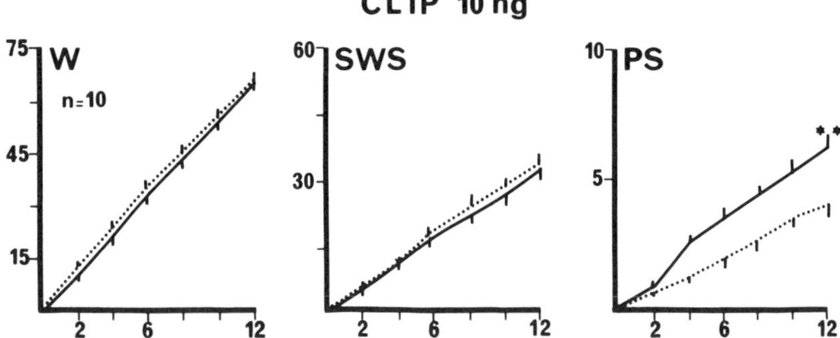

FIGURE 6. Effects of CLIP (10 ng i.c.v.) on wakefulness (W), SWS, and PS during the 12-h nocturnal period in rats. Values in the ordinates represent the percentage of the total recording time expressed cumulatively (mean ± SEM). The abscissae indicate the time in hours after injection. $**p < 0.02$. (From Chastrette, N., Clement, H. W., Prevautel, H., and Cespuglio, R., in *Sleep Peptides: Basic and Clinical Approaches,* Inoué, S. and Schneider-Helmert, D., Eds., Japan Scientific Societies Press, Tokyo/Springer-Verlag, Berlin, 1988, 27. With permission.)

pothalamo-hypophysectomized pontine cats up to 10 d. Similarly, s.c. injections of bovine posterior and intermediate lobe extract (PIL, 0.5 to 1 unit/kg) increased PS up to 350min/d and survival. In addition, PIL (1.5 unit s.c.; 1 milliunit into the ventral medulla) was able to restore PS in PCPA-pretreated insomniac cats. Although the active materials in PIL are unknown, the above-mentioned studies of Chastrette and Cespuglio[51-54] indicate that CLIP may be involved in the PS enhancement in Jouvet's studies.

E. Arginine Vasotocin

Contrary to the first report that arginine vasotocin (AVT), a pineal hormone, induced SWS in cats,[56] Coculescu et al.[57-59] demonstrated the REM sleep-enhancing capacity of AVT in humans. Two s.c. injections of AVT (each 1 to 2 μg) at night induced REM sleep at the expense of non-REM sleep in healthy humans (see Figure 7). However, diurnal administration of AVT through continuous i.v. perfusion of 2 to 12 μg in the early morning (between 04.00 and 08.00 h) or in the afternoon (between 16.00 and 19.00 h), resulted in no sleep-modulatory effect. In contrast, the same Romanian research team reported a tremendous increase in the total amount of nocturnal sleep, both non-REM sleep and REM sleep, after intranasal administration of AVT (10 μg) in healthy subjects and insomniacs.[60]

Tobler and Borbély,[61] on the other hand, found no somnogenic effect from an i.c.v. infusion of AVT at doses of 0.01 to 1 fmol in freely behaving rats. The treatment instead exerted a reduction of EEG delta activity. In contrast, Krueger et al.[62] observed a transient increase in EEG delta activity and excess SWS in rabbits within the first hour following i.c.v. infusion of 25 pmol of AVT. Mendelson et al.[63] found that an i.p. injection of AVT at a dose of 50μg/kg but not 0.5 μg/kg reduced REM latency by 65% and improved REM efficiency in rats without affecting any other sleep parameters. On the contrary, Riou et al.[64] reported a complete suppression of PS in rats during an 8-h diurnal period following an i.p. injection of AVT (10 and 50μg/kg). Pavel and Goldstein[65] reported that an i.c.v administration of AVT antigen caused a reduction of PS latency in cats.

Apart from these controversial results on the sleep-modulatory effects of AVT, there is some associated information. Goldstein and Psatta[66] suggested that the site of action of AVT is located in the raphe dorsalis nucleus (RDN). Their assumption is based on their experimental findings that RDN lesions produced a decrease in SWS and a transient increase in PS and narcoleptic-like alternations of sleep, and that AVT could not prevent the occurrence of such RDN lesion-induced symptoms. Goldstein[67] further speculated that AVT plays a

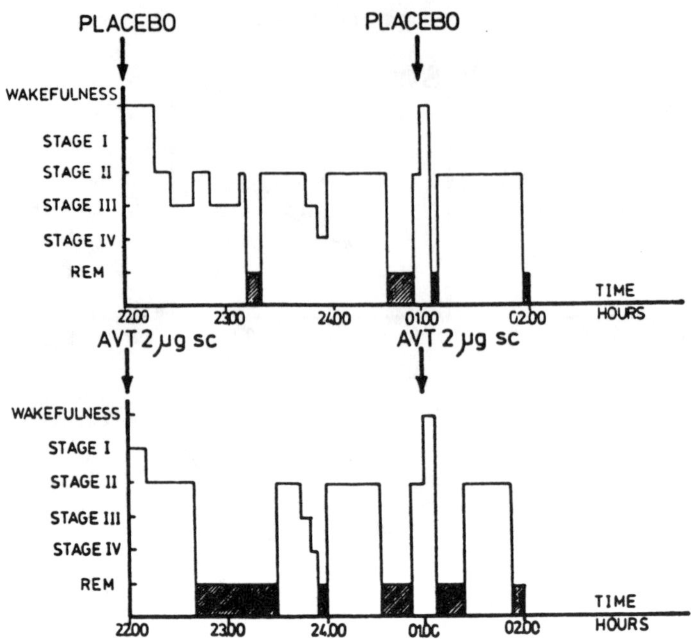

FIGURE 7. A 4-h hypnogram in a healthy subject receiving two s.c. injections of either a placebo (top) or AVT (bottom) at night. (From Coculescu, M., Serbanescu, A., and Temeli, E., *Waking Sleeping*, 3, 273, 1979. With permission.)

role in inhibiting the maturation of the brain, since daily i.p. injections of AVT (1 ng/day) in neonatal kittens resulted in an increase in the total amount and intensity of active sleep (the precursor of PS), a decrease in locomotor and investigative activities, a decrease in the total lipid levels in the brain, and a retardation of the eye opening. Normanton and Gent[68] found that locally applied AVT by microiontophoresis induced long-lasting excitement with a profound desensitization in the neurons of the nucleus gigantocellularis of the brainstem in rabbits and rats. Prechel et al.[69] on the other hand, suggested that AVT may be involved in the time-keeping mechanism of annual cycles, since the pineal AVT immunoactivity showed a marked yearly variation with a dramatic elevation in mid August in golden hamsters and rats.

F. Deoxycytidine

We have recently found that a 10-h nocturnal i.c.v. infusion of deoxycytidine, a pyrimidine nucleoside, dose-related enhanced PS without affecting SWS in rats[70] (see Figure 8). The dose of 10 pmol significantly increased the amount of PS, which was due to an elevated occurrence of PS episodes. For detailed experimental data and the chemical structure of deoxycytidine, see Table 2 and Figure 15 of Chapter 4, respectively.

III. PS-ASSOCIATED MATERIALS

A. PS-Suppressive Substances

Gillin et al.[71] found that an i.v. infusion of ACTH (40 IU) for 8 h beginning either in the morning or in the afternoon induced a suppression and delay of REM sleep in normal humans. Riou et al.[64] reported several endogenous peptides, such as angiotensin II (100 ng i.c.v), renin (0.01 units i.c.v.), and substance P (100 ng i.c.v.), selectively decreased PS in rats. Jouvet et al.[55] noted that arginine vasopressin (AVP, 10 milliunits s.c.) suppressed

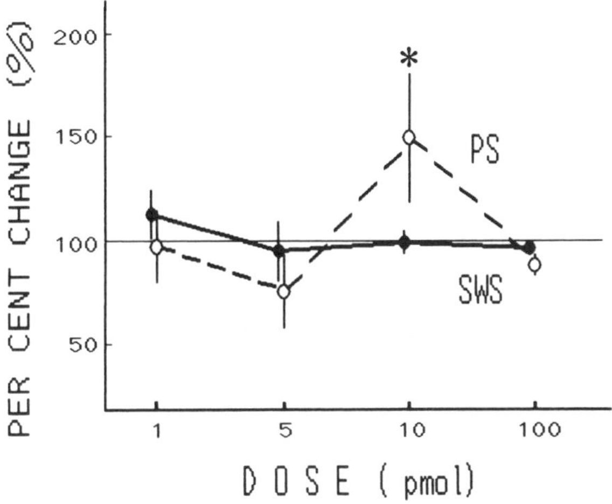

FIGURE 8. Dose-response relationship of the sleep modulatory effects of deoxycytidine, which was i.c.v. infused for 10 h during the dark period in freely behaving rats. *$p < 0.05$.

PS in pontine cats. In this connection, it should be mentioned here that Rees et al.[72] reported hyperactivity in mice after the administration of lysine vasopressin. Krejci and Kupková[73] demonstrated that an AVP analog, deamino-6-carba-8-vasopressin (10 to 20 μg/kg s.c.) dose-dependently induced behavioral sleep. Bibene et al.[74] demonstrated that a marked increase in wakefulness with a concomitant reduction of EEG sleep were induced during continuous i.c.v. infusion of AVP (2 μ*M*/0.5 μl/h) by an osmotic minipump, whereas i.c.v. infusion of anti-AVP antibodies significantly decreased wakefulness, especially during the light period. Finally it is reported that DSIP may sometimes suppress PS in rats (7 nmol/kg i.c.v.)[75] and in cats (30 nmol/kg i.p.)[76]

B. Lecithin

A cholinomimetic agent, lecithin or phosphatidyl choline, is known to be widely distributed in living organisms. Ishizuka et al.[77,78] reported that a daily uptake of lecithin (30 g/d) for 2 weeks resulted in a significant reduction of the latency to the first REM sleep from 87.5 min of baseline nights to 63.9 min in healthy humans. No other sleep parameters were affected by the lecithin uptake. The results suggest that lecithin may increase REM sleep propensity in the early night.

REFERENCES

1. **Yanagisawa, I. and Yoshikawa, H.,** A bromine compounds isolated from human cerebrospinal fluid, *Biochim. Biophys. Acta,* 329, 283, 1973.
2. **Jouvet, M., Cier, A., Mounier, D., and Valatx, J. L.,** Effets du 4-butyrolactone et du 4-hydroxybutyrate de sodium sur l'EEG et le comportement du chat, *C. R. Seances Soc. Biol. Paris,* 155, 1313, 1961.
3. **Matsuzaki, S., Takagi, H., and Tokizane, T.,** Paradoxical phase of sleep: its artificial induction in the cat by sodium butyrate, *Science,* 146, 1328, 1964.
4. **Yanagisawa, I., Yamane, M., Torii, S., and Inubushi, S.,** Biosynthesis and behavior of 1-methylheptyl γ-bromoacetoacetate in animal body, *Bull. Jpn. Neurochem. Soc.,* 14, 120, 1975.
5. **Yamane, M., Maeda, K., and Yanagisawa, I.,** Organic bromine compound in animal organs, *Sleep Res.,* 9, 60, 1980.

6. **Yamane, M. and Yanagisawa, I.,** Microdetermination of 1-methylheptyl γ-bromoacetoacetate in human blood, *J. Biochem.,* 92, 2009, 1982.
7. **Okudaira, N., Koike, H., Inubushi, S., Abe, A., Yamane, M., Yanagisawa, I., and Torii, S.,** Organic bromine substance: an endogenous REM sleep modulator?, in *Endogenous Sleep Substances and Sleep Regulation,* Inoué, S. and Borbély, A. A., Eds., Japan Scientific Societies Press, Tokyo/VNU Sience Press BV, Utrecht, 1985, 237.
8. **Torii, S., Mitsumori, K., Inubushi, S., and Yanagisawa, I.,** The REM sleep-inducing action of a naturally occurring organic bromine compound in the encéphale isolé cat, *Psychopharmacologia,* 29, 65, 1973.
9. **Yamane, M., Abe, A., and Yanagisawa, I.,** Anticholinesterase action of a bromine compound isolated from human cerebrospinal fluid, *J. Neurochem.,* 42, 1650, 1984.
10. **Okudaira, N., Hagino, K., Kudoh, K., Ohshige, H., Koike, H., Inubushi, S., Yamane, M., and Torii, S.,** MHBAA and REM sleep, *J. Physiol. Soc. Jpn.,* 49, 438, 1987.
11. **Okudaira, N., Inubushi, S., Koike, H., Hagino K., Torii, S., and Yamane, M.,** Sleep induction by MHBAA, *Annu. Rep. Natl. Inst. Physiol. Sci., Okazaki,* 8, 450, 1987.
12. **Torii, S., Inubushi, S., and Sakura, A.,** Effects of related compound of naturally occurring bromo-substance on the sleep-wakefulness cycle in the cat, in *Advances in Sleep Research,* Vol. 2, Weitzman, E. D., Ed., Spectrum, New York, 1975, 155.
13. **Okudaira, N., Torii, S., and Endo, S.,** The effect of butoctamide hydrogen succinate on nocturnal sleep: all-night polygraphical studies, *Psychopharmacology,* 70, 117, 1980.
14. **Okudaira, N., Kanemoto, H., Koike, H., Suenaga, K., Iguchi, Y., and Torii, S.,** Butoctamide hydrogen succinate (BAHS) and human delta wave, *Sleep Res.,* 16, 116, 1987.
15. **Hayashi, Y., Otomo, E., Okudaira, N., and Endo, S.,** Drug-increased REM sleep in aged subjects: butoctamide hydrogen succinate (BAHS), *Psychopharmacology,* 77, 367, 1982.
16. **Gruber, J. C., Gigli, G. L., Colognola, R. M., Ferri, R., Musumeci, S. A., and Bergonzi, P.,** Sleep patterns of Down's syndrome children: effects of butoctamide hydrogen succinate (BAHS) administration, *Psychopharmacology,* 90, 119, 1986.
17. **Colognola, R. M., Gruber, J. C., Gigli, G. L., Ferri, R., Petrella, M. A., Musumeci, S. A., and Bergonzi, P.,** Effects of acute and chronic administration of butoctamide hydrogen succinate (BAHS) on the nocturnal sleep of mental retardates (MR), *Sleep Res.,* 16, 82, 1987.
18. **Honda, K., Yanagisawa, I., and Inoué, S.,** Sleep-inducing action of a γ-Br derivative, *Annu. Rep. Natl. Inst. Physiol. Sci., Okazaki,* 8, 449, 1987.

18a. **Miura, M., Honda, K., Yanagisawa, I., and Inoué, S.,** unpublished.

19. **Oswald, I.,** Human brain proteins, drugs, and dreams, *Nature,* 223, 893, 1969.
20. **Drucker-Colín, R. and Ververde-R., C.,** Endocrine and peptide functions in the sleep-waking cycle, in *Sleep — Clinical and Experimental Aspects,* Ganten, D. and Pfaff, D., Eds., Springer-Verlag, Berlin, 1982, 37.
21. **Drucker-Colín, R., Aguilar-Roblero, R., and Arankowsky-Sandoval, G.,** Sleep factors released from brain of unrestrained cats: a critical appraisal, *Ann. N.Y. Acad. Sci.,* 473, 449, 1986.
22. **Drucker-Colín, R.,** Neuroproteins, brain excitability and REM sleep, in *Sleep, Dreams and Memory,* Fishbein, W., Ed., MTP Press, Falcon House, Lancaster, England, 1981, 73.
23. **Pegram, V., Hammond, D., and Bridgers, W.,** The effects of protein synthesis inhibition in sleep in mice, *Behav. Biol.,* 9, 377, 1973.
24. **Drucker-Colín, R. R., Spanis, C. W., Hunyadi, J., Sassin, J. F., and McGaugh, J. L.,** Growth hormone effects on sleep and wakefulness in the rat, *Neuroendocrinology,* 18, 1, 1975.
25. **Kitahama, K. and Valatx, J. L.,** Effet du chloramphenicol sur le sommeil de la souris, *C. R. Seances Soc. Biol. Paris,* 169, 1522, 1975.
26. **Rojas-Ramirez, J. A., Anhuilar-Jiménez, E., Posadas-Andrews, A., Bernal-Pedraza, J., and Drucker-Colín, R. R.,** The effect of various protein synthesis inhibitors on the sleep-waking cycle of rats, *Psychopharmacology,* 53, 147, 1977.
27. **Petitjean, F., Buda, C., Janin, M., David, M., and Jouvet, M.,** Effets du chloramphénicol sur le sommeil du chat — comparaison avec le thiamphénicol, l'érythromycine et l'oxytétracycline, *Psychopharmacology,* 66, 147, 1979.
28. **Matsumoto, J., Uezu, E., Seno, H., Sano, A., and Morita, Y.,** Reevaluation of humoral factors in sleep mechanism, *Proc. Jpn. Acad.,* 56, 492, 1980.
29. **Drucker-Colín, R., Zamora, J., Bernal-Pedraza, J., and Sosa, B.,** Modification of REM sleep and associated phasic activities of protein synthesis inhibitors, *Exp. Neurol.,* 63, 458, 1979.
30. **Drucker-Colín, R., Bowersox, S. S., and McGinty, D. J.,** Sleep and medial reticular unit responses to protein synthesis inhibitors: effects of cloramphenicol and thiamphenicol, *Brain Res.,* 252, 117, 1982.
31. **Drucker-Colín, R., Tuena de Gómez-Puyou, M., Gutiérrez, M. C., and Dreyfus-Cortés, G.,** Immunological approach to the study of neurohumoral sleep factors: effects on REM sleep of antibodies to brain stem proteins, *Exp. Neurol.,* 69, 563, 1980.

32. **Sastre, J. P., Sallanon, M., Buda, C., and Jouvet, M.,** La bacitracine, un inhibiteur des peptidases, augumente le sommeil paradoxal chez le chat, *C. R. Acad. Sci.*, 296, 965, 1983.
33. **Sallanon, M., Buda, C., Janin, M., and Jouvet, M.,** Restoration of paradoxical sleep by cerebrospinal fluid transfer to PCPA-treated cats, *Brain Res.*, 251, 137, 1982.
34. **Prospéro-Garcia, O., Morales, M., Arankowsky-Sandoval, G., and Drucker-Colín, R.,** Vasocative intestinal polypeptide (VIP) and cerebrospinal fluid (CSF) of sleep-deprived cats restores REM sleep in insomniac recipients, *Brain Res.*, 385, 169, 1986.
35. **Adrien, J. and Dugovic, C.,** Presence of a paradoxical sleep (PS) factor in the cerebrospinal fluid of PS-deprived rats, *Eur. J. Pharmacol.*, 100, 223, 1984.
36. **Adrien, J. and Dugovic, C.,** PS-inducing factors and the noradrenergic system, in *Endogenous Sleep Substances and Sleep Regulation,* Inoué, S. and Borbély, A. A., Eds., Japan Scientific Societies Press, Tokyo/VNU Science Press BV, Utrecht, 1985, 227.
37. **Drucker-Colín, R., Prospéro-García, G., Arankowsky-Sandoval, G., and Pérez-Montfort, R.,** Gastropancreatic peptides and sensory stimuli as REM sleep factors, in *Sleep Peptides: Basic and Clincal Approaches,* Inoué, S. and Schneider-Helmert, D., Eds., Japan Scientific Societies Press, Tokyo/Springer-Verlag, Berlin, 1988, 73.
38. **Takahashi, Y., Kipnis, D. M., and Daughaday, W. H.,** Growth hormone secretion during sleep, *J. Clin. Invest.*, 47, 2079, 1968.
39. **Takahashi, Y., Ebihara, S., Nakamura, Y., and Takahashi, K.,** A model of human sleep-related growth hormone secretion in dogs: effects of 3, 6, and 12 hours of forced wakefulness on plasma growth hormone, cortisol, and sleep stages, *Endocrinology,* 109, 262, 1981.
40. **Takahashi, Y., Kato, N., Nakamura, Y., Ebihara, S., Tsuji, A., and Takahashi, K.,** Delta sleep inducing peptide (DSIP): a preliminary report of effects of intraventricular infusion on sleep and GH secretion, and of plasma DSIP concentrations in the dog and rat, in *Integrative Control Functions of the Brain,* Vol. 2, Ito, M., Tsukahara, N., Kubota, K., and Yagi, K., Eds., Kodansha, Tokyo, 1980, 339.
41. **Iyer, K. S. and McCann, S. M.,** Delta sleep-inducing peptide (DSIP) stimulates growth hormone (GH) release in the rat by hypothalamic and pituitary actions, *Peptides,* 8, 45, 1987.
42. **Stern, W. C., Jalowiec, J. E., Shabshelowitz, H., and Morgane, P. J.,** Effects of growth hormone on sleep-waking patterns in cats, *Horm. Behav.*, 6, 189, 1975.
43. **Mendelson, W. B., Slater, S., Gold, P., and Gillin, J. C.,** The effect of growth hormone administration on human sleep: dose-dependent response study, *Biol. Psychiatr.*, 15, 613, 1980.
44. **Mendelson, W. B., Gillin, J. C., and Wyatt, R. J.,** The search for circulating sleep-promoting factors, in *Advances in Pharmacology and Therapeutics II,* Vol. 1, Yoshida, H., Hagiwara, Y., and Ebashi, S., Eds., Pergamon Press, Oxford, 1982, 227.
45. **Mendelson, W. B., Wyatt, R. J., and Gillin, J. C.,** Whither the sleep factors?, in *Sleep Disorders: Basic and Clinical Research,* Chase, M. H. and Weitzman, E. D., Eds., Spectrum, New York, 1983, 281.
46. **Danguir, J.,** Intracerebroventricular infusion of somatostatin selectively increases paradoxical sleep in rats, *Brain Res.*, 367, 26, 1986.
47. **Danguir, J.,** Internal milieu and sleep homeostasis, in *Sleep Peptides: Basic and Clinical Approaches,* Inoué, S. and Schneider-Helmert, D., Eds., Japan Scientific Societies Press, Tokyo/Springer-Verlag, Berlin, 1988, 53.
48. **Havlicek, V., Rezek, M., and Friesen, H.,** Somatostatin and thyrotropin releasing hormone: central effect on sleep and motor system, *Pharmacol. Biochem. Behav.*, 4, 455, 1976.
49. **Rezek, M., Havlick, V., Hughes, K. H., and Friesen, H.,** Cortical administration of somatostatin (SRIF): effect on sleep and motor behavior, *Pharmacol. Biochem. Behav.*, 5, 73, 1976.
50. **Takahasi, Y., Usui, S., Ebihara, S., and Honda, Y.,** Effects of continuous infusions of TRH and GIF on circadian rhythm parameters of sleep, ambulation, eating, and drinking in rats, in *Endogenous Sleep Substances and Sleep Regulation,* Inoué, S. and Borbély, A. A., Eds., Japan Scientific Societies Press, Tokyo/VNU Science Press BV, Utrecht, 1985, 101.
51. **Chastrette, N. and Cespuglio, R.,** Effets hypnogènes de la des-acetyl-α-MSH et du CLIP (ACTH 18-39) chez le rat, *C. R. Acad. Sci.*, 301, 527, 1985.
52. **Chastrette, N. and Cespuglio, R.,** Influence of proopiomelanocortin-derived peptides on the sleep-waking cycle of the rat, *Neurosci. Lett.*, 62, 365, 1985.
53. **Chastrette, N., Lin, Y. L., Faradji, H., and Cespuglio, R.,** Hypnogenic properties of desacetyl-α-MSH and CLIP (ACTH 18-39), in *Sleep '86,* Koella, W. P., Obál, F., Schulz, H., and Visser, P., Eds., Gustav Fischer Verlag, Stuttgart, 1988, 165.
54. **Chastrette, N., Clement, H. W., Prevautel, H., and Cespuglio, R.,** Proopiomelanocortin components: differential sleep-waking regulation?, in *Sleep Peptides: Basic and Clinical Appraoches,* Inoué, S. and Schneider-Helmert, D., Eds., Japan Scientific Societies Press, Tokyo/Springer-Verlag, Berlin, 1988, 27.
55. **Jouvet, M., Buda, C., Cespuglio, R., Chastrette, N., Denoyer, M., Sallanon, M., and Sastre, J. P.,** Hypnogenic effects of some hypothalamo-pituitary peptides, *Clin. Neuropharmacol.*, 9(Suppl. 4), 465, 1986.

56. **Pavel, S., Psatta, D., and Goldstein, R.,** Slow-wave sleep induced in cats by extremely small amounts of synthetic and pineal vasotocin injected into the third ventricle of the brain, *Brain Res. Bull.*, 2, 251, 1977.
57. **Coculescu, M., Serbanescu, A., and Temeli, E.,** Influence of arginine vasotocin administration on nocturnal sleep of human subjects, *Waking Sleeping*, 3, 273, 1979.
58. **Coculescu, M., Simionescu, L., Cristoveanu, A., Servanescu, A., Matulevicius, V., Grigorescu, F., and Temeli, E.,** Administration of arginine-vasotocin (AVT) to human subjects, *Prog. Brain Res.*, 52, 535, 1979.
59. **Coculescu, M., Serbanescu, A., and Temeli, E.,** Influence of arginine vasotocin administration on nocturnal sleep of human subjects, in *Sleep 1978*, Popoviciu, L., Asigian, B., and Badiu, G., Eds., S. Karger, Basel, 1980, 315.
60. **Popoviciu, L., Tudosie, M., Corfariu, O., and Pavel, S.,** Sleep organization peculiarities in normal subjects and insomniacs after arginine vasotocin administration, *Abstr. 6th Eur. Congr. Sleep Res.*, 1982, 119.
61. **Tobler, I. and Borbély, A. A.,** Effect of delta sleep inducing peptide (DSIP) and arginine vasotocin (AVT) on sleep and locomotor activity in the rat, *Waking Sleeping*, 4, 139, 1980.
62. **Krueger, J. M., Bacsik, J., and García-Arrarás, J.,** Sleep-promoting material from human urine and its relation to factor S from brain, *Am. J. Physiol.*, 238, E116, 1980.
63. **Mendelson, W. B., Gillin, J. C., Pisner, G., and Wyatt, R. J.,** Arginine vasotocin and sleep in the rat, *Brain Res.*, 182, 246, 1980.
64. **Riou, F., Cespuglio, R., and Jouvet, M.,** Endogenous peptides and sleep in the rats. I. Peptides decreasing paradoxical sleep, *Neuropeptides*, 2, 243, 1982.
65. **Pavel, S. and Goldstein, R.,** Narcoleptic-like alterations of the sleep cycle in cats induced by a specific vasotocin antiserum, *Brain Res. Bull.*, 7, 453, 1981.
66. **Goldstein, R. and Psatta, D.,** Sleep in the cat: raphe dorsalis and vasotocin, *Sleep*, 7, 373, 1984.
67. **Goldstein, R.,** The involvement of arginine vasotocin in the maturation of the kitten brain, *Peptides*, 5, 25, 1984.
68. **Normanton, J. R. and Gent, J. P.,** Comparison of the effect of two 'sleep' peptides, delta sleep-inducing peptide and arginine-vasotocin, on single neurons in the rat and rabbit brain stem, *Neuroscience*, 8, 107, 1983.
69. **Prechel, M. M., Audhya, T. K., and Simmons, W. H.,** Pineal arginine vasotocin activity increases 200-fold during August in adult rats and hamsters, *J. Pineal Res.*, 1, 175, 1984.
70. **Naito, K., Honda, K., Komoda, Y., and Inoué, S.,** Sleep-promoting effects of deoxycytidine in unrestrained rats, *Jpn. J. Psychiatr.*, 42, 128, 1988.
71. **Gillin, J. C., Jacobs, L. S., Snyder, F., and Henkin, R. I.,** Effects of ACTH on the sleep of normal subjects and patients with Addison's disease, *Neuroendocrinology*, 15, 21, 1974.
72. **Rees, H. D., Dunn, A. J., and Iuvone, P. M.,** Behavioral and biochemical responses of mice to the intraventricular administration of ACTH analogs and lysine vasopressin, *Life Sci.*, 18, 1333, 1976.
73. **Krejci, I. and Kupková, B.,** Sleep-inducing effect of a vasopressin analog, deamino-6-carba-ornithine-8-vasopressin (DCOV) in rats, *Act. Nerv. Super. (Praha)*, 20, 60, 1978.
74. **Bibene, V., Arnauld, E., Meynard, J., Rodriguez, F., Poncet, C., and Vincent, J.-D.,** Influence of vasopressin on circadian sleep and wakefulness, *Sleep*, 16, 43, 1987.
75. **Obál, F., Jr., Török, A., Alföldi, P., Sáry, S., Hajós, M., and Penke, B.,** Effects of intracerebroventricular injection of delta sleep-inducing peptide (DSIP) and an analogue on sleep and brain temperature in rats at night, *Pharmacol. Biochem. Behav.*, 23, 953, 1985.
76. **Sommerfelt, L.,** Reduced sleep in cats after intraperitoneal injection of δ-sleep-inducing peptide, *Neurosci. Lett.*, 58, 73, 1985.
77. **Ishizuka, Y., Atsumi, Y., Uchiyama, M., Tanaka, K., Iijima, M., and Kojima, T.,** The effect of lecithin on human sleep, *Folia Psychiatr. Neurol. Jpn.*, 39, 213, 1985.
78. **Ishizuka, Y., Atsumi, Y., Uchida, S., Uchiyama, M., Hibino, H., Kojima, T., and Fukuzawa, H.,** The effect of lecithin on human sleep and body temperature: the cholinomimetic agent and biological rhythm, *Sleep Res.*, 16, 216, 1987.

Chapter 8

OTHER SLEEP MODULATORS

I. PIPERIDINE AS A HYPNOGENIC SUBSTANCE

A. Natural Occurrence

Piperidine (see Figure 1) is a biogenic amine and a normal constituent of the central nervous system of both invertebrates and vertebrates, and regarded as a neuromodulator or a hypnogenic substance.[1] This substance is also widely distributed in the peripheral organs. Kasé et al.[2] demonstrated that piperidine intraperitoneally (i.p.) injected at doses of 40 and 80 mg/kg induced pronounced sedation and prolongation of hexobarbital-induced narcosis in mice. They also suggested that piperidine may act as a stimulator of the nicotinic cholinergic receptor. Stepita-Klauco et al.[3] found that brain concentrations of piperidine increased to 36.6 nmol/g during dormancy from 2.0 nmol/g during the active state in mice. The brain content of piperidine also increased during dormancy or hibernation in snails.[4] Miyata et al.,[5,6,9-12] Okano et al.,[7,8] and Aisaka et al.[13] studied extensively the biological and pharmacological properties of piperidine, such as its hypnogenic activity;[5] its concentrations in the brain,[6,7] blood,[8] and urine;[8] its changes in brain levels depending on seasonal activity,[9] anesthesia,[10-12] or sleep deprivation;[12] and its vasodilating activity.[13]

B. Experiments in Cats

Miyata et al.[5] first demonstrated the electroencelphalographic (EEG)-sleep enhancing activity of piperidine. Microinjection of piperidine at doses of 20 to 50 μg into the hippocampus induced high-amplitude slow waves in the cortical EEG and concomitant slow-wave sleep (SWS) about 15 min after the treatment in freely behaving cats. Higher doses (300 to 500 μg) caused spike seizure discharges in the hippocampal EEG, which occurred about 5 min after the treatment and appeared repeatedly during the 20- to 30-min postinjection period, followed by fast seizure discharges for more than 20 min. Marked behavioral changes such as attention, exploratory reaction, and stereotypy during the 20- to 30-min postinjection period were followed by grooming and sexual behaviors. Microinjection of piperidine into the amygdala at the same low and high doses caused hypnogenic and behavioral changes largely similar to those induced by the hippocampal administration. In contrast, no effect was induced after microinjection of low doses of piperidine at 50 to 100 μg into the caudate nucleus, whereas behavioral sleep and a potentiation of electrically induced caudate-spindles were induced after the administration of high doses (300 to 700 μg).

Microinjection of piperidine at low doses of 10 to 20 μg into the pontine reticular formation induced a significant enhancement of both SWS and paradoxical sleep (PS) at the expense of wakefulness (see Figure 2). During a 420-min postinjection period, the total time of SWS and PS rose to 180 and 68 min, respectively, from 148 and 33 min, respectively, of the saline control. The number of PS episodes rose from 9.0 to 18.0, whereas the duration of each episode remained unchanged. On the other hand, higher doses of piperidine at 100 to 300 μg produced an arousal pattern in cortical EEG. Microinjection of piperidine at both low and high doses into the cerebellum resulted in an enhancement of SWS within 10 min.

Thus, it appears likely that piperidine microinjected at relatively low doses can provoke a definite sleep-enhancing activity in cats. The locus of microinjection seems to be crucial for the induction of either SWS or PS. The observations of Miyata et al.[5] were confirmed by Drucker-Colín and Giacobini[14] in cats. Using the push-pull perfusion technique, the Mexican investigators found a significant increase in the total amount of PS and a significant decrease in the latency to SWS and PS onset after the administration of piperidine (333 μg for 15 min) in the cat midbrain reticular formation.

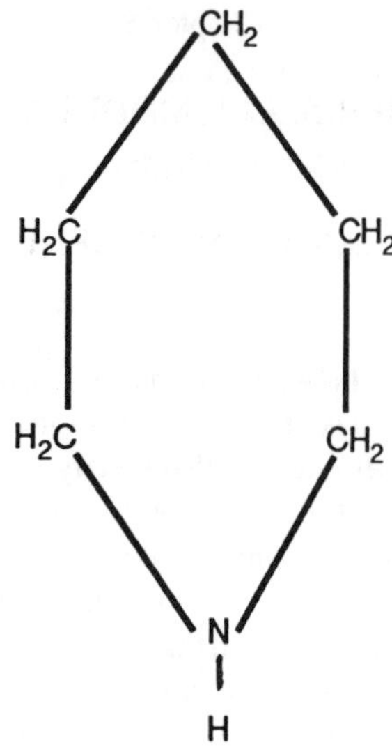

FIGURE 1. Chemical structure of piperidine.

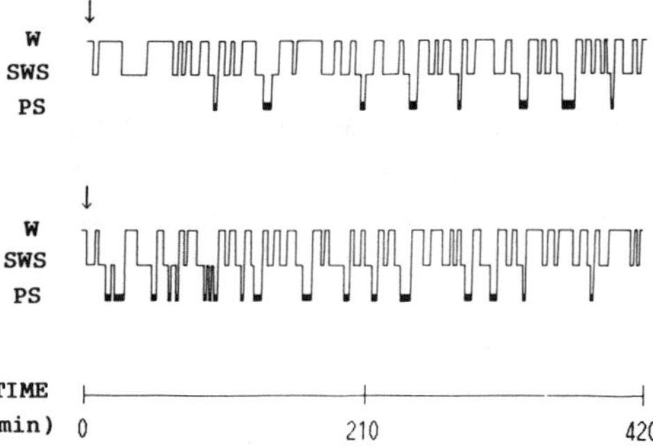

FIGURE 2. Hypnogram of freely behaving cats microinjected with saline (top) and 20 μg of piperidine (bottom) into the pontine reticular formation. The arrows indicate the time of injection. PS: paradoxical sleep; SWS: slow-wave sleep; W: wakefulness. (From Miyata, T., Kamata, K., Nishikibe, M., Kasé, Y., Takahama, K., and Okano, Y., *Life Sci.*, 15, 1135, 1974. With permission.)

C. Experiments in Rats

On the contrary, Nixon and Karnovsky[15] reported that an intracerebroventricular (i.c.v.) infusion of piperidine into the lateral cerebral ventricle at doses of 0.1 to 20 μg over 7 min dose-dependently resulted in a significant reduction of both SWS and PS during 20 min in rats. Although the reduced level of SWS recovered to that of the baseline value during the subsequent period (a 20- to 140-min postinfusion period), PS did not recover to the baseline level.

Miyata and Okano[12] have recently measured the content of piperidine in the brainstem of rats after anesthesia and PS deprivation. They found that piperidine was distributed at high levels in the reticular formation of the midbrain, the pons, and the medulla. The content of piperidine in these regions increased after urethane anesthesia. The content of piperidine in the reticular formation of the midbrain and the anterior pons decreased significantly during PS deprivation, whereas it increased significantly during PS rebound.

D. Experiments in Humans

Mendelson et al.[16] reported that an intravenous (i.v.) infusion of piperidine at a dose of 100 mg for 30 min starting at sleep onset in normal young adults did not change any sleep parameters during the subsequent nocturnal sleep.

II. NUTRIENTS — ISCHYMETRIC SLEEP THEORY

A. Food and Sleep

Since Aristotle's theory on sleep (see Chapter 1), the relationship between food and sleep has been discussed repeatedly. In 1980, Danguir in collaboration with Nicolaidis[17] published a paper on an ischymetric sleep-regulation hypothesis, suggesting that sleep is related to both the nature and the degree of utilization of circulating metabolites. This theory is based on the following experimental findings in rats.

1. A continuous i.v. infusion of high-energy nutrients induced a significant increase in both SWS and PS.
2. A continuous i.v. infusion of amino acids brought about a significant increase in PS, whereas SWS remained unchanged.
3. Although a continuous i.v. infusion of glucose or lipids did not affect the daily sleep quotas, glucose, in combination with insulin (6 units for 24 h), significantly increased both SWS and PS.
4. Obese rats with ventromedial hypothalamic lesions exhibited hypersomnia during the dynamic phase of hyperphasia.[18]
5. Food deprivation in lean rats, but not in obese rats, caused a decrease in both SWS and PS, whereas the normal amount of sleep was recovered with the availability of food.[19] (In this connection, it should be noted that suckling enhanced SWS, and the gut load of milk enhanced PS in milk-deprived rat pups.)[20]
6. There were significant correlations between meal size and the amount of time spent in SWS and PS in intermeal interval that followed during the activity phase, i.e., the dark period, in rats.[21]

On the other hand, there is another interesting and inverse feature of sleep. It should not be overlooked that the amount of sleep may become increased at the time of food inavailability, since sleep has an adaptative function to save energy expenditure.[22]

B. Insulin and SWS

Insulin is a 51-amino acid residue polypeptide originally known as a pancreatic hormone. Insulin is also regarded to be distributed in the brain, although its functional role in the brain is still an open question. Subsequent to the observation of Danguir and Nicolaidis,[17] Sangiah et al.[23] reported that an i.p. injection of a single nonconvulsive dose of insulin (1 unit/kg = 45.5 μg/kg = 7.6 nmol/kg) completely suppressed PS during the first 3-h postinjection period, returning to the normal level by the sixth hour without gross behavioral changes in rats. The administration of insulin concomitantly increased SWS over 5 h.

Danguir and Nicolaidis[24,25] and Danguir[26] further undertook a series of studies on the sleep-enhancing activity of centrally administered insulin. A continuous i.c.v. infusion of insulin for 3 consecutive days (200 μU/d) in rats resulted in a profound increase in SWS but not in PS (see Figure 3). The increased SWS was essentially due to a prolongation of episodes, particularly during the dark phase of the circadian cycle. In order to ascertain the direct effect of peripherally liberated insulin on the cerebral mechanism via the blood-brain barrier, the inactivation of insulin in the brain or in the peripheries was conducted by the administration of insulin antiserum. Both i.c.v and i.v. infusions of anti-insulin serum in

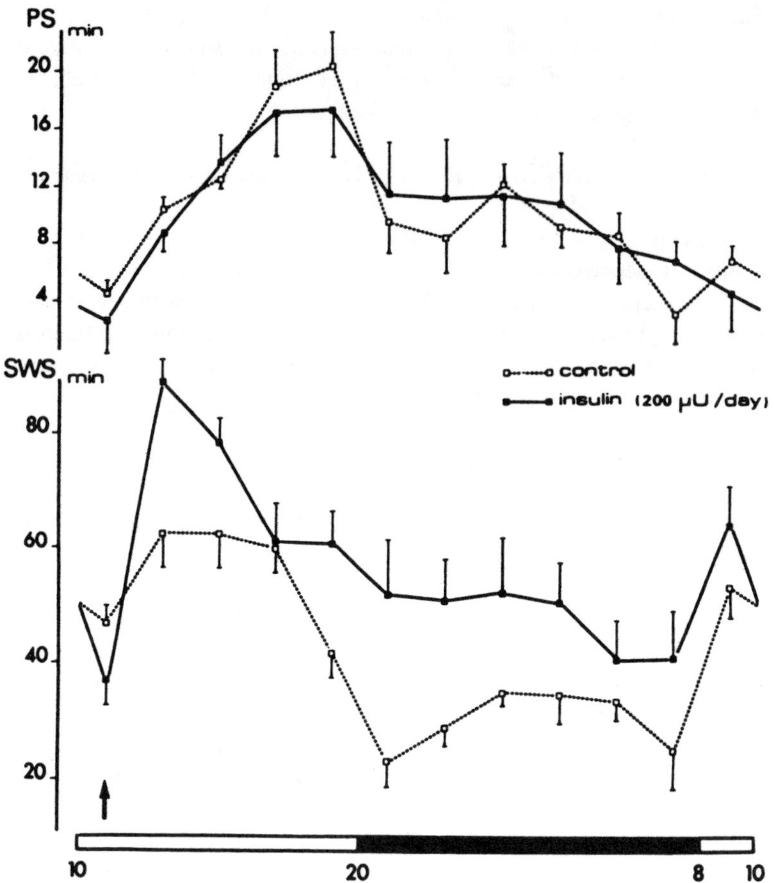

FIGURE 3. Effects of continuous i.c.v. infusion of insulin in rats.[23] Time spent in PS (top) and SWS (bottom) are expressed in minutes (mean ± SEM) on a 2-h period base during infusion of insulin (200 µU/d; solid lines) and during infusion of artificial CSF (dashed lines). The arrow indicates the initiation of the infusion. The abscissa represents the time of day, in which the open and closed bars stand for the light and the dark periods, respectively. (From Danguir, J. and Nicolaidis, S., *Brain Res.*, 306, 97, 1984. With permission.)

rats brought about a significant decrease in SWS but not in PS. This was caused by a shortening of the mean duration of SWS episodes, during both the light and the dark periods. Thus, it seems likely that the brain insulin, originating from the periphery, is responsible for the enhancement of SWS.

The involvement of insulin in the modulation of SWS was also confirmed by investigations of sleep in diabetic rats.[27] Streptozotocin-induced diabetic rats exhibited a continued decrease in SWS with a transient reduction of PS. The supply of exogenous insulin (50 to 200 µU, i.v. infused) dose-dependently recovered SWS in these diabetic rats.

In this connection, some insulin-related information should be documented here. According to Bernstein et al.,[28] an i.c.v. injection of a very small dose of insulin (0.05 fmol) did not influence the duration of hexabarbital-induced narcosis in rats. In addition, the metabolic effect of insulin and glucagon is temperature-dependent, since Castex and Hoo-Paris[29] reported that blood glucose, insulin, and glucagon were low in the blood of the hedgehog and the edible dormouse during hibernation. The secretion of both hormones was not stimulated by glucose, and blood glucose was not regulated by insulin under lowered body temperature during hibernation. After arousal and temperature rise, the effect of glucagon and insulin on glucose increased markedly. Thus, the SWS-inducing mechanism seems to be different from the hibernation-maintaining mechanism so far as the activity of insulin is concerned.

C. Somatostatin and PS

As already mentioned in a previous chapter, growth hormone (GH) enhances PS in cats, humans, and rats. The phenomena have been elucidated as a result of the elevated production of proteins in the brain. The reduction of PS by protein synthesis inhibitors appears to support such speculation. Furthermore, an elevated level of GH may trigger the release of somatostatin (SRIF) via a negative-feedback mechanism and it is the activity of SRIF that may be responsible for the enhancement of PS.[30] It is interesting that cerebral protein synthesis is correlated closely with SWS in rats:[31] higher rates of cerebral protein synthesis are associated with higher levels of SWS, whereas the level of PS is not linked to the rate of cerebral protein synthesis.

D. Food-Dependent Differential Sleep Regulation

Danguir[32] recently reported that the ''cafeteria'' choice of several palatable, high-energy diets including white bread, chocolate, biscuits, and rat chow resulted in a significant increase in daily amounts of both SWS and PS in rats. SWS increased throughout the light and the dark periods, while PS increased only during the light period. The increase was due mainly to a significant prolongation of the respective episodes. After the withdrawal of the cafeteria diet, the elevated levels of SWS and PS lasted for 3 and 1 d, respectively, and returned to the normal level.

It is established that the secretion of insulin and SRIF is related closely to both the quantity and the quality of food. Carbohydrates are known to be particularly potent inducers of insulin secretion. Protein diets can stimulate SRIF secretion, which can induce PS. Taking all these facts into consideration, Danguir[33,34] has recently formulated the following interesting causal relationship:

Carbohydrate diet	—	(herbivors)	—	Insulin release	—	Increase in SWS
Protein diet	—	(carnivors)	—	SRIF release	—	Increase in PS
Composite diet	—	(omnivors)	—	Insulin/SRIF release	—	Increase in SWS/PS

III. SLEEP-MODULATORY NEUROPEPTIDES

A. Vasoactive Intestinal Polypeptide

Vasoactive intestinal polypeptide (VIP) is a 28-amino acid residue polypeptide originally known as a gastrointestinal hormone. Today, VIP is known to occur widely in both the central and the peripheral nervous systems as a neuromodulator, exerting a wide variety of biological activities such as induction of hyperthermia, stimulation of release of prolactin (PRL), GH, and adrenocorticotropic hormone (ACTH), enhancement of glucose utilization, vasodilation in cerebral vessels, etc. It is of particular interest that VIP is distributed in the suprachiasmatic nucleus and the arcuate nucleus, since both the hypothalamic nuclei play an important role as neuroendocrine information transducers.

Riou et al.[35,36] first reported the marked sleep-enhancing activity of VIP in rats. The effects of VIP were dependent on the dosage and the time of day. A single i.c.v. injection of VIP in the morning at a dose of 100 ng induced a significant increase in PS at the expense of wakefulness during an 8-h, diurnal postinjection period. The 15% increase in the amount of PS was due to an increase in the number of PS episodes. A lower dose (10 ng) of VIP increased diurnal wakefulness 17%, but the difference was not significant. A higher dose (1 μg) of VIP exerted no sleep-modulatory effect immediately, but it significantly increased diurnal PS and decreased diurnal wakefulness on the next day. It is interesting that the i.c.v.

administration of 100 ng of VIP at the dark period onset resulted in a significant increase in both SWS (18%) and PS (65%), and a significant decrease in wakefulness (12%) during the subsequent 12-h dark period. The increase in SWS was due to a prolongation of SWS episodes, while the increase in PS was caused by an increased occurrence of PS episodes and their prolongation. The administration of VIP antiserum brought about arousal characterized by an increase in wakefulness, a decrease in PS, and no change in SWS. It was also found that VIP (100 ng i.c.v.) restored sleep, both SWS and PS, in parachlorophenylaranine (PCPA)- or chloramphenicol (CAP)-pretreated, partially insomniac rats.

Obál et al.[37] and Obál[38] reported that an acute i.c.v. injection of VIP (100 ng = 30 pmol) shortly before the dark period onset increased significantly the amount of SWS in rats for the first 6 h in the dark period and the first 3 h in the subsequent light period. PS also increased for the first 3-h postinjection period. Brain temperature was little affected by the treatment. According to the recent studies of Kruisbrink et al.,[39] a constant delivery of VIP into the lateral cerebral ventricle via an Accurel®/collodion implant containing 10 μg VIP enhanced both diurnal and nocturnal sleep at the expense of wakefulness in rats. The increment of SWS in the light and dark periods was 9 and 23%, respectively, while that of PS was 38 and 36%, respectively. The increase in PS was caused by more frequent and longer periods of PS. Nakagaki and Takahashi[40] reported that an i.c.v. injection of 1 nmol/kg VIP shortly before the onset of the dark period induced a significant increase in nocturnal sleep, both SWS and PS. The enhancement of SWS lasted over the second day. A lower dose (0.1 nmol/kg) resulted in a prolongation of PS latency without modulating other sleep parameters.

In contrast to the somnogenic activity of the relatively low dosage of VIP, Itoh et al.[41] reported that an i.c.v. injection of VIP dose-dependently shortened the duration of pentobarbital-induced narcosis at doses of 0.5 to 10 μg and increased spontaneous motility at doses of 10 to 20 μg in rats.

VIP can modify sleep-waking states in other species. In PCPA-pretreated cats, 10 ng of VIP, but not 100 ng or 1 μg, restored PS selectively during the light period.[35] This observation has been recently confirmed by Prospéro-García et al.[42] and Drucker-Colín et al.,[43] who reported that 200 ng of VIP restored PS in PCPA- and CAP-pretreated insomniac cats. In normal cats, the i.c.v. administration of VIP at a dose of 100 ng, but not 10 ng, selectively induced a slight but significant increase in PS.[44] In rabbits, either an i.c.v. or intracarotid infusion of VIP (10 to 20 ng/min) for 1 h induced an increase in EEG spindle activity.[45]

Some additional information should be referred to here. VIP is regarded as an energy-mobilizing hormone. The plasma VIP level is known to increase during and after muscular exercise.[46] Plasma VIP significantly increased after a 30-min ergometer exercise test in young men who had undergone 5-d sleep deprivation in combination with prolonged strain and energy deficiency.[47] The feeling of sleepiness or an elevated demand for sleep after exercise might be partly accounted for by the rise in the plasma level of VIP, if VIP can cross the blood-brain barrier and affect the neural activity in the central nervous system. Shimatsu et al.[48] reported that the release of VIP was stimulated by the presence of prostaglandin D_2, a potent sleep inducer (see Chapter 6). A definite circadian rhythm was found in the cerebrospinal fluid (CSF) level of VIP in monkeys, with a peak in the early dark period and a nadir in the early light period.[49] Finally, VIP is reportedly correlated to the immune system,[50] suggesting interactions with immunomodulatory somnogens.[51]

B. Cholecystokinin

Cholecystokinin (CCK) is distributed in the brain and gut in mammals and a 33-amino acid residue peptide (CCK-33). However, the C-terminal octapeptide fragment of the original CCK has a full biological activity. Hence, the octapeptide, Asp-Tyr(SO_3)-Met-Gly-Trp-Met-Asp-Phe-NH_2 (CCK-8), has been widely used for studies on the biological activities of CCK. Here, the abbreviation CCK represents CCK-8, but not CCK-33.

Riou et al.[52] reported that no sleep-modulatory effect was induced after the injection of CCK (0.1 or 1 μg i.c.v.; 2μg/kg i.p.) in normal rats during the light period. However, they found that CCK (0.1 μg i.c.v.) restored sleep, both SWS and PS, in CAP-pretreated, partially insomniac rats, but not in PCPA-pretreated rats. Interestingly, using a different species, Prospéro-García et al.[53] demonstrated that CCK (0.1 μg i.c.v. for 5 min) restored PS during an 8-h diurnal period in PCPA-pretreated cats.

Mansbach and Lorenz[54] observed that an i.p. injection of CCK at doses of 5 to 80 Ivy Dog Units/kg dose-dependently increased behavioral rest and EEG sleep, both SWS and PS, in rats who had been deprived of food for 17 h, then fed 10 min before the CCK administration. Itoh et al.[41] reported that an i.c.v. injection of CCK at doses of 0.2 to 2 μg dose-dependently suppressed the VIP-induced hypermotility in rats. In addition to a depression of the respiratory rate during SWS and PS, DeMesoquita and Haney[55] detected a significant increase in the total number of rapid-eye-movement periods occurring during SWS in the light period in rats under chronic i.c.v. infusion of sulfated CCK (1.1 ng/min for 5 d) by an osmotic minipump. However, no other sleep parameters were affected by the CCK infusion.

Recently, Obál[38] and Kapás and associates[56] reported that an acute i.p. injection of CCK (9 and 45 nmol) in normal rats shortly before the dark period onset increased significantly the amount of SWS at the expense of wakefulness for the first postinjection hour. In contrast to the previous studies, the Hungarian researchers could not observe a significant enhancement of PS. It is suggested that the SWS-promoting effect of CCK may belong to the behavioral sequence elicited by the peptide, i.e., a reduction in motor activity and food intake, which is often attributed to satiety.

Thus, the sleep-modulatory activity of CCK is currently not convincing. Since it is suggested that the central production of CCK occurs predominantly during PS,[57] further studies are required to prove the involvement of CCK in sleep regulation.

C. Growth Hormone-Releasing Factor

Growth hormone-releasing factor (GRF), or somatocrinin, is a carboxy-terminal amidated, 44-residue peptide along with two biologically active amino-terminal fragments of 37 and 40 residues, respectively. There are slight structural differences in amino acid sequences among GRFs in different mammalian species. The primary structures of various GRFs are closely related to glucagon and VIP. GRF stimulates the release of hypophyseal GH.

Ehlers et al.[58] demonstrated that 2 nmol of GRF acutely injected into the lateral cerebral ventricle of rats induced a reduction in sleep latency, an increase in the total amount of SWS for a 30-min postinjection period, an increase in the power density of an EEG delta band (1 to 2 Hz), and a reduced amount of locomotor activity. A higher (5 nmol) and a lower (0.2 nmol) dose of GRF exerted little effect. According to the American researchers, the effects of GRF may be ascribed to an action of reducing the activity or release of hypothalamic CRF or a production of an indirect effect on other sleep modulators. Obál[38] also reported that an acute i.c.v. injection of GRF (0.03 nmol) in rats increased significantly the amount of both SWS and PS for the first postinjection hour. In rabbits, GRF (0.04 to 4 nmol i.c.v.) enhanced SWS in a dose-dependent manner for 1 h.[51]

Nistico et al.[59] reported that differential sleep-modulatory effects were induced by i.c.v., intrahippocampal, and intracaudate microinjections of human pancreatic GRF (hpGRF) in rats. The administration of 50 to 100 ng hpGRF into the third cerebral ventricle produced a dose-dependent behavioral sedation accompanied by SWS and an increase in the total voltage power of EEG (predominantly in lower frequency bands), which occurred within 10 min after injection and lasted for 2 to 4 h. The i.c.v. injection of smaller doses were ineffective. The dose of 100 ng given into the hippocampus produced effects similar to those in rats with 100 ng hpGRF into the third cerebral ventricle. The intrahippocampal microinjection of smaller doses were ineffective. However, if 75 ng of hpGRF were injected into the head of the caudate nucleus, an increase in locomotor activity and stereotyped behaviors such as licking, grooming, chewing, and gnawing occurred within 15 min and lasted for about 50 min. No significant changes were observed in the EEG spectrum power. From these results, it is speculated that an interaction between hpGRF and specific receptors (presumably monoaminergic receptors) located in the periventricular region of the hypothalamus was involved in the induction of behavioral and EEG sleep, whereas the brain structures, including the caudate, were related to the dopaminergic mechanism for behavioral and locomotor stimulation.

D. Corticotropin-Releasing Factor

Corticotropin-releasing factor (CRF) is a 41-residue amidated peptide. Although CRF is regarded originally as a hypophysiotropic neurohormone stimulating the secretion of ACTH, it is also known to exhibit behavior-modulatory properties. Ehlers et al.[58] reported that an i.c.v. injection of CRF at doses of 1.5 to 150 pmol produced increases in cortical EEG fast-wave activity associated with behavioral arousal. CRF at higher doses (1.5 to 3.75 nmol i.c.v.) induced convulsive seizures over a period of 3 to 7 h in rats. These EEG and behavioral findings are consistent with the behaviors frequently correlated with the known circadian timing of the release of GH and ACTH/corticosteroids during the sleep-waking cycle in rats and humans. However, it is uncertain whether CRF as well as GRF (see above) actually participate in the normal regulation of the sleep-waking cycle under physiological conditions. It is rather likely that the sleep-modulatory effects of CRF are mediated by nonspecific, indirect actions on the regulatory mechanisms of sleep, presumably analogous to the conditions of stress. In this connection, it should be noted here that CRF, either i.c.v. injected (0.5 nmol) or directly microinjected, can activate noradrenergic neurons in the locus coeruleus to increase the discharge rates.[60]

Recently, considering the reciprocal interaction between GRF and CRF, Ehlers and Kupfer[61] hypothesized that GRF and CRF are responsible for the sleep-dependent process S and the sleep-independent circadian process C, respectively, of Borbély's two-process model.[62] Hence the ratio of GRF to CRF seems to be crucial for the induction and/or maintenance of sleep.

E. Thyrotropin-Releasing Factor

Thyrotropin-releasing factor (TRF or TRH; see Figure 4) is a tripeptide isolated from the mammalian hypothalamus. Apart from the main physiological function of stimulating the secretion of thyrotropic hormone (TSH) from the anterior hypophysis, TRF is known to elicit some effects in modulating behaviors. Havlicek et al.[63] reported that a marked suppression of sleep and an increase in exploratory and motor behavior were induced after i.c.v., i.p., and supracortical infusions of TRF (10 μg) over 10 to 15 min during the first postinfusion period in normal and hypophysectomized rats. Takahashi et al.[64] demonstrated that a continuous i.c.v. infusion of TRF (0.1 or 1 μg/h) for 7.5 consecutive days by an osmotic minipump significantly enhanced the daily amount of wakefulness in rats at the expense of SWS and PS in a dose-dependent manner. A slight but significant phase advancement was observed in the acrophase of total sleep, both SWS and PS, although no phase shift was induced by 0.1 μg/h TRF in PS acrophase. In this regard, it should be mentioned here that there is evidence of a sleep-related inhibition in the release of TSH in humans.[65]

F. Neuropeptide Y

Neuropeptide Y (NPY), a hexatriacontapeptide amide, is a natural constituent of the mammalian brain and structurally belongs to the pancreatic polypeptide family.[66] Fuxe et al.[67] observed that an i.c.v. injection of NPY at a dose of 1.25 nmol brought about an

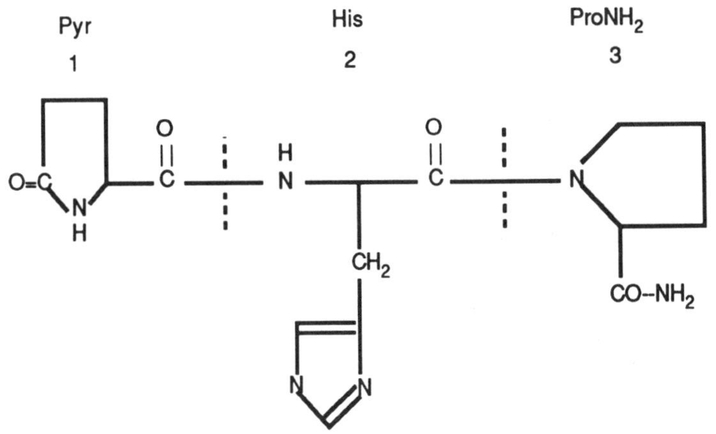

Pyr
1
His
2
ProNH$_2$
3

FIGURE 4. Chemical structure of thyrotropin-releasing factor (TRF or TRH).

increase in EEG synchronization and a decrease in EEG desynchronization in rats during a 90-min recording period. Zini et al.[68] further demonstrated circadian and strain differences in the effect of an i.c.v. injection of NPY (1.25 nmol) on EEG synchronization in rats. In Sprague-Dawley rats, the NPY administration between 10.00 and 11.00 h (morning session) decreased significantly desynchronized EEG activity and the latency to SWS, and increased synchronized and mixed EEG activity. No significant change was induced after the administration of the same dose of NPY between 19.30 and 20.30 h (evening session) except for a significant decrease in the latency to SWS. In Wistar-Kyoto rats, NPY administered in the morning session was ineffective, while in the evening session it increased synchronized EEG activity without modifying other parameters. In spontaneous hypertensive rats of the Wistar-Kyoto strain, NPY administration in both the morning and evening sessions induced an increase in the latency to SWS and desynchronized EEG activity. NPY administration in the morning session also resulted in a significant decrease in mixed EEG activity. These differences seem to be attributed to the difference in the activity of α2-adrenoceptors in the brain.

G. Neurotensin

Neurotensin (NT) is a tridecapeptide (*p*Glu-Leu-Tyr-Glu-Asn-Lys-Pro-Arg-Arg-Pro-Tyr-Ile-Leu-OH), which is known to be present at a high concentration in the hypothalamus, the amygdala, the preoptic area (POA), the nucleus acumbens, and the olfactory tubercle. NT has been regarded as an important neuroregulatory peptide in the mammalian central nervous system.[69] Riou et al.[52] reported that no sleep-modulatory effect was induced after an i.c.v. injection of NT (0.3 or 2 μg) in rats during the light period.

On the contrary, Kalivas and Taylor[70] reported that a bilateral microinjection of NT (1 μg/side) for 5 consecutive days into the rat ventral tegmental area induced a significant decrease in behavioral sleep and a significant increase in motor activity during a 90-min postinjection period. The increased motor response to NT was considerably long-lasting and still present 8 d after the last daily injection with NT. Since the enhancement of motor activity was produced only by NT injection into the dopamine region, the NT-induced behavioral changes seem to be mediated by the dopaminergic system.

H. Luteinizing Hormone-Releasing Hormone

Luteinizing hormone-releasing hormone (LHRH) is a hypothalamic decapeptide, which stimulates hypophyseal gonadotropins, luteinizing hormone (LH), and follicle-stimulating hormone. Although sleep-related LH secretion is well known to occur around puberty in humans, LHRH seems to have no sleep-modulatory activity. Mendelson et al.[71] observed that a nocturnal i.v. infusion of LHRH at a dose of 100 μg for 8 h resulted in no change in sleep parameters in adult humans.

I. Substance P

Substance P is an unadecapeptide of the tachykinin family having the amino acid sequence Arg-Pro-Lys-Pro-Gln-Gln-Phe-Phe-Gly-Leu-Met-NH_2. This peptide is known to be distributed widely in the central nervous system and subserves multiple physiological and behavioral roles. Riou et al.[72] reported that substance P (100 ng i.c.v.) selectively decreased diurnal PS in rats. In contrast, Wachtel et al.[73] demonstrated that an i.p. injection of substance P (250 μg/kg/d) for 4 consecutive days restored sleep, both SWS and PS, in stress-induced, insomniac rats. An analog of substance P, eledoisine-hexapeptide, is reported to have a similar effect.[74]

IV. PROOPIOMELANOCORTIN-RELATED PITUITARY PEPTIDES

A. ACTH and its Derivatives

It is well known that proopiomelanocortin (POMC) is the prohormone of β-lipotropic hormone (β-LPH), which is cleaved into γ-LPH and β-endorphin, and ACTH, a 39-amino acid residue peptide, which is cleaved into α-melanophore-stimulating hormone (α-MSH; = $ACTH_{1-13}$) and corticotropin-like intermediate lobe peptide (CLIP; = $ACTH_{18-39}$). ACTH primarily stimulates the release of corticosteroids from the adrenal gland, but it exerts a wide spectrum of behavioral effects. α-MSH and CLIP may undergo some modifications, such as acetylation and phosphorylation. Desacetyl-α-MSH (des-α-MSH) is known to exist in a natural form of α-MSH. These POMC-derived peptides have been regarded to be closely related to various kinds of mental activities, such as motivation, vigilance, fear, stress, learning, and memory retrieval. Circadian variations of these hormones are well known. α-MSH can enhance EEG delta waves.[75] Recently, it has been reported that the α-MSH content exhibits a definite circadian rhythm in the medial basal hypothalamus and the POA in rats, being high during the dark period, i.e., the active phase of the animal species.[76] α-MSH seems to play a role in temperature regulation as an antipyretic neuropeptide.[77]

As mentioned in the previous chapter, Gillin et al.[78] observed that an i.v. infusion of ACTH (40 IU) for 8 h, beginning either in the morning or in the afternoon, induced a suppression and delay of REM sleep in normal humans. Chastrette and Cespuglio[79-81] reported that doses of 1 to 100 ng of ACTH were ineffective in modulating sleep-wakefulness, when i.c.v. injected in rats at the end of the light period. Recently, however, Chastrette et al.[82] demonstrated that a larger dose of ACTH (1 μg i.c.v.) significantly enhanced nocturnal wakefulness at the expense of SWS and PS in rats. This effect was characterized by an increase in the frequency and duration of waking episodes. The French researchers have further found that des-α-MSH (1 ng i.c.v.) selectively enhanced SWS (see Figure 5), whereas α-MSH (1 to 100 ng i.c.v.) had no hypnogenic effect. The SWS-enhancing effect of des-α-MSH was due to an increase in the frequency and duration of SWS episodes. The selective PS-enhancing effect of CLIP is dealt with in the previous chapter. It is interesting that stress causes an elevated release of the POMC-derived neuropeptides. Hence, it is not excluded that an analogous situation, both temporally and functionally, may occur in the brain for these peptides differentially involved in the physiological regulation of sleep and wakefulness.

In this connection, it is worth mentioning that ACTH and its analog $ACTH_{1-24}$ inhibit the incorporation of uridine into brain or brainstem RNA in rats and mice.[83,84] Hence, it might be speculated that the enhancement of arousal after ACTH administration is caused by the suppression of the somnogenic activity of uridine (see Chapter 4). In addition, delta-sleep-

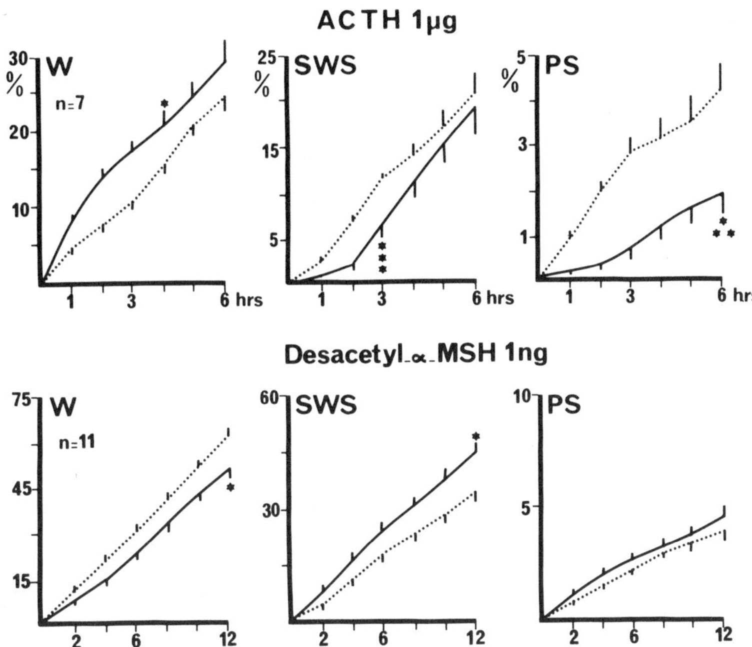

FIGURE 5. Effects of ACTH (1 µg i.c.v.; top) and desacetyl-α-MSH (1 ng i.c.v.; bottom) on wakefulness (W), SWS, and PS during a 12-h nocturnal period in rats. The administration was done shortly before the dark period onset. The values in the ordinates represent the percentage of the total recording time expressed cumulatively (mean ± SEM). The abscissae indicate the time in hours after injection. $^{*}p < 0.05$, $^{***}p < 0.01$. (From Chastrette, N., Clement, H. W., Prevautel, H., and Cespuglio, R., in *Sleep Peptides: Basic and Clinical Approaches*, Inoué, S. and Schneider-Helmert, D., Eds., Japan Scientific Societies Press, Tokyo/Springer-Verlag, Berlin, 1988, 27. With permission.)

inducing peptide (DSIP) is reported to inhibit CRF-induced ACTH secretion from the rat anterior hypophysis *in vitro*.[85] Finally, it should not be overlooked that the wakefulness-enhancing property of ACTH is not consistent with a previous report; subcutaneous (s.c.) injection of ACTH could restore PS to approximately control levels in hypophysectomized rats.[86]

B. LPH and Endorphin

Riou et al.[52] reported that no sleep-modulatory effect was observed after i.c.v. injection of β-endorphin (1 µg) in rats during the light period. Recently, Chastrette et al.[82] demonstrated that an i.c.v. injection of 1 to 100 ng of β-LPH and α- and β-endorphins resulted in no enhancement of nocturnal sleep in rats.

V. BENZODIAZEPINE-RELATED SUBSTANCES

A. Benzodiazepine Receptor Agonists and Antagonists

Benzodiazepines (BZ) are a group of well-established medicaments having strong anxiolytic, anticonvulsant, muscle-relaxant, sedative, and hypnotic properties with low toxicity. Such characteristics of BZs may imply some similarity to endogenous compounds. Actually, in addition to synthetic agonists and antagonists, such as β-carboline derivatives and the imidazobenzodiazepine Ro 15-1788, a number of endogenous ligands of BZ receptors have been identified from the central nervous system.[87] Currently, the following substances have been documented as endogenous ligands existing in the brain: β-carboline-3-carboxylic acid

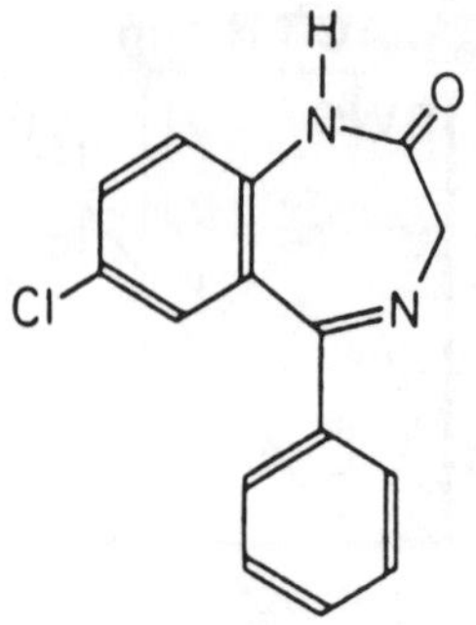

FIGURE 6. Chemical structure of nordiazepam.

ethyl ester,[88] inosine,[89,90] hypoxanthine,[89,90] *N*-butyl β-carboline-3-carboxylate,[91] nicotinamide,[92] and tribulin (monoamine oxidase inhibitor).[93] They are classified into agonists, partial agonists, competitive antagonists, partial inverse agonists, and inverse agonists according to their biological activities.[94] Specific protein compounds that bind BZs exist in the brain and in the periphery. At the binding sites, BZs can enhance the ability of the γ-aminobutyric acid (GABA) receptor to open the chloride channel. Today, the BZ receptor is regarded to form a GABA-BZ-barbiturate-picrotoxin-receptor complex.

Recently, Alho et al.[95] demonstrated that a neuropeptide termed diazepam-binding inhibitor (DBI) was identified in the rat brain. This neuropeptide can act as an inverse agonist of BZs. DBI has 104 amino acid residues and contains two identical octadecaneuropeptide (ODN), which might act as an endogenous ligand of BZ-recognition sites. Ferrero et al.[96] further demonstrated that ODN is more potent than the parent compound in the displacement of a specifically bound β-carboline derivative. The amino acid sequence of ODN is

Gln-Ala-Thr-Val-Gly-Asp-Val-Asn-Thr-Asp-Arg-Pro-Gly-Leu-Leu-Asp-Leu-Lys

In addition, the following three synthetic peptides containing the COOH-terminal segment of ODN were active:

Octapeptide:	Arg-Pro-Gly-Leu-Leu-Asp-Leu-Lys
Heptapeptide:	Pro-Gly-Leu-Leu-Asp-Leu-Lys
Hexapeptide:	Gly-Leu-Leu-Asp-Leu-Lys

On the other hand, De Blas and Sangameswaran[97] and Sangameswaran et al.[98] succeeded in detecting a known BZ, *N*-desmethyldiazepam (nordiazepam; see Figure 6), from bovine, human, and rat brains. It is uncertain whether the detected nordiazepam originated from the natural product of the brain itself or from the synthetic product contaminated during the experimental procedures. However, it is suggested that nordiazepam could be taken from the diet as such or in the form of a precursor that could be converted to nordiazepam in the animal body.

B. Melatonin

Melatonin, or 5-methoxy-*N*-acetyltryptamine (see Figure 7), is the major hormone of the mammalian pineal gland and is regarded to be involved in the mechanisms that cause circadian rhythms in sleep. The blood level of melatonin exhibits a clear, light-dependent rhythmicity. This hormone is also well known as a regulator of photoperiodic reproductive and behavioral rhythms. Recently, it has been established that, in humans, the seasonal affective disorder syndrome with depressions and hypersomnia is associated closely with a phase delay of the circadian rhythm of melatonin secretion.[99,100]

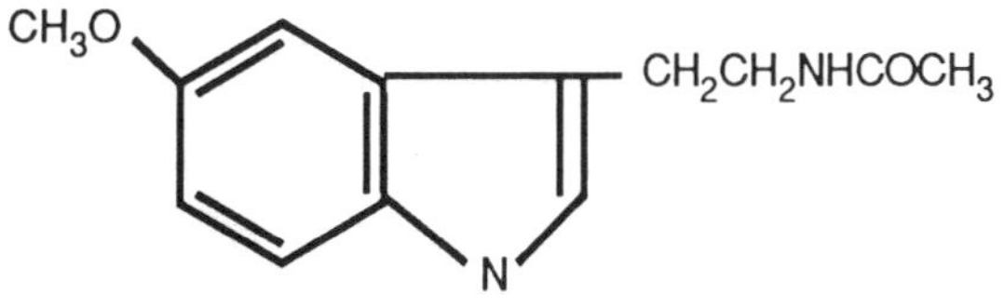

FIGURE 7. Chemical structure of melatonin.

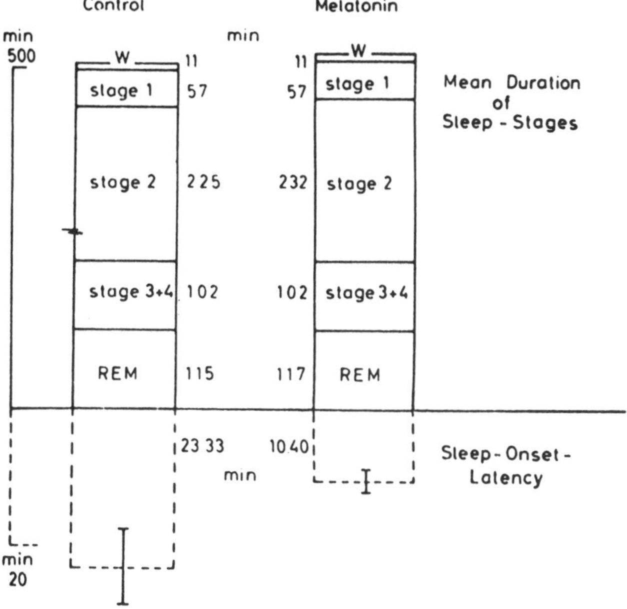

FIGURE 8. Effect of melatonin (50 mg i.v.) on nocturnal sleep in healthy humans. A marked decrease in sleep-onset latency is shown, but no change in other sleep parameters are demonstrated. (From Cramer, H., Rudolph, J., Consbruch, U., and Kendel, K., *Adv. Biochem. Psychopharmacol.*, 11, 187, 1974. With permission.)

In 1964, Marczynski et al.[101] reported that a bilateral microinjection of crystalline melatonin (15 to 30 μg) into the POA of alert cats resulted in EEG synchronization with a markedly increased occurrence of delta waves appearing 15 to 20 min later and lasting for 2 to 3 h. The cats did not react to acoustic stimuli, and curled up and slept with bradycardia and bradypnea. The administration of melatonin to the nucleus centralis medialis induced similar behavioral and EEG changes to a lesser extent, whereas an administration to the brainstem reticular formation did not. Hishikawa et al.[102] observed a powerful sedative and hypnotic effect of melatonin (10 to 60 ng/kg i.p.) in young chicks with enhanced EEG delta activity. The administration of melatonin concomitantly decreased the latency and the amount of PS.

In healthy young humans, Antón-Tay et al.[103] reported that the administration of melatonin (0.25 to 1.25 mg/kg i.v.) at 16.00 h induced an elevation of EEG alpha activity and sleep. Also, Cramer et al.[104] reported that an i.v. injection of 50 mg of melatonin in the daytime resulted in enhanced stage 3 + 4 sleep. Furthermore, the same dose of melatonin i.v. injected at 21.30 h shortened significantly the latency to sleep onset, without affecting the amount of nocturnal sleep (see Figure 8). Interestingly, the administration of melatonin (50 mg i.v.) in the afternoon markedly enhanced the secretion of GH, but not β-MSH, during afternoon sleep or wakefulness.[105]

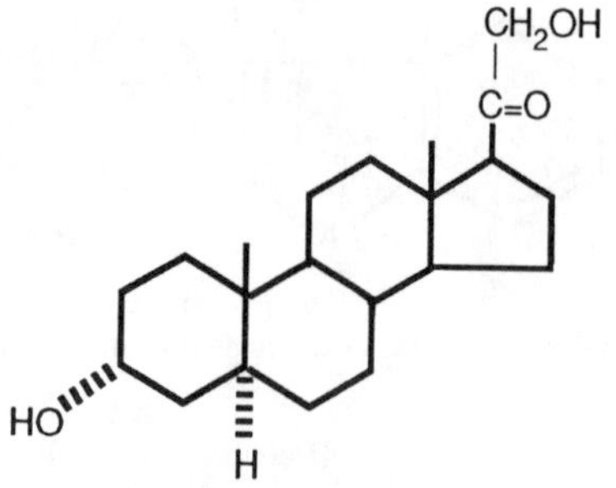

FIGURE 9. Chemical structure of 3-α, 5-α-tetrahydrodeoxycorticosterone (THDOC).

Holmes and Sugden[106] demonstrated that melatonin (2.5 or 10 mg/kg i.p.) dose-dependently decreased the latency to sleep and wakefulness and increased SWS and PS in rats. Since melatonin shares the muscle relaxant and anticonvulsant properties with BZs in addition to the sleep-promoting activity, the British investigators suggested that melatonin or a melatonin metabolite may act as an agonist at the BZ receptor. In contrast, Mendelson et al.[107] found a significant decrease in SWS after an i.p. injection of melatonin (833 μg/kg) at the light period onset, while finding no effect after the administration at the dark period onset. Recently, Mirmiran and Pévet[108] observed a long-term effect of melatonin continuously liberated into the systemic circulation through s.c. implanted silastic capsules (25 to 30 μg/day) for 1 to 3 months in rats. The treated animals exhibited increased levels of both SWS and PS regardless of the light-dark cycles, but the circadian sleep-waking rhythmicity persisted intact.

Although there are controversial discussions on the modulatory action of melatonin on the circadian sleep-waking rhythm,[109] it should be mentioned here that Arendt et al.[110] successfully controlled jet lag by the timely use of melatonin, and that Cassone et al.[111] found an inhibition of metabolic activity of melatonin on the rat suprachiasmatic nucleus, the pacemaker of the circadian rhythm.

C. Steroid Hormones

Mendelson et al.[112] reported that an adrenal mineralocorticosteroid metabolite, 3-α,5-α-tetrahydrodeoxycorticosterone (THDOC, 10 mg/kg; see Figure 9) i.p. injected in the light period significantly reduced sleep latency and enhanced SWS for a 2-h postinjection, recording period in male rats. No influence was found on PS by the treatment. Since THDOC has barbiturate-like properties *in vitro*[113] and anxiolytic activities *in vivo*, the hypnotic effect may be mediated by actions at the GABA-BZ-barbiturate-receptor complex. It is suggested that THDOC or similar steroids may play some role in sleep regulation.

The other steroids, especially the sex steroids so far investigated, seem to have no direct sleep-regulatory actions. Branchey et al.[114] reported that s.c. injections (5 μg/d) of estradiol benzoate (E_2B) in the light period for 6 consecutive days brought about a slight decrease in the amount of nocturnal PS in ovariectomized female rats. A similar reduction in PS was obtained by the local implantation of crystalline E_2B into the medial POA in ovariectomized rats.[115] Two daily s.c. injections with E_2B (5 μg/d) followed by progesterone (0.5 mg s.c.), which may mimic the hormonal state of the proestrus, resulted in significant reductions in both SWS and PS in the subsequent dark period in ovariectomized female rats[114] and feminized male rats which had been neonatally castrated.[116] In this connection, Yamaoka[117] detected a significant decrease in the amount of nocturnal SWS and PS at proestrus in intact, cyclically estrous female rats. In contrast, male rats castrated in adulthood did not respond to the E_2B-progesterone treatment. Thus, the sexual dimorphism is apparent in response to gonadal steroids. However, it must be noted that these steroids affect mainly the hypothalamic-pituitary axis to trigger the ovulatory surge of gonadotropins, which eventually accompany activated sexual arousal and the reduction of sleep.

Under short light-dark cycles (30 min light and 30 min dark), no effect on PS was observed after the s.c. administration and E_2B (5 μg/d) and progesterone (40 to 500 μg/d) in male and ovariectomized female rats.[118] Neither adrenalectomy nor the administration of cortisol (10 mg/kg/d s.c.) in these rats altered the total daily amount of PS. Peder[119] detected that castration in male rats did not modify the PS rebound occurring after PS deprivation. This fact suggests that the depletion of testosterone, an anabolic hormone, is not essential for the triggering of PS, which is regarded to be influenced by protein synthesis (see Chapter 7).

VI. AROUSAL-ENHANCING SUBSTANCES

A. Lactate

Koenigsberg et al.[120] demonstrated that an i.v. infusion of 0.5 *M* sodium lactate solution (flow rate: 0.5 ml/kg/min) during stage 3 or 4 sleep caused arousal 5.7 min after the initiation of the lactate infusion in healthy volunteers having a panic disorder. Although lactate infusion is known to precipitate panic episodes in patients with a panic disorder, but not in normal subjects, the authors observed neither vocalization, night terror, nor panic symptoms and found no difference from spontaneous arousal. It was suggested that sodium lactate may activate the central nervous system involved in arousal.

B. Neurohormones, Hormones, and Putative Sleep Substances

As already mentioned in the previous chapters, there are several endogenous substances which may induce enhancement of wakefulness depending on the dosage, the time of day, the animal species, interactions with other factors, and so on. It is largely unknown how these substances partipicate in the physiological regulation of sleep-waking mechanisms in the brain. However, the function of wakefulness-promoting substances should be considered in the context with that of sleep-promoting substances. Table 1 shows the list of substances that are reported to have an awaking activity.

Table 1
LIST OF WAKEFULNESS-ENHANCING SUBSTANCES

Substance	Effective dose and route	Test animal	Effect on SWS	Effect on PS	Ref.
Adrenocorticotropic hormone (ACTH)	40 IU i.v.	Human	+	+	78
	1 μg i.c.v.	Rat	+	+	82
Angiotensin II	100 ng i.c.v.	Rat	−	+	72
Arginine vasopressin (AVP)	2 μ*M*/0.5 μl/h i.c.v.	Rat	±	±	122
	20 μg *i.c.v.	Rat	+	−	40
Arginine vasotocin (AVT)	10—50 μg/kg i.p.	Rat	−	+	72
Corticotropin-releasing factor (CRF)	1.5—15 pmol	Rat	+	±	58
Delta-sleep-inducing peptide (DSIP)	7 nmol/kg i.c.v.	Rat	+	+	123
	30 nmol/kg i.p.	Cat	+	+	124
	—	Human	±	±	125
Lactate	0.5 *M*/0.5 ml/kg/min	Human	+	+	120
Neurotensin (NT)	2 μg/d	Rat	±	±	70
Prostaglandin E_2 (PGE_2)	0.25—2500 pmol intra-POA	Rat	+	+	126
	0.36—36 nmol i.c.v.	Rat	+	+	127
	0.36—36 nmol i.c.v.	Rhesus monkey	+	+	128
Renin	0.01 unit i.c.v.	Rat	−	+	71
Somatostatin (SRIF)	10 μg i.c.v.	Rat	+	+	61
	10 μg supracortical	Rat	+	+	129
	10 μg i.p.	Rat	+	+	129
	1—10 nmol/h i.c.v.	Rat	+	−	62
Thyrotropin-releasing factor (TRF)	10 μg i.c.v.	Rat	+	+	61
	0.1—1 μg/h i.c.v.	Rat	+	+	62

Note: −, Not affected; ±, Uncertain; +, Suppressed; *, an i.c.v. implant.

REFERENCES

1. **Giacobini, E.,** Piperidine: a new neuromodulator or a hypnogenic substance?, *Adv. Biochem. Psychopharmacol.,* 15, 17, 1976.
2. **Kasé, Y., Miyata, T., Kamikawa, Y., and Kataoka, M.,** Pharmacological studies on alicyclic amines. II. Central actions of piperidine, pyrrolidine and piperazine, *Jpn. J. Pharmacol.,* 19, 300, 1969.
3. **Stepita-Klauco, M., Dolezalova, H., and Fairweather, R.,** Piperidine increase in the brain of dormant mice, *Science,* 183, 536, 1974.
4. **Dolezalova, H., Stepita-Klauco, M., and Fairweather, R.,** The accumulation of piperidine in the central ganglia of dormant snails, *Brain Res.,* 72, 115, 1974.
5. **Miyata, T., Kamata, K., Nishikibe, M., Kasé, Y., Takahama, K., and Okano, Y.,** Effects of intracerebral administration of piperidine on EEG and behavior, *Life Sci.,* 15, 1135, 1974.
6. **Miyata, T., Okano, Y., Fukunaga, K., Takahama, K., and Kasé, Y.,** Analysis of regional concentrations of piperidine in the brain by mass fragmentography, *Brain Res.,* 188, 291, 1980.
7. **Okano, Y., Miyata, T., Fukunaga, K., Takahama, K., Hitoshi, T., and Kasé, Y.,** Mass fragmentographic analysis of piperidine levels in tissues of rats during development, *J. Pharm. Dyn.,* 4, 197, 1981.
8. **Okano, Y., Kadota, T., Naka, M., Nagata, J., Ijima, S., Matsuda, A., Iwamura, H., Hitoshi, T., Takahama, K., and Miyata, T.,** Imino acid and related alicyclic amine levels in biological fluid, *J. Pharmacobio.-Dyn.,* 8, 487, 1984.
9. **Miyata, T., Okano, Y., Fukunaga, K., Takahama, K., and Kasé, Y.,** Seasonal and activity-dependent variations of piperidine levels in the amphibian brain, *Brain Res.,* 197, 279, 1980.
10. **Miyata, T., Okano, Y., Fukunaga, K., Takahama, K., Hitoshi, T., and Kasé, Y.,** Effects of anesthetics on piperidine levels in mouse brain, *Eur. J. Pharmacol.,* 71, 79, 1981.
11. **Miyata, T., Okano, Y., Iwasaki, K., Takahama, K., Hitoshi, T., and Kasé, Y.,** Changes in brain piperidine levels under anesthesia: mass fragmentographic analysis, *Eur. J. Pharmacol.,* 78, 457, 1982.

12. **Miyata, T. and Okano, Y.,** Sleep-inducing activity of piperidine and its mechanism, *Annu. Rep. Natl. Inst. Physiol. Sci., Okazaki.,* 8, 445, 1987.
13. **Aisaka, K., Hattori, Y., Ishihara, T., Morita, M., Kasé, Y., and Miyata, T.,** The effects of piperidine and its related substances on blood vessels, *Jpn. J. Pharmacol.,* 37, 345, 1985.
14. **Drucker-Colín, R. and Giacobini, E.,** Sleep-inducing effect of piperidine, *Brain Res.,* 88, 186, 1975.
15. **Nixon, R. A. and Karnovsky, M. L.,** Uptake and metabolism of intraventricularly administered piperidine and its effect on sleep and wakefulness in the rat, *Brain Res.,* 134, 501, 1977.
16. **Mendelson, W. B., Lantigua, R. A., Wyatt, R. J., Gillin, J. C., and Jacobs, L. S.,** Piperidine enhances sleep-related and insulin-induced growth hormone secretion: further evidence for a cholinergic secretory mechanism, *J. Clin. Endocrinol. Metab.,* 52, 409, 1981.
17. **Danguir, J. and Nicolaidis, S.,** Intravenous infusions of nutrients and sleep in the rat: an ischymetric sleep regulation hypothesis, *Am. J. Physiol.,* 238, E307, 1980.
18. **Danguir, J. and Nicolaidis, S.,** Sleep and feeding patterns in the ventromedial hypothalamic lesioned rat, *Physiol. Behav.,* 21, 769, 1978.
19. **Danguir, J. and Nicolaidis, S.,** Dependence of sleep on nutrients' availability, *Physiol. Behav.,* 22, 735, 1979.
20. **Lorenz, D. N.,** Alimentary sleep satiety in suckling rats, *Physiol. Behav.,* 38, 557, 1986.,
21. **Danguir, J. and Nicolaidis, S.,** Relations between feeding and sleep patterns in the rat, *J. Comp. Physiol. Psychol.,* 93, 820, 1979.
22. **Phillips, N. H. and Berger, R. J.,** Effects of diurnal glucose infusion on sleep, metabolism and body temperature in the fasting pigeon, *Sleep Res.,* 16, 54, 1987.
23. **Sangiah, S., Caldwell, D. F., Villeneuve, M. J., and Clancy, J. J.,** Sleep: sequential deduction of paradoxical (REM) and elevation of slow-wave (NREM) sleep by a non-convulsive dose of insulin in rats, *Life Sci.,* 31, 763, 1982.
24. **Danguir, J. and Nicolaidis, S.,** Chronic intracerebroventricular infusion of insulin causes increase of slow wave sleep in rats, *Brain Res.,* 306, 97, 1984.
25. **Danguir, J. and Nicolaidis, S.,** Feeding, metabolism, and sleep: peripheral and central mechanisms of their interaction, in *Brain Mechanisms of Sleep,* McGinty, D. J., Drucker-Colín, R., Morrison, A., and Parmeggiani, P. L., Eds., Raven Press, New York, 1985, 321.
26. **Danguir, J.,** Insulin as a hypnogenic factor, in *Sleep '84,* Koella, W. P., Rüther, E., and Schulz, H., Eds., Gustav Fischer Verlag, Stuttgart, 1985, 210.
27. **Danguir, J.,** Sleep deficits in diabetic rats: restoration following chronic intravenous or intracerebroventricular infusions of insulin, *Brain Res. Bull.,* 12, 641, 1984.
28. **Bernstein, H.-G., Schwarzberg, H., Reiser, M., Günter, O., and Dorn, A.,** Intracerebroventricular infusion of insulin alters the behavior of rats not related to food intake, *Endocrinol. Exp.,* 20, 387, 1986.
29. **Castex, C. and Hoo-Paris, R.,** Régulation des sécrétions du pancréas endocrine (insuline et glucagon) au cours cycle léthargie-réveil périodique du mammifère hibernant, *Diabete Metab. (Paris),* 13, 176, 1987.
30. **Danguir, J.,** Intracerebroventricular infusion of somatostatin selectively increases paradoxical sleep in rats, *Brain Res.,* 367, 26, 1986.
31. **Ramm, P. and Smith, C. T.,** Cerebral protein synthesis during slow wave and REM sleep in the rat, *Sleep Res.,* 16, 67, 1987.
32. **Danguir, J.,** Cafeteria diet promotes sleep in rats, *Appetite,* 8, 49, 1987.
33. **Danguir, J.,** Insulin and somatostatin as sleep-inducing hormones, *Sleep Res.,* 16, 44, 1987.
34. **Danguir, J.,** Internal milieu and sleep homeostasis, in *Sleep Peptides: Basic and Clinical Approaches,* Inoué, S. and Schneider-Helmert, D., Eds., Japan Scientific Societies Press, Tokyo/Springer-Verlag, Berlin, 1988, 53.
35. **Riou, F., Cespuglio, R., and Jouvet, M.,** Endogenous peptides and sleep in the rat. III. The hypnogenic properties of vasoactive intestinal polypeptide, *Neuropeptides,* 2, 265, 1982.
36. **Riou, F., Cespuglio, R., and Jouvet, M.,** Sleep-facilitating effect of vasoactive intestinal polypeptide in the rat, in *Sleep 1982,* Koella, W. P., Ed., S. Karger, Basel, 1983, 114.
37. **Obál, F., Jr., Sáry, G., Alföldi, P., Rubiscek, G., and Obál, F.,** Vasoactive intestinal polypeptide promotes sleep without effects on brain temperature in rats at night, *Neurosci. Lett.,* 64, 236, 1986.
38. **Obál, F., Jr.,** Effects of peptides (DSIP, DSIP analogues, VIP, GRF and CCK) on sleep in the rat, *Clin. Neuropharmacol.,* 9(Suppl. 4), 459, 1986.
39. **Kruisbrink, J., Mirmiran, M., Van der Woude, T. P., and Boer, G. J.,** Effects of enhanced cerebrospinal fluid levels of vasopressin, vasopressin antagonists or vasoactive intestinal polypeptide on circadian sleep-wake rhythm in the rat, *Brain Res.,* 419, 76, 1987.
40. **Nakagaki, K. and Takahashi, Y.,** Sleep-inducing activity of DSIP-P and VIP, *Annu. Rep. Natl. Inst. Physiol. Sci., Okazaki,* 8, 447, 1987.
41. **Itoh, S., Katsura, G., and Yoshikawa, K.,** Hypermotility induced by vasoactive intestinal peptide in the rat: its reciprocal action to cholecystokinin octapeptide, *Peptides,* 6, 53, 1985.

42. **Prospéro-García, O., Morales, M., Arankowsky-Sandoval, G., and Drucker-Colín, R.,** Vasoactive intestinal polypeptide (VIP) and cerebrospinal fluid (CSF) of sleep-deprived cats restores REM sleep in insomniac recipients *Brain Res.*, 385, 169, 1986.
43. **Drucker-Colín, R., Prospéro-García, G., Arankowsky-Sandoval, G., and Pérez-Montfort, R.,** Gastropancreatic peptides and sensory stimuli as REM sleep factors, in *Sleep Peptides: Basic and Clinical Approaches,* Inoué, S. and Schneider-Helmert, D., Eds., Japan Scientific Societies Press, Tokyo/Springer-Verlag, Berlin, 1988, 73.
44. **Drucker-Colín, R., Bernal-Pedraza, J., Fernandez-Cacino, F. and Oksenberg, A.,** Is vasoactive intestinal polypeptide (VIP) a sleep factor?, *Peptides,* 5, 837, 1984.
45. **Tartara, A., Bo, P., Savoldi, F., and Said, S. I.,** Centrally administered VIP increases spindle activity in unanaesthetized rabbits, *Pharmacol. Res. Commun.*, 15, 307, 1983.
46. **Woie, L., Kaada, B., and Opstad, P. K.,** Increase in plasma vasoactive intestinal peptide (VIP) in muscular exercise in humans, *Gen. Pharmacol.*, 3, 321, 1986.
47. **Opstad, P. K.,** The plasma vasoactive intestinal peptide (VIP) response to exercise is increased after prolonged strain, sleep and energy deficiency and extinguished by glucose infusion, *Peptides,* 8, 175, 1987.
48. **Shimatsu, A., Kato, Y., Matsushita, N., Ohta, H., Kabayama, Y., Yanaihara, N., and Imura, H.,** Prostaglandin D_2 stimulates vasoactive intestinal peptide release into rat hypophysial portal blood, *Peptides,* 5, 395, 1984.
49. **Sharpless, N. S., Thal, L. J., Perlow, M. J., Tabaddor, K., Waltz, J. M., Shapiro, K. N., Amin, I. M., Engel, J., Jr., and Cranndall, P. H.,** Vasoactive intestinal peptide in cerebrospinal fluid, *Peptides,* 5, 429, 1984.
50. **Werner, G. H., Floc'h, F., Migliore-Samour, D., and Jollès, P.,** Immunomodulating peptides, *Experientia,* 42, 521, 1986.
51. **Krueger, J. M., Toth, L. A., Cady, A. B., Johanssen, L., and Obál, F., Jr.,** Immunomodulation and sleep, in *Sleep Peptides: Basic and Clinical Approaches,* Inoué, S. and Schneider-Helmert, D., Eds., Japan Scientific Societies Press, Tokyo/Springer-Verlag, Berlin, 1988, 95.
52. **Riou, F., Cespuglio, R., and Jouvet, M.,** Endogenous peptides and sleep in the rats. II. Peptides without significant effect on the sleep-waking cycle, *Neuropeptides,* 2, 255, 1982.
53. **Prospéro-García, O., Ott, T., and Drucker-Colín, R.,** Cerebroventricular infusion of cholecystokinin (CCK-8) restores REM sleep in parachlorophenylalanine (PCPA)-pretreated cats, *Neurosci. Lett.*, 78, 205, 1987.
54. **Mansbach, R. S. and Lorenz, D. N.,** Cholecystokinin (CCK-8) elicits prandial sleep in rats, *Physiol. Behav.*, 30, 179, 1983.
55. **DeMesoquita, S. and Haney, W. H.,** Effect of chronic intracerebroventricular infusion of cholecystokinin on respiration and sleep, *Brain Res.*, 378, 127, 1986.
56. **Kapás, L., Obál, F., Jr., Alföldi, P., Rubiscek, G., Penke, B., and Obál, F.,** Effects of nocturnal intraperitoneal administration of cholecystokinin in rats: simultaneous increase in sleep, increase in EEG slow-wave activity, reduction of motor activity, suppression of eating, and decrease in brain temperature, *Brain Res.*, 438, 155, 1988.
57. **Elomaa, E.,** Is the action of peptide hormones of gastrointestinal origin on the hypothalamic satiety center modulated by REM sleep-dependent release of their central nervous analogues?, *Med. Hypotheses,* 967, 1980.
58. **Ehlers, C. L., Reed, T. K., and Henriksen, S. J.,** Effects of corticotropin-releasing factor and growth hormone-releasing factor on sleep and activity in rats, *Neuroendocrinology,* 42, 467, 1986.
59. **Nistico, G., De Sarro, G. B., Bagetta, G., and Müller, E. E.,** Behaviural and electrocortical spectrum power effects of growth hormone releasing factor in rats, *Neuropharmacology,* 26, 75, 1987.
60. **Valentino, R. J., Foote, S. L., and Aston-Jones, G.,** Corticotropin-releasing factor activates noradrenergic neurons of the locus coeruleus, *Brain Res.*, 270, 363, 1983.
61. **Ehlers, C. L. and Kupfer, D. J.,** Hypothalamic peptide modulation of EEG sleep in depression: a further application of the S-process hypothesis, *Biol. Psychiatry,* 22, 513, 1987.
62. **Borbély, A. A.,** A two process model of sleep regulation, *Human Neurobiol.*, 1, 195, 1982.
63. **Havlicek, V., Rezek, M., and Friesen, H.,** Somatostatin and thyrotropin releasing hormone: central effect on sleep and motor system, *Pharmacol. Biochem. Behav.*, 4, 455, 1976.
64. **Takahashi, Y., Usui, S., Ebihara, S., and Honda, Y.,** Effects of continuous infusions of TRH and GIF on circadian rhythm parameters of sleep, ambulation, eating, and drinking in rats, in *Endogenous Sleep Substances and Sleep Regulation,* Inoué, S. and Borbély, A. A., Eds., Japan Scientific Societies Press, Tokyo/VNU Science Press BV, Utrecht, 1985, 101.
65. **Parker, D. C., Rossman, L. G., Perkary, A. E., and Hershman, J. M.,** Effect of 64-hour sleep deprivation on the circadian waveform of thyrotropin (TSH): further evidence of sleep-related inhibition of TSH release, *J. Clin. Endocrinol. Metab.*, 64, 157, 1987.
66. **Tatemoto, K., Carlquist, M., and Mutt, V.,** Neuropeptide Y — a novel brain peptide with structural similarities to peptide YY and pancreatic polypeptide, *Nature,* 296, 659, 1982.

67. **Fuxe, K., Agnati, L., Härfstrand, A., Zini, I., Tatemoto, K., Pich, E. M., Hökfelt, T., Mutt, V., and Terenius, L.,** Central administration of neuropeptide Y induced hypotension, bradypnea and EEG synchronization in the rat, *Acta Physiol. Scand.*, 118, 189, 1983.
68. **Zini, I., Pich, E. M., Fuxe, K., Lenzi, P. L., Agnati, L. F., Harfstrand, A., Mutt, V., Tatemoto, K., and Moscara, M.,** Actions of centrally administered neuropeptide Y on EEG activity in different rat strains and in different phases of their circadian cycle, *Acta Physiol. Scand.*, 122, 71, 1984.
69. **Nemeroff, C. B.,** Neurotensin, in *Encyclopedia of Neuroscience,* Vol. 2, Adelman, G., Ed., Birkhäuser, Boston, 1987, 851.
70. **Kalivas, P. W. and Taylor, S.,** Behavioral and neurochemical effect of daily injection with neurotensin into the ventral tegmental area, *Brain Res.*, 358, 70, 1985.
71. **Mendelson, W. B., Gold, P. W., Slater, S., Gillin, J. C., and Goodwin, F. K.,** LH-RH administration and human sleep, *Sleep Res.*, 7, 122, 1978.
72. **Riou, F., Cespuglio, R., and Jouvet, M.,** Endogenous peptides and sleep in the rats. I. Peptides decreasing paradoxical sleep, *Neuropeptides,* 2, 243, 1982.
73. **Wachtel, E., Koplik, E., Kolometsewa, I. A., Balzer, H.-U., Hecht, K., Oehme, P., and Ivanov, V. T.,** Vergleichende Untersuchungen zur Wirkung von DSIP und SP_{1-11} auf streßinduzierte chronische Schlafstörungen der Ratte, *Pharmazie,* 42, 188, 1987.
74. **Hecht, K., Kolometsewa, I. A., Ljowschina, I. P., Oehme, P., Poppei, M., Airapetjanz, M. G., and Wachtel, E.,** The influence of a substance P analogue on the sleep disturbances of stressed rats, in *Sleep 1978,* Popoviciu, L., Asigian, B., and Badiu, G., Eds., S. Karger, Basel, 1980, 508.
75. **Miller, L. H., Kastin, A. J., Hayes, M., Sterste, A., Garcia, J., and Coy, D. H.,** Inverse relationship between onset and duration of EEG effects of six peripherally administered peptides, *Pharmacol. Biochem. Behav.*, 15, 845, 1981.
76. **Scimonelli, T., Celis, M. E., and Eberle, A. N.,** Changes in alpha-melanocyte-stimulating hormone content in discrete hypothalamic areas of the male rat during a twenty-four-hour period, *Hormone Res.*, 27, 78, 1987.
77. **Bell, R. C. and Lipton, J. M.,** Pulsatile release of antipyretic neuropeptide α-MSH from septum of rabbit during fever, *Am. J. Physiol.*, 252, R1152, 1987.
78. **Gillin, J. C., Jacobs, L. S., Snyder, F., and Henkin, R. I.,** Effects of ACTH on the sleep of normal subjects and patients with Addison's disease, *Neuroendocrinology,* 15, 21, 1974.
79. **Chastrette, N. and Cespuglio, R.,** Effets hypnogènes de la des-acetyl-α-MSH et du CLIP (ACTH 18-39) chez le rat, *C. R. Acad. Sci. Paris,* 301, 527, 1985.
80. **Chastrette, N. and Cespuglio, R.,** Influence of proopiomelanocortin-derived peptides on the sleep-waking cycle of the rat, *Neurosci. Lett.*, 62, 365, 1985.
81. **Chastrette, N., Lin, Y. L., Faradji, H., and Cespuglio, R.,** Hypnogenic properties of desacetyl-α-MSH and CLIP (ACTH 18-39), in *Sleep '86,* Koella, W. P., Obál, F., Schulz, H., and Visser, P., Eds., Gustav Fischer Verlag, Stuttgart, 1988, 165.
82. **Chastrette, N., Clement, H. W., Prevautel, H., and Cespuglio, R.,** Proopiomelanocortin components: differential sleep-waking regulation?, in *Sleep Peptides: Basic and Clinical Approaches,* Inoué, S. and Schneider-Helmert, D., Eds., Japan Scientific Societies Press, Tokyo/Springer-Verlag, Berlin, 1988, 27.
83. **Jakoubek, B., Buresová, M., Hájek, I., Etrychová, J., Pavík, A., and Dedicová, A.,** Effect on ACTH on the synthesis of rapidly labelled RNA in the nervous system of mice, *Brain Res.*, 43, 417, 1972.
84. **Gispen, W. H., Reith, M. E. A., Schotman, P., Wiegant, V. M., Zwiers, H., and de Wied, D.,** CNS and ACTH-like peptides: neurochemical response and interaction with opiates, in *Neuropeptide Influences on the Brain and Behavior,* Miller, L. H., Sandman, C. A., and Kastin, A. J., Eds., Raven Press, New York, 1977, 61.
85. **Okajima, T. and Hertting, G.,** Delta-sleep-inducing peptide (DSIP) inhibited CRF-induced ACTH secretion from rat anterior pituitary gland in vitro, *Horm. Metab. Res.*, 18, 497, 1986.
86. **Valatx, J. L., Chouvet, G., and Jouvet, M.,** Sleep-waking cycle of the hypophysectomized rat, *Prog. Brain Res.*, 42, 115, 1975.
87. **Skolnick, P., Mendelson, W. B., and Paul, S. M.,** Benzodiazepine receptors in the central nervous system, in *Psychobiology of Sleep,* Wheatley, D., Ed., Raven Press, New York, 1981, 117.
88. **Braestrup, C., Nielsen, M., and Olsen, C. E.,** Urinary and brain β-carboline-3-carboxylates as potent inhibitors of brain benzodiazepine receptors, *Proc. Natl. Acad. Sci. U.S.A.*, 77, 2288, 1980.
89. **Skolnick, P., Marangos, P. J., Goodwin, F. K., Edwards, M., and Paul, S.,** Identification of inosine and hypoxanthine as endogenous inhibitiors of [^{3}H] diazepam binding in the central nervous system, *Life Sci.*, 23, 1473, 1978.
90. **Asano, T. and Spector, S.,** Identification of inosine and hypoxanthine as endogenous ligands for the brain benzodiazepine-binding sites, *Proc. Natl. Acad. Sci. U.S.A.*, 77, 2288, 1979.
91. **Pena, C., Medina, J. H., Novas, M. L., Paladini, A. C., and De Robertis, E.,** Isolation and identification in bovine cerebral cortex of *n*-butyl β-carboline-3-carboxylate, a potent benzodiazepine binding inhibitor, *Proc. Natl. Acad. Sci. U.S.A.*, 83, 4952, 1986.

92. **Möhler, H., Polc, P., Cumin, R., Pieri, L., and Kettler, R.,** Nicotinamide is a brain constituent with benzodiazepine-like actions, *Nature,* 278, 563, 1979.
93. **Armando, I., Glover, V., and Sandler, M.,** Distribution of endogenous benzodiazepine receptor ligand-monoamine oxidase inhibitory activity (tribulin) in tissues, *Life Sci.,* 38, 2063, 1986.
94. **Pieri, L.,** The mechanism of action, in *Benzodiazepines: An Update. Where do we go from here?,* Rafaelsen, O. J. and Ward, J., Eds., Editiones Roche, Basel, 1986, 17.
95. **Alho, H., Costa, E., Ferrero, P., Fujimoto, M., Cosenza-Merphy, D., and Guidotti, A.,** Diazepam-binding inhibitor: a neuropeptide located in selected neural populations of rat brain, *Science,* 182, 179, 1985.
96. **Ferrero, P., Santi, M. R., Conti-Tronconi, B., Costa, E., and Guidotti, A.,** Study of an octadecaneuropeptide derived from diazepam binding inhibitor (DBI): biological activity and presence in rat brain, *Proc. Natl. Acad. Sci. U.S.A.,* 83, 827, 1986.
97. **De Blas, A. and Sangameswaran, L.,** Demonstration and purification of an endogenous benzodiazepine from the mammalian brain with a monoclonal antibody of benzodiazepines, *Life Sci.,* 39, 1927, 1986.
98. **Sangameswaran, L., Fales, H. M., Friedrich, P., and De Blas, A.,** Purification of a benzodiazepine from bovine brain and detection of benzodiazepine-like immunoreactivity in human brain, *Proc. Natl. Acad. Sci. U.S.A.,* 83, 9236, 1986.
99. **Lewy, A. J., Sack, R. L., and Singer, C. M.,** Melatonin, light and chronobiological disorders, *Ciba Found. Symp.,* 117, 231, 1985.
100. **Lewy, A. J., Sack, R. L., Miller, L. S., Hoban, T. M., Singer, C. M., Samples, J. R., and Krauss, G. L.,** The use of plasma melatonin levels and light in the assessment and treatment of chronobiologic sleep and mood disorders, *J. Neural Transm.,* Suppl. 21, 311, 1986.
101. **Marczynski, T. J., Yamaguchi, N., Ling, G. M., and Grodzinska, L.,** Sleep induced by the administration of melatonin (5-methoxy-*N*-acetyltryptamine) to the hypothalamus in unrestrained cats, *Experientia,* 20, 435, 1964.
102. **Hishikawa, Y., Cramer, H., and Kuhlo, W.,** Natural and melatonin-induced sleep in young chickens — a behavioral and electrographic study, *Exp. Brain Res.,* 7, 84, 1969.
103. **Antón-Tay, F., Díaz, J. L., and Fernández-Guardiola, A.,** On the effect of melatonin upon human brain. Its possible therapeutic implications, *Life Sci.,* 10, 841, 1971.
104. **Cramer, H., Rudolph, J., Consbruch, U., and Kendel, K.,** On the effects of melatonin on sleep and behavior in man, *Adv. Biochem. Psychopharmacol.,* 11, 187, 1974.
105. **Cramer, H., Böhme, W., Kendel, K., and Donnaudieu, M.,** Freisetzung von Wachstumshormon und von Melanozyten stimulierendem Hormon im durch Malatonin gebahnten Schlaf beim Menschen, *Arzneim.-Forsch.,* 26, 1076, 1976.
106. **Holmes, S. W. and Sugden, D.,** Effects of melatonin on sleep and neurochemistry in the rat, *Br. J. Pharmacol.,* 76, 95, 1982.
107. **Mendelson, W. G., Gillin, J. C., and Wyatt, R. J.,** The search for circulating sleep-promoting factors, in *Advances in Pharmacology and Therapeutics II,* Vol. 1, Yoshida, H., Hagiwara, Y., and Ebashi, S., Eds., Pergamon Press, Oxford, 1982, 227.
108. **Mirmiran, M. and Pévet, P.,** Effects of melatonin and 5-methozytryptamine on sleep-wake patterns in the male rat, *J. Pineal Res.,* 3, 135, 1986.
109. **Wurtman, R. J., and Lieberman, H. R.,** Melatonin secretion as a mediator of circadian variations in sleep and sleepiness, *J. Pineal Res.,* 2, 301, 1985.
110. **Arendt, J., Aldhous, M., and Marks, V.,** Alleviation of jet lag by melatonin: preliminary results of controlled double blind trial, *Br. Med. J.,* 292, 1170, 1986.
111. **Cassone, V. M., Roberts, M. H., and Moore, R. Y.,** Melatonin inhibitis metabolic activity in the rat suprachiasmatic nuclei, *Neurosci. Lett.,* 81, 29, 1987.
112. **Mendelson, W. B., Martin, J. V., Wagner, R. R., Perlis, M., Majewska, M., and Paul, S. M.,** Hypnotic properties of an endogenous corticosteroid, *Sleep Res.,* 16, 108, 1987.
113. **Majewska, M., Harrison, N. L., Schwartz, R. D., Barker, J. L., and Paul, S. M.,** Steroid hormone metabolites are barbiturate-like modulators of the GABA receptor, *Science,* 232, 1004, 1986.
114. **Branchey, M., Branchey, L., and Nadler, R. D.,** Effects of estrogen and progesterone on sleep patterns of female rats, *Physiol. Behav.,* 6, 743, 1971.
115. **Matsushima, M. and Takeuchi, M.,** Effect of sex hormone on sleep-wakefulness circadian rhythm in ovariectomized rat, *Neurosciences,* 12, 182, 1986.
116. **Branchey, M., Branchey, L., and Nadler, R. D.,** Effects of sex hormones on sleep patterns of male rats gonadectomized in adulthood and in the neonatal period, *Physiol. Behav.,* 11, 609, 1973.
117. **Yamaoka, S.,** Participation of limbic-hypothalamic structures in circadian rhythm of slow wave sleep and paradoxical sleep in the rat, *Brain Res.,* 151, 255, 1978.
118. **Johnson, J. H. and Sawyer, C. H.,** Adrenal steroids and the maintenance of a circadian distribution of paradoxical sleep in rats, *Endocrinology,* 89, 507, 1971.

119. **Peder, M.,** Rapid eye movement sleep deprivation affects sleep similarly in castrated and noncastrated rats, *Behav. Neural Biol.*, 47, 186, 1987.
120. **Koenigsberg, H. W., Pollak, C., and Sullivan, T.,** The sleep lactate infusion: arousal and the panic mechanism, *Biol. Psychiatr.*, 22, 786, 1987.
121. **Jouvet, M., Buda, C., Cespuglio, R., Chastrette, N., Denoyer, M., Sallanon, M., and Sastre, J. P.,** Hypnogenic effects of some hypothalamopituitary peptides, *Clin. Neuropharmacol.*, 9(Suppl. 4), 465, 1986.
122. **Bibene, V., Arnauld, E., Meynard, J., Rodriguez, F., Poncet, C., and Vincent, J.-D.,** Influence of vasopressin on circadian sleep and wakefulness, *Sleep Res.*, 16, 43, 1987.
123. **Obál, F., Jr., Török, A., Alföldi, P., Sáry, S., Hajós, M., and Penke, B.,** Effects of intracerebroventricular injection of delta sleep-inducing peptide (DSIP) and an analogue on sleep and brain temperature in rats at night, *Pharmacol. Biochem. Behav.*, 23, 953, 1985.
124. **Sommerfelt, L.,** Reduced sleep in cats after intraperitoneal injection of δ-sleep-inducing peptide, *Neurosci. Lett.*, 58, 73, 1985.
125. **Ernst, A. and Schoenenberg, G. A.,** DSIP: basic findings in human beings, in *Sleep Peptides: Basic and Clinical Approaches,* Inoué, S. and Schneider-Helmert, D., Eds., Japan Scientific Societies Press, Tokyo/Springer-Verlag, Berlin, 1988, 131.
126. **Matsumura, H., Goh, Y., Ueno, R., Sakai, T., and Hayaishi, O.,** Awaking effect of PGE_2 microinjected into the preoptic area of rats, *Brain Res.*, 444, 265, 1988.
127. **Matsumura, H., Honda, K., Goh, Y., Ueno, R., Sakai, T., Inoué, S., and Hayaishi, O.,** Awaking effect of prostaglandin E_2 in freely moving rats, *Brain Res.*, in press.
128. **Onoe, H., Ueno, R., Fujita, I., Nishino, H., Oomura, Y., and Hayaishi, O.,** Prostaglandins D_2, a cerebral sleep-inducing substance in monkey, *Proc. Natl. Acad. Sci. USA,* 85, 4082, 1988.
129. **Rezek, M., Havlicek, V., Hughes, K. H., and Friesen, H.,** Cortical administration of somatostatin (SRIF): effect on sleep and motor behavior, *Pharmacol. Biochem. Behav.*, 5, 73, 1976.

Chapter 9

INTERACTIONS OF SLEEP SUBSTANCES

I. EXISTENCE OF MULTIPLE SLEEP FACTORS

A. Possible Interrelations

It is generally known that a highly developed integrative function in the organism is regulated by multiple factors, both neural and humoral. The complementary relationships between the neural and the endocrine regulatory systems are established in the homeothermal animal.[1] Hence, it appears likely that sleep, one of the highest order biological functions, may be controlled through both neural and humoral information. It has been well established that neural circuits in the basal forebrain, hypothalamus, and brainstem are involved in the regulation of sleep. In addition, as dealt with in the previous chapters, there are a wide variety of sleep-related endogenous factors, which can more or less enhance sleep.

The existence of such a large number of sleep substances in the brain puzzles us and may lead to the question why so many substances are involved in the regulation of sleep and wakefulness, and how they interact in the process of sleep induction, maintenance, and termination[2,3] (see Figure 1). Each investigator has been busy isolating and identifying his or her own material, and has been concerned mostly with proving the validity of a single substance and its related compounds. Little attempt has been made to analyze the mutual relationship of the putative sleep substances. However, it has been pointed out, as mentioned in each chapter, that the somnogenic activity of some putative sleep substances is more or less influenced by the presence of other substances. The knowledge so far found in the literature is summarized in Table 1.

However, as far as experimental studies are concerned, no systematic approach had been made to analyze interactions occurring among coexisting sleep substances before our attempt since 1985. In most studies on sleep factors, analyses have been individually performed on each substance which was singly administered. Hence, synergistic or antagonistic interactions, if any, among several substances could hardly be detected. Moreover, bioassay techniques were originally developed by different researchers using different recipient animals, different timings and routes of administration, different dosage, different recording durations, and so on. Thus, comparison of experimental results from different laboratories are extremely difficult.

B. Differential Somnogenic Properties of Different Sleep Substances

From the above reasons, we have attempted to compare the somnogenic activity of different putative sleep substances by means of our routine nocturnal 10-h i.c.v. assay in freely behaving rats (see Chapter 2), and found differential sleep-enhancing properties in several substances.[5,6] As summarized in Figure 2, five different sleep substances exhibited a compound-dependent somnogenic activity, each different from the others in the time course and quantity of sleep modulation. Differential effects were also apparent between the modulation of slow-wave sleep (SWS) and that of paradoxical sleep (PS). The results suggest that sleep is regulated by multiple humoral factors that play a specific role in the dynamic regulatory process of sleeping and waking.

Furthermore, there is a marked circadian difference in the sleep-modulatory effect between nocturnal and diurnal administration:[6,7] the same dose of the substances that caused excess sleep at night (the active phase of rats) during the nocturnal administration period failed to do so at daytime during the diurnal infusion period (see Figure 3). Hence, as repeatedly mentioned before, it appears likely that a homeostatic mechanism is operative to maintain

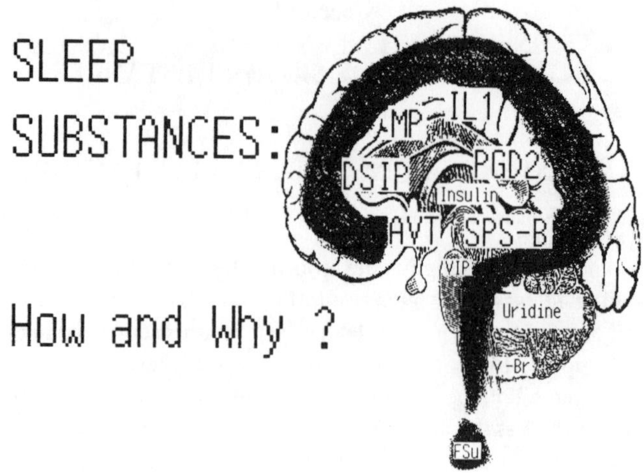

FIGURE 1. The mystery of sleep substances. (From Inoué, S., *Nemuri no Sei o Motomete*, Dobutsu-sha, Tokyo, 1986. With permission.)

the normal level of diurnal sleep (the resting phase of rats), and eventually to cancel the activity of exogenously supplemented sleep substances. In addition, it should be recalled that the diurnal administration of prostaglandin (PG) synthesis inhibitors could suppress diurnal sleep to an extent, but not completely in rats[8,9] (see Chapter 6): the treated rats could still sleep in spite of PG depletion. Partial sleep suppression has also been reported after the administration of antibodies to insulin[10,11] and vasoactive intestinal polypeptide (VIP).[12] These facts may also suggest that sleep is under the interacting control of multiple sleep substances.

II. INTERACTIONS AMONG COEXISTING SLEEP SUBSTANCES

A. Simultaneous Administration Studies

Among a number of putative sleep substances, three compounds, such as delta-sleep-inducing peptide (DSIP), uridine, and muramyl peptides (MP) were isolated directly from the body fluids or the cerebral tissues of sleeping or sleepy animals. DSIP was originally found and identified in the blood of sleeping rabbits.[13] Uridine was identified as one of the active components of sleep-promoting substance (SPS), which was extracted from the brain-stems of sleep-deprived rats.[14] Muramyl dipeptide (MDP) is the simplest analog of those MPs which were first extracted as Factor S from the cerebrospinal fluid (CSF) of sleep-deprived goats and finally identified in human urine.[15]

If these substances actually participate in the regulation of sleep *in vivo*, what is the specific role played by each of them? Although the compound-dependent differential somnogenic effects are demonstrated as above, the analyses were individually done on each substance, which was singly administered, and no information about their interactions were available. Hence, a series of studies were conducted to analyze the sleep-modulatory effect of simultaneously administered DSIP, MDP, and uridine.[2,3,16-19] Since it was demonstrated that 2.5 nmol of DSIP, 2 nmol of MDP, and 10 pmol of uridine could induce a maximal sleep enhancement if nocturnally infused into the third cerebral ventricle for 10 h in freely moving rats, the same dosage was adopted for the combined infusion.

B. DSIP-MDP Combination

The simultaneous intracerebroventricular (i.c.v.) infusion of DSIP and MDP resulted in a profound increase in SWS in the middle part of the 10-h infusion period (see Figure 4,

Table 1
INTERRELATIONSHIPS BETWEEN SLEEP SUBSTANCES IN SLEEP REGULATION

Description[a]	Chapter
ACTH suppresses incorporation of uridine in brain mRNA	4,8
CRF actions may be mediated by ACTH	8
CRF and GRF exert a reciprocal activity	8
DSIP suppresses ACTH release	3,8
DSIP suppresses corticosterone, PRL, TRF, and TSH release	3
DSIP stimulates GH release	3,7
DSIP antagonizes P-DSIP	3
Endotoxin stimulates HLA-DR2, IL1, and PGD_2 production	5
GH stimulates SRIF release	8
HLA-DR2 potentiates cell sensitivity to IL1	5
IFN stimulates the production of PGD_2	5
IL1 alters IFN metabolism	5
Lipid A stimulates the production of IFN and IL1	5
MDP stimulates PGD_2 production	6
Melatonin acts as a BZ receptor agonist	8
MPs stimulate the production of HLA-DR2, IFN, IL1, and PGD_2	5
PGD_2 and PGE_2 exert a reciprocal activity	6
PGD_2 stimulates VIP release	5
THDOC acts as a BZ receptor agonist	8
TNF stimulates the production of IFN, IL1, and PGD_2	5
VIP may interact with immunomodulatory somnogens	8

[a] Abbreviations — ACTH: adrenocorticotropic hormone; BZ: benzodiazepine; CRF: corticotropin-releasing factor; GH: growth hormone; GRF: growth hormone-releasing factor; DSIP: delta-sleep-inducing peptide; HLA-DR2: a certain histocompatibility antigen; IFN: interferon; IL1: interleukin-1; MDP: muramyl dipeptide; MPs: muramyl peptides; P-DSIP: phosphorylated DSIP; PGD_2: prostaglandin D_2; PGE_2: prostaglandin E_2; PRL: prolactin; TRF: thyrotropin-releasing factor; TSH: thyrotropin; THDOC: 3-α,5-α-tetrahydrodeoxycorticosterone; TNF: tumor necrosis factor; VIP: vasoactive intestinal polypeptide.

bottom). In contrast, no difference from the baseline was found in the early and late parts of the infusion period even though the two sleep substances were steadily supplied. In our previous studies[5,6] on the similar nocturnal i.c.v. administration of the single substance, DSIP showed a prompt but short-lasting SWS enhancement, while MDP exhibited a slowly rising SWS-promoting effect, which accompanied a peak in the middle infusion period. Although the DSIP sample used here (from Peptide Institute, Inc., Osaka) has a delayed SWS-promoting property[20] (see Figure 4, top), the little effect at the early phase of the DSIP-MDP administration period could not be attributed to the DSIP characteristics alone. Hourly amounts were significantly higher than the baseline values between 21.00 and 24.00 h. In contrast, no remarkable change in PS was induced by the treatment (see Figure 5, bottom). The excess SWS between 21.00 and 01.00 h, which was significantly different from the baseline, was due largely to the elevated number of SWS episodes (42.1%) but not to the prolongation of their duration. Summated SWS and PS amounts in the 12-h dark period differed from those of the baseline by 48.7 (a 23.5% increment) and −6.3 min (an 18.3% decrement), respectively, but the differences were statistically insignificant. Sleep on the recovery day was characterized by a notable but insignificant decrease in the total time of diurnal SWS and PS.

C. DSIP-Uridine Combination

The simultaneous i.c.v. infusion of DSIP and uridine induced transient increases in both SWS and PS in the early and late parts of the infusion period. Hourly amounts of SWS

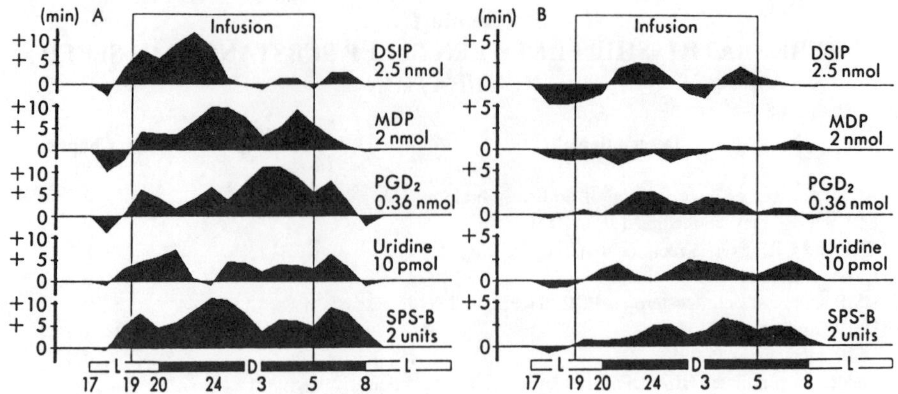

FIGURE 2. A comparison of the somnogenic effects of five sleep substances individually infused into the third cerebral ventricle of freely moving rats between 19.00 and 05.00 h (indicated by a frame). The ordinate shows the difference from the baseline under saline infusion; the abscissa indicates the time of day, where L and D represent the light and dark periods, respectively. DSIP: delta-sleep-inducing peptide; MDP: muramyl dipeptide; PGD_2: prostaglandin D_2; SPS-B: an active fraction of sleep-promoting substance (SPS). (From Inoué, S., in *Endogenous Sleep Substances and Sleep Regulation,* Inoué, S. and Borbély, A. A., Eds., Japan Scientific Societies Press, Tokyo/VNU Science Press BV, Utrecht, 1985, 3. With permission.)

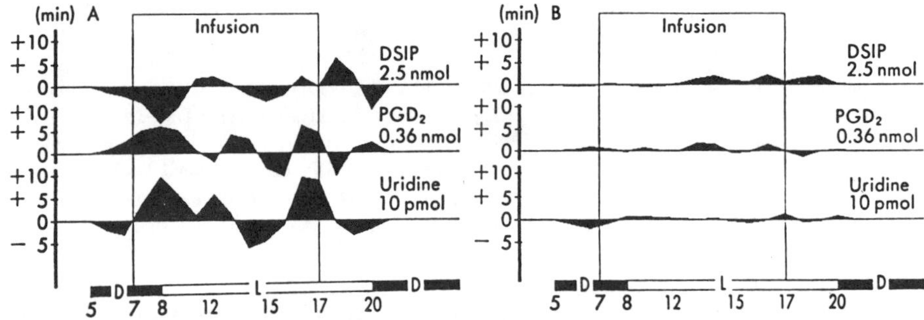

FIGURE 3. A comparison of the somnogenic effects of three sleep substances individually infused into the third cerebral ventricle of freely moving rats between 07.00 and 17.00 h (indicated by a frame). For other explanations, see Figure 2. (From Inoué, S., in *Endogenous Sleep Substances and Sleep Regulation,* Inoué, S. and Borbély, A. A., Eds., Japan Scientific Societies Press, Tokyo/VNU Science, Press BV, Utrecht, 1985, 3. With permission.)

were significantly higher than the baseline values between 02.00 and 04.00 h (see Figure 6, bottom). The largest increment was observed between 03.00 and 04.00 h. The excess SWS between 01.00 and 04.00 h, which was significantly different from the baseline, was due largely to the elevated number of SWS episodes (40.6%), but not to the prolongation of their duration. No significant change in PS was induced by the combined infusion (see Figure 7, bottom). Furthermore, the total increment of SWS and PS in the 12-h dark period was 3.2 (1.4%) and 9.3 min (25.0%) from the baseline, respectively. The DSIP-uridine combination exerted a slight SWS-promoting effect, which was quite different from and far less than the continued SWS-PS enhancement by the single administration of uridine.[5,21,22]

D. MDP-Uridine Combination

The simultaneous infusion of MDP and uridine resulted in a rapid and profound increase in both SWS and PS, which lasted over the early and middle parts of the infusion period

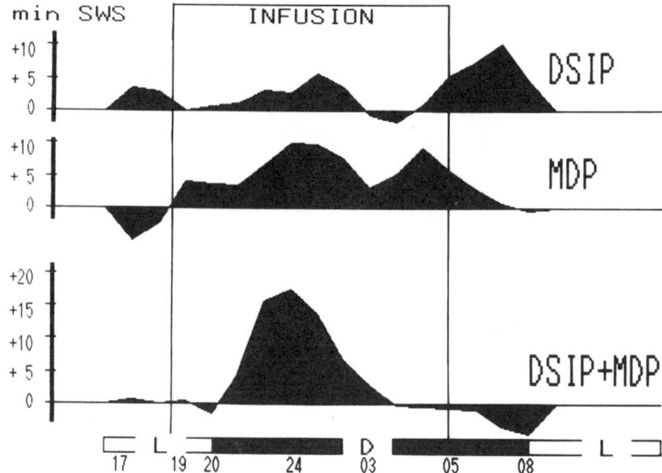

FIGURE 4. A comparison of the somnogenic effects of two sleep substances (2.5 nmol DSIP and 2 nmol MDP) on SWS singly or simultaneously i.c.v. infused in freely moving rats. For other explanations, see Figure 2.

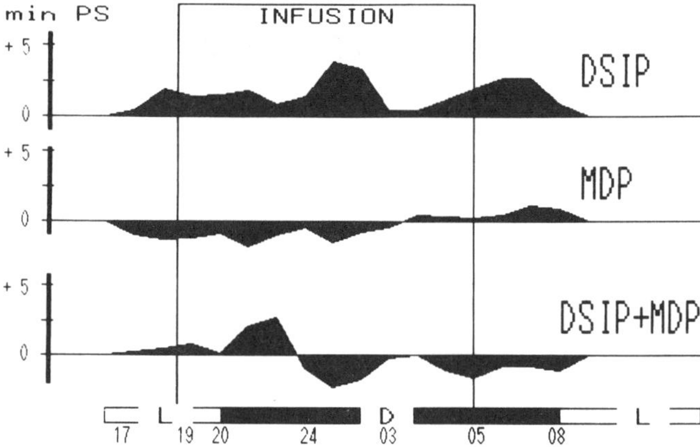

FIGURE 5. A comparison of the somnogenic effects of two sleep substances (2.5 nmol DSIP and 2 nmol MDP) on PS singly or simultaneously i.c.v. infused in freely moving rats. For other explanations, see Figure 2.

(see Figures 8 and 9, bottom). The largest increment of SWS and PS, which was observed between 24.00 and 01.00 h, exceeded the baseline by 23.6 and 3.9 min, respectively. The excess SWS between 19.00 and 01.00 h, which was significantly different from the baseline, was due largely to the elevated number of SWS episodes (36.9%), but not to the prolongation of their duration. Similarly, the excess PS between 19.00 and 22.00 h, which was significantly different from the baseline, was due largely to the elevated number of PS episodes (93.7%). Although the total increment of SWS and PS in the 12-h dark period was 62.7 (30.9%) and 8.4 min (35.0%) from the baseline, respectively, they were statistically insignificant.

E. DSIP-MDP-Uridine Combination

The simultaneous infusion of DSIP, MDP, and uridine induced an extremely large increase in SWS, also from the early to the middle phase of the infusion period (see Figure 10,

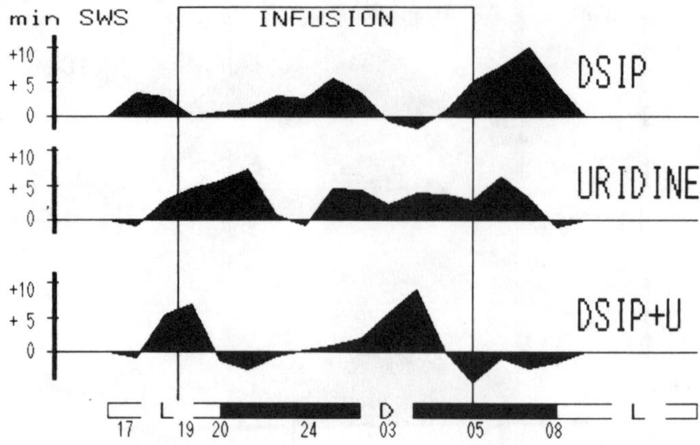

FIGURE 6. A comparison of the somnogenic effects of two sleep substances (2.5 nmol DSIP and 10 pmol uridine) on SWS singly or simultaneously i.c.v. infused in freely moving rats. For other explanations, see Figure 2.

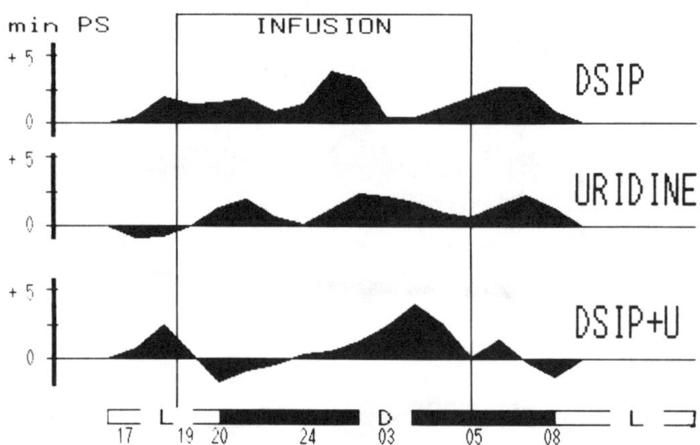

FIGURE 7. A comparison of the somnogenic effects of two sleep substances (2.5 nmol DSIP and 10 pmol uridine) on PS singly or simultaneously i.c.v. infused in freely moving rats. For other explanations, see Figure 2.

bottom). However, PS increased significantly only at the early administration period (see Figure 11, bottom). Interestingly, no difference from the baseline was observed at the late part of the infusion period. Hourly amounts were significantly higher than the baseline values between 22.00 and 24.00 h in SWS and between 20.00 and 21.00 h in PS. The largest increment of SWS and PS was observed between 23.00 and 24.00 h and between 20.00 and 21.00 h, respectively. The excess SWS between 19.00 and 02.00 h, which was significantly different from the baseline, was due largely to the elevated number of SWS episodes (59.4%), but not to the prolongation of their duration. Similarly, the excess PS between 20.00 and 22.00 h, which was significantly different from the baseline, was due largely to the elevated number of PS episodes (171.1%). The difference in nocturnal SWS and PS amounts was a significant increment by 70.4 (30.8%) and an insignificant decrement by 1.2 min (−2.3%), respectively, as compared with the baseline.

F. Possible Mechanisms

The simultaneous nocturnal infusion of DSIP, MDP, and/or uridine induced a significant, combination-dependent change in sleep-waking dynamics. Sleep enhancement at a certain

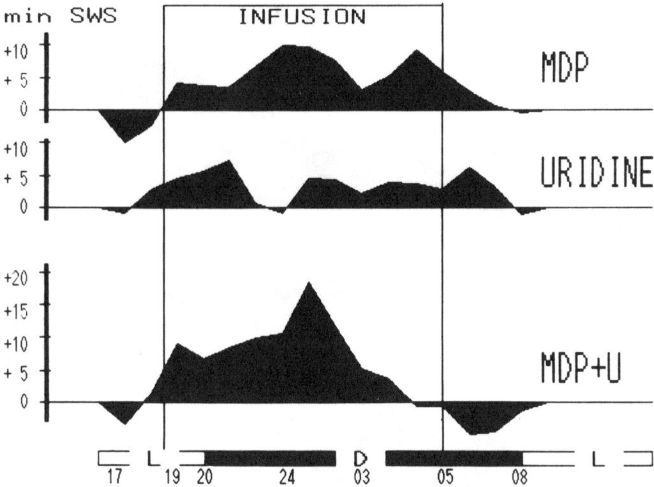

FIGURE 8. A comparison of the somnogenic effects of two sleep substances (2 nmol MDP and 10 pmol uridine) on SWS singly or simultaneously i.c.v. infused in freely moving rats. For other explanations, see Figure 2.

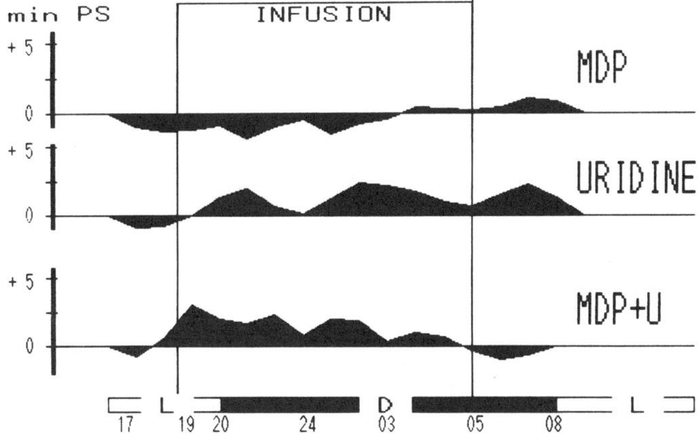

FIGURE 9. A comparison of the somnogenic effects of two sleep substances (2 nmol MDP and 10 pmol uridine) on PS singly or simultaneously i.c.v. infused in freely moving rats. For other explanations, see Figure 2.

part of the infusion period was due largely to the increased number of sleep episodes, but not to the prolongation of their duration. This indicates that the enhanced sleep was of a normal nature.[5,20]

The time-course characteristics and the grade of excess sleep were quite different from those which might be deduced by the superimposition of sleep-promoting effects of each substance, as presented in the top and middle graphs of Figures 4 through 11. Since the dosage for each substance was optimal to induce excess sleep when individually administered, the combination of two or three substances may indicate the doubled or tripled sleep-enhancing potency. However, the actual increments of nocturnal sleep (1.4 to 30.9% for SWS and -18.3 to 35.0% for PS) were less than those values obtained by summating the increments due to the single administration[5] (DSIP: 18.4% for SWS and 19.0% for PS; MDP: 28.0% for SWS and -5.1% for PS; uridine: 21.0% for SWS and 68.1% for PS). This may suggest that some negative interactions occur among the substances. Such a

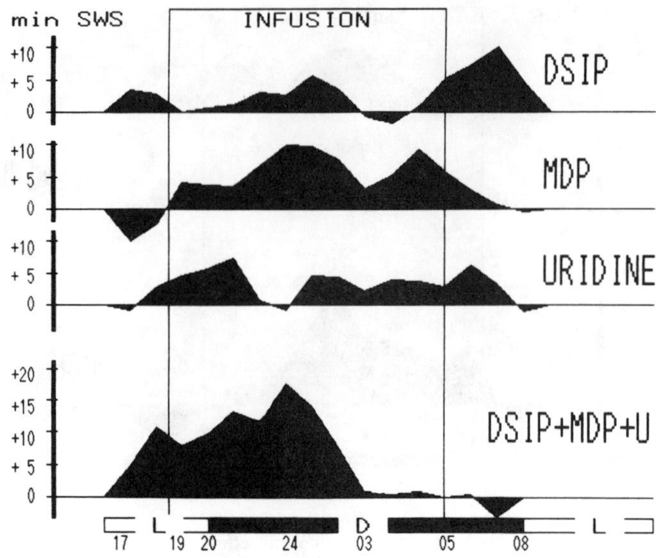

FIGURE 10. A comparison of the somnogenic effects of three sleep substances (2.5 nmol DSIP, 2 nmol MDP, and 10 pmol uridine) on SWS singly or simultaneously i.c.v. infused in freely moving rats. For other explanations, see Figure 2.

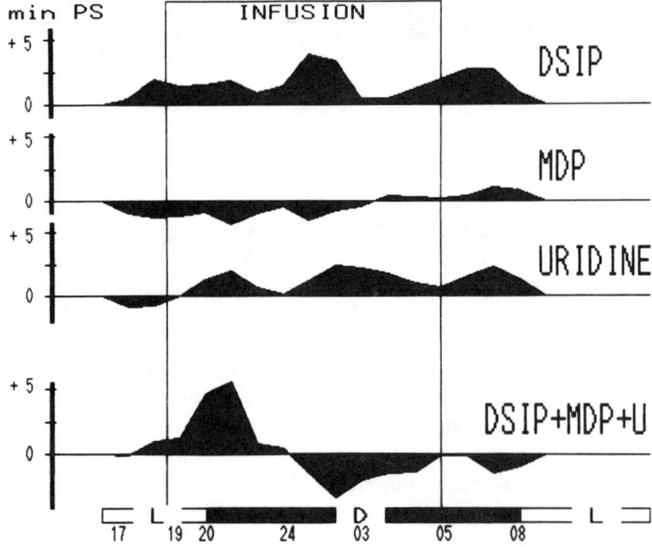

FIGURE 11. A comparison of the somnogenic effects of three sleep substances (2.5 nmol DSIP, 2 nmol MDP, and 10 pmol uridine) on PS singly or simultaneously i.c.v. infused in freely moving rats. For other explanations, see Figure 2.

nonlinear feature of sleep modulation by the multiple sleep factors may also suggest that some interfering relationships could be operative among the substances.[2,3]

The time-course pattern of sleep enhancement shows that if SWS was increased rapidly, no more excess SWS followed at the later period even under the supply of sleep substances as in the case of the MDP-uridine and the DSIP-MDP-uridine combination (see Figure 12,

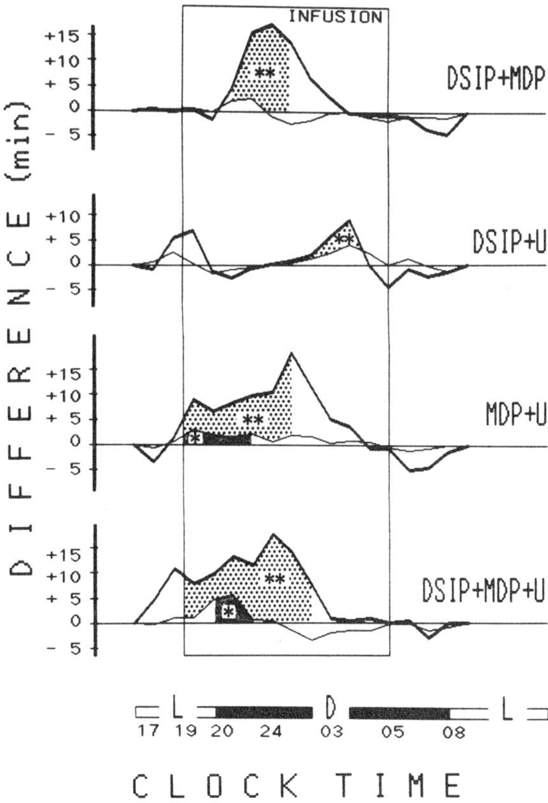

FIGURE 12. A comparison of the somnogenic effects of four combinations of three sleep substances (2.5 nmol DSIP, 2 nmol MDP, and 10 pmol uridine) i.c.v. infused in freely moving rats. The curves show hourly increments expressed as the difference from the baseline value. The thick and thin curves represent SWS and PS, respectivly. The hatched areas indicate that the difference from the baseline is statistically significant at $p < 0.05$ (*) and $p < 0.01$ (**). For other explanations, see Figure 2. (From Kimura, M., Honda, K., Komoda, Y., and Inoué, S., *Neurosci. Res.*, 5, 157, 1987. With permission.)

the two bottom graphs). If the effect appeared slowly, then suddenly as in the case of the DSIP-MDP combination, the effect also rapidly faded away (see Figure 12, the first graph). However, if the sleep enhancement appeared still later, the effect was not remarkable as in the case ofthe DSIP-uridine combination (see Figure 12, the second graph). It seems plausible that the induction rate of "sleep pressure" may depend on the combination of sleep substances and that the clearance mode of the pressure following the response to the substances may affect the time-course of sleep modulation.[2,3] Hence, the integration of antagonistic and synergistic interactions seems to determine nonspecifically the sleep-waking dynamics.[2]

III. INTERACTIONS WITH PREEXISTING SLEEP SUBSTANCES

A. Serial Administration of Two Sleep Substances

The time sequence interaction of coexisting, multiple sleep substances is suggested in the case of the MP-mediated excess sleep[23-26] (see Table 1). The reciprocal sleep-modulatory activity between corticotropin-releasing factor (CRF) and growth hormone-releasing factor (GRF),[27] between DSIP and phosphorylated DSIP (P-DSIP),[28] and between prostaglandin

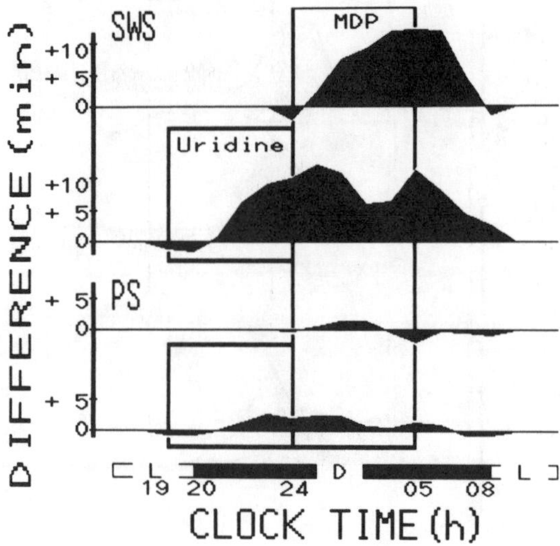

FIGURE 13. The effects of uridine pretreatment on the somnogenic activity of MDP in unrestrained rats. A 5-h i.c.v. infusion of uridine (5 pmol between 19.00 and 24.00 h, indicated by a frame) enhanced the occurrence of increased SWS and PS during the subsequent 5-h period (between 00.00 and 05.00 h, indicated by a frame) of an i.c.v. infusion of MDP (1 nmol). The first graph shows changes in the hourly amount of SWS when MDP was infused without uridine pretreatment. The second graph shows changes in the hourly amount of SWS when MDP was infused after uridine pretreatment. The third graph shows changes in the hourly amount of PS when MDP was infused without uridine pretreatment. The fourth graph shows changes in the hourly amount of PS when MDP was infused after uridine pretreatment.

D_2 (PGD_2) and prostaglandin E_2 (PGE_2) (see Chapter 6) may also suggest a time-sequential interrelationship. Yet no experimental approach has been made to study the effects of serially administered sleep substances. Hence we attempted to nocturnally infuse the third cerebral ventricle of freely behaving rats with 1.0 nmol MDP for 5 h and then with 5 pmol uridine for 5 h, and vice versa.[19,29,30]

B. Uridine Pretreatment and the Somnogenic Effect of MDP

A nocturnal i.c.v. infusion of MDP for 5 h (between 00.00 and 05.00 h) induced a gradual increase in SWS but not PS in rats, typical of the somnogenic activity of this substance (see Figure 13, the first and third graphs).The 5-h uridine pretreatment had brought about an increased level of SWS and PS before the 5-h MDP administration was initiated (see Figure 13, the second graph). The elevated level of SWS due to the uridine administration was maintained throughout the subsequent 5-h period of MDP infusion, although it was not amplified further. However, an increasing tendency of PS gradually ceased after the MDP administration (see Figure 13, the fourth graph).

C. MDP Pretreatment and the Somnogenic Effect of Uridine

In contrast, the MDP/uridine sequence was characterized by a quite different time course of sleep-waking dynamics. A nocturnal 5-h i.c.v. infusion of uridine between 00.00 and 05.00 h exhibited a less marked SWS enhancement (see Figure 14, the first graph) than that between 19.00 and 24.00 h (see Figure 13, the second graph), due presumably to the circadian

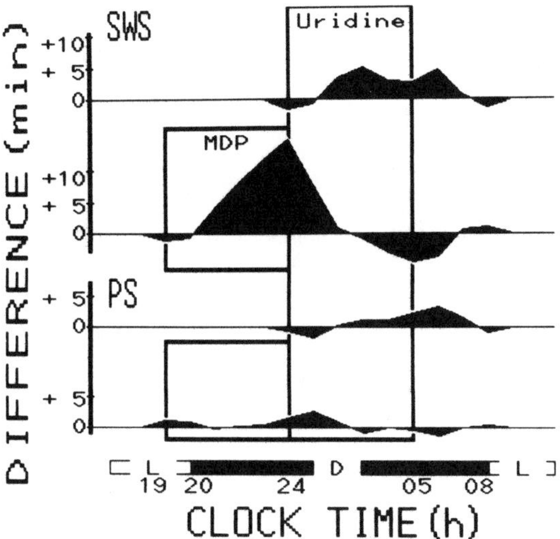

FIGURE 14. The effects of MDP pretreatment on the somnogenic activity of uridine in unrestrained rats. A 5-h i.c.v. infusion of MDP (1 nmol, between 19.00 and 24.00 h, indicated by a frame) suppressed the occurrence of SWS and PS during the subsequent 5-h period (between 00.00 and 05.00 h, indicated by a frame) of an i.c.v. infusion of uridine (5 pmol). The first graph shows changes in the hourly amount of SWS when uridine was infused without MDP pretreatment. The second graph shows changes in the hourly amount of SWS when uridine was infused after MDP pretreatment. The third graph shows changes in the hourly amount of PS when uridine was infused without MDP pretreatment. The fourth graph shows changes in the hourly amount of PS when uridine was infused after MDP pretreatment.

rhythm-dependent changes in the reactivity to uridine.[31,32] PS showed a tendency of gradual enhancement by the uridine administration (see Figure 14, the third graph). The MDP pretreatment had elevated the level of SWS but not PS when the uridine infusion was started (see Figure 14, the second and fourth graphs). Interestingly, uridine, after the MDP pretreatment, exerted a negative or rather sleep-suppressive effect on the subsequent state of vigilance in rats. The elevated level of SWS dropped to the baseline level within 2 h after the uridine administration. The level of PS was little affected by the uridine administration.

D. Sequence-Dependent Interactions

The sequential i.c.v. infusion of the above two sleep substances resulted in a significant, sequence-dependent change in sleep-waking dynamics, which was quite different from the time-course sleep promotion induced by the single or combined administration of each substance.

Thus, the preceding activity of a sleep substance during the latest phase of rest and the early phase of activity in rats brought about differential changes in the subsequent sleep modulation by another different substance. The preexistence of uridine profoundly enhanced the SWS-promoting effect of MDP, whereas the preexistence of MDP negatively affected the action of uridine. It seems likely that some programming mechanism (see Chapter 3) may be involved in these differential reactions to sequentially administered sleep substances, and that further investigation is required to elucidate the phenomenon. It should be noted that the occurrence of SWS and PS was not synchronized. This fact clearly indicates that the regulatory process of SWS and PS is independently operated.

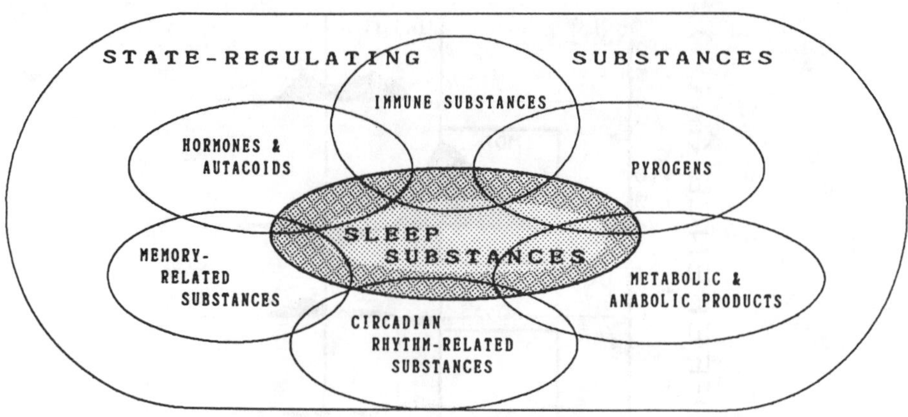

FIGURE 15. A schematic representation of the involvement of multiple humoral factors in the regulation of sleep and wakefulness.

IV. SLEEP REGULATION BY MULTIPLE SLEEP FACTORS

From our experimental analyses, it may be concluded that the sleep modulation by sleep substances depends on the following functions: (1) a specific sleep-enhancing action of each substance, (2) the time of day, (3) the interactions of the coexisting and/or preexisting multiple factors, and (4) the feedback regulation of the sleep-waking states.

As regards items (1) and (2), a considerably large quantity of information has been accumulated. In the past decade, the literature concerning sleep factors has been published amazingly often. The number of candidate substances, which are chemically identified, have increased enormously. Most of them possess sleep-unrelated biological properties, since sleep per se is closely related to various aspects of bodily functions, such as the circadian clock mechanism, the length of wakefulness, temperature regulation, food intake, immunoreaction, etc. Consequently, it has become unlikely that a specific sleep factor regulates all aspects of sleep, but that multiple, nonspecific factors interact dynamically to modulate some aspects of sleep (see Figure 15).

In this connection, we should be challenged to unveil the mechanism related to items (3) and (4) above. To our present knowledge, the total system of humoral sleep regulation is obscurely understood, as shown in Figure 16. The interrelationship between the neural and the neuroendocrine activities also awaits further investigation. Finally, it should be mentioned here that an approach to the evolutionary aspects is informative to understanding the role sleep substances play in sleep regulation, since the evoluation of sleep might be supplemented by multiple humoral factors.[6,33] These aspects will be dealt with in the next chapter.

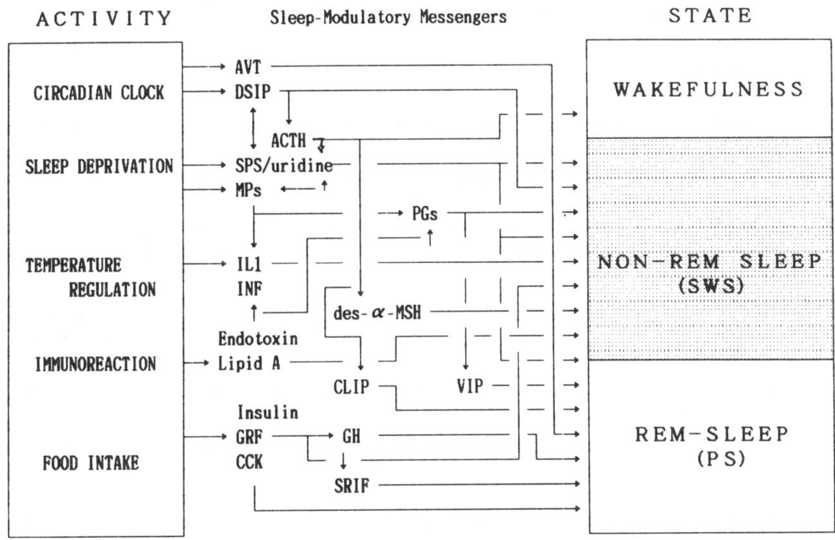

FIGURE 16. The possible interrelationships among physiological activity, sleep substances, and sleep-waking states.

REFERENCES

1. **Inoué, S. and Sekiguchi, T.,** Neuroendocrine control systems and systems analyses of the brain-hypophyseal-gonadal mechanism, in *Neuroendocrine Control,* Yagi, K. and Yoshida, S., Eds., University of Tokyo Press, Tokyo, 1973, 361.
2. **Inoué, S.,** Multifactorial humoral regulation of sleep, *Clin. Neuropharmacol.,* 9 (Suppl. 4), 470, 1986.
3. **Inoué, S.,** Interference among sleep substances simultaneously infused in unrestrained rats, in *Sleep '86,* Koella, W. P., Obál, F., Schulz, H., and Visser, P., Eds., Gustav Fischer Verlag, Stuttgart, 1988, 175.
4. **Inoué, S.,** *Nemuri no Sei o Motomete,* Dobutsu-sha, Tokyo, 1986.
5. **Inoué, S., Honda, K., Komoda, Y., Uchizono, K., Ueno, R., and Hayaishi, O.,** Differential sleep-promoting effects of five sleep substances nocturnally infused in unrestrained rats, *Proc. Natl. Acad. Sci. U.S.A.,* 81, 6240, 1984.
6. **Inoué, S.,** Sleep substances: their roles and evolution, in *Endogenous Sleep Substances and Sleep Regulation,* Inoué, S. and Borbély, A. A., Eds., Japan Scientific Societies Press, Tokyo/VNU Science Press BV, Utrecht, 1985, 3.
7. **Inoué, S., Honda, K., Komoda, Y., Uchizono, K., Ueno, R., and Hayaishi, O.,** Little sleep-promoting effect of three sleep substances diurnally infused in unrestrained rats, *Neurosci. Lett.,* 49, 207, 1984.
8. **Naito, K., Ueno, R., Hayaishi, O., Honda, K., and Inoué, S.,** Suppression of sleep by prostaglandin synthesis inhibitors in unrestrained rats, *Sleep Res.,* 16, 64, 1987.
9. **Naito, K., Osama, H., Ueno, R., Hayaishi, O., Honda, K., and Inoué, S.,** Suppression of sleep by prostaglandin synthesis inhibitors in unrestrained rats, *Brain Res.,* 453, 329, 1988.
10. **Danguir, J.,** Intracerebroventricular infusion of somatostatin selectively increases paradoxical sleep in rats, *Brain Res.,* 367, 26, 1986.
11. **Danguir, J.,** Internal milieu and sleep homeostasis, in *Sleep Peptides: Basic and Clinical Approaches,* Inoué, S. and Schneider-Helmert, D., Eds., Japan Scientific Societies Press, Todyo/Springer-Verlag, Berlin, 1988, 53.
12. **Riou, F., Cespuglio, R., and Jouvet, M.,** Endogenous peptides and sleep in the rat. III. The hypogenic properties of vasoactive intestinal polypeptide, *Neuropeptides,* 2, 265, 1982.
13. **Schoenenberger, G. A., Maier, P. F., Tobler, H. J., and Monnier, M.,** A naturally occurring delta-EEG enhancing nonapeptide in rabbits. X. Final isolation, characterization and activity test, *Pfluegers Arch.,* 369, 99, 1977.
14. **Komoda, Y., Ishikawa, M., Nagasaki, H., Iriki, M., Honda, K., Inoué, S., Higashi, A., and Uchizono, K.,** Uridine, a sleep-promoting substance from brainstems of sleep-deprived rats, *Biomed. Res.,* 4 (Suppl.), 223, 1983.

15. **Krueger, J. M., Karnovsky, M. L., Martin, S. A., Pappenheimer, J. R., Watler, J., and Biemann, K.,** Peptideglycans as promoters of slow-wave sleep. II. Somnogenic and pyrogenic activities of some naturally occurring muramyl peptides; correlations with mass spectrometric structure determination, *J. Biol. Chem.*, 259, 12659, 1984.
16. **Kimura, M., Honda, K., Okano, Y., Komoda, Y., and Inoué, S.,** Effects of simultaneously infused sleep substances on nocturnal sleep in rats, *Zool. Sci.*, 2, 989, 1985.
17. **Kimura, M., Honda, K., Okano, Y., Komoda, Y., and Inoué, S.,** Interacting sleep-modulatory effects of simultaneously administered sleep substances in rats, *J. Physiol. Soc. Jpn.*, 48, 280, 1986.
18. **Kimura, M., Honda, K., Komoda, Y., and Inoué, S.,** Interacting sleep-modulatory effects of simultaneously administered delta-sleep-inducing peptide (DSIP), muramyl dipeptide (MDP) and uridine in unrestrained rats, *Neurosci. Res.*, 5, 157, 1987.
19. **Inoué, S., Kimura, M., Honda, K., and Komoda, Y.,** Sleep peptides: general and comparative aspects, in *Sleep Peptides: Basic and Clinical Approaches,* Inoué, S. and Schneider-Helmert, D., Eds., Japan Scientific Societies Press, Tokyo/Springer-Verlag, Berlin, 1988, 1.
20. **Kimura, M., Honda, K., and Inoué, S.,** Effects of intracerebroventricularly infused DSIP of different sources on nocturnal sleep in unrestrained rats, *Zool. Sci.*, 3, 1088, 1986.
21. **Honda, K., Komoda, Y., Nishida, S., Nagasaki, H., Higashi, A., Uchizono, K., and Inoué, S.,** Uridine as an active component of sleep-promoting substance: its effects on the nocturnal sleep in rats, *Neurosci. Res.*, 1, 243, 1984.
22. **Honda, K., Komoda, Y., and Inoué, S.,** Effects of sleep-promoting substances on the rat circadian sleep-waking cycles, in *Endogenous Sleep Substances and Sleep Regulation,* Inoué, S. and Borbély, A. A., Eds., Japan Scientific Societies Press, Tokyo/VNU Science Press BV, Utrecht, 1985, 203.
23. **Krueger, J. M., Kubillus, S., Shoham, S., and Davenne, D.,** Enhancement of slow wave sleep by endotoxin and lipid A, *Am. J. Physiol.*, 251, R591, 1986.
24. **Krueger, J. M., Toth, L. A., Cady, A. B., Johanssen, L., and Obál, F., Jr.,** Immunomodulation and sleep, in *Sleep Peptides: Basic and Clinical Approaches,* Inoué, S. and Schneider-Helmert, D., Eds., Japan Scientific Societies Press, Tokyo/Springer-Verlag, Berlin, 1988, 95.
25. **Shoham, S., Davenne, D., Cady, A. B., Dinarello, C. A., and Krueger, J. M.,** Recombinant tumor necrosis factor and interleukin-1 enhance slow-wave sleep in rabbits, *Am. J. Physiol.*, 253, R142, 1987.
26. **Krueger, J. M., Dinarello, C. A., Shoham, S., Davenne, D., Walter, J. and Kubillus, S.,** Interferon alpha-2 enhances slow-wave sleep in rabbits, *Int. J. Immunopharmacol.*, 9, 23, 1987.
27. **Ehlers, C. L. and Kupfer, D. J.,** Hypothalamic peptide modulation of EEG sleep in depression: a further application of the S-process hypothesis, *Biol. Psychiatry,* 22, 513, 1987.
28. **Ernst, A. and Schoenenberger, G. A.,** DSIP: basic findings in human beings, in *Sleep Peptides: Basic and Clinical Approaches,* Inoué, S. and Schneider-Helmert, D., Eds., Japan Scientific Societies Press, Tokyo/Springer-Verlag, Berlin, 1988, 131.
29. **Kimura, M., Honda, K., Komoda, Y., and Inoué, S.,** Effects of serially administered MDP and uridine on nocturnal sleep in rats, *Zool. Sci.*, 4, 1088, 1987.
30. **Kimura, M. and Inoué, S.,** Simultaneous and serial intracerebroventricular administration of MDP and uridine: differential effects on nocturnal sleep in rats, *Neurosci. Res.*, Suppl. 7, S192, 1988.
31. **Honda, K., Okano, Y., Komoda, Y., and Inoué, S.,** Sleep-promoting effects of intraperitoneally administered uridine in unrestrained rats, *Neurosci. Lett.*, 62, 137, 1985.
32. **Honda, K. and Inoué, S.,** Sleep substances and sleep-related rhythms, in *Comparative Aspects of Circadian Clocks,* Hiroshige, T. and Honma, K., Eds., Hokkaido University Press, Sapporo, 1988, 157.
33. **Inoué, S.,** Sleep substances: their roles and evolution, *Seikagaku,* 55, 445, 1983.

Chapter 10

EVOLUTIONARY ASPECTS

I. ORIGIN AND DIFFERENTIATION OF SLEEP SUBSTANCES

A. Evolutionary Point of View

The varying features of sleep substance do not allow us to obtain a unitary view of their mode of action. The differential sleep-modulating properties of the various candidate substances may indicate that sleep is regulated as a combined action of multiple humoral factors as repeatedly mentioned. However, the question still remains open as to why so many sleep-related factors are required, how they participate in the regulation of sleep, and what the specific role of each of them is. The circadian variations in the somnogenic property lead to the further question about the relationship between the sleep-regulating mechanism and the circadian clock mechanism which is so common in biological organism. Moreover, the problem must be addressed as to why sleep substances modify rest-activity rhythms in lower animals, as will be described below. The relationship between temperature regulation and sleep or between pyrogenic and somnogenic substances indicates that the homeothermic state in higher animals is closely correlated. The state of vigilance is closely interwoven by the nutritional state. However, taking into consideration the evolutionary aspects of the neuroendocrine control system in higher animals may provide a possible explanation.[1,2]

The higher vertebrates are evolved, the more humoral messengers are involved in the elaborate control of the body. The circulatory system of the body fluid is utilized as a common route of information transfer to mediate such substances, which play a complementary role to neural regulation[3] (see Figure 1A). The humoral circulatory route is obviously more economical and efficient than the cabled nervous route for the purpose of simultaneous, long-lasting communication to multiple component systems of the body, as long as the receiver components have specific receptors for a certain chemical messenger. It may be postulated that sleep-specific messengers are also in similar ways endogenously produced and distributed to the entire central nervous system via the cerebrospinal fluid (CSF), since sleep requires a synchronized transfer of vigilance states in the total body from arousal to rest. Such a neurohumoral regulatory system of sleep may have evolved from the basic circadian and/or ultradian rest-activity mechanism in parallel with the development of sleep, which represents a later acquired function in higher animals (see Figure 1B).

The evolution of sleep has differentiated the sleep-like state in lower vertebrates into slow-wave sleep (SWS) and paradoxical sleep (PS) in homeotherms.[4] The immobilization and remobilization processes of PS seem to be closely related to the archaic type of sleep. Deep sleep, especially the deep stages of nonrapid-eye-movement (non-REM) sleep, might be required to restitute the highly developed cerebrum as a specially differentiated pattern of cerebral rest.[5] Since higher vertebrates have established warm-bloodedness, rest-activity rhythms may not be absolutely dependent on the environmental temperature. Rather, the rhythms should be controlled by the internal environmental situation depending on the various physiological, psychological, and ecological conditions. Since homeotherms cannot drastically lower their body temperature, temperature regulation, coupled with sleep regulation, has become vitally important for controlling energy expenditure. Thus, the function of sleep has become more and more specialized and diversely differentiated. It seems likely that the multivariate functions of sleep may require multiple humoral factors in addition to the hierarchical organization of the central nervous system.

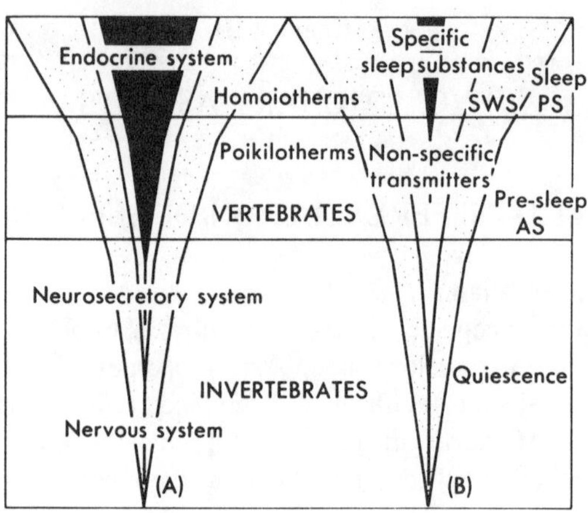

FIGURE 1. The evolution of the neuroendocrine system and the sleep-regulating system. The higher stage of evolution requires the involvement of the larger number of differentiated humoral factors. (From Inoué, S., *Seikagaku,* 55, 445, 1983. With permission.)

B. Two Facts — Evolution of Benzodiazepine Receptors and Bacterial Products

The above speculation is substantiated by two findings. Nielsen et al.[6] investigated the distribution of ^{3}H-diazepam (*N*-methyl-^{3}H-diazepam) receptors in the nerve tissues (head ganglia, whole head, or spinal cord) of invertebrate species and in the forebrain of 18 vertebrate species. As summarized in Table 1, the benzodiazepine (BZ) receptors were detected in higher animals, but not in animals lower than jawless fish.

Krueger[7] similarly speculated from the viewpoint of the relationship between temperature regulation and SWS. He found that muramyl peptides (MP), i.e., bacterial products, and interleukin-1 (IL1) are pyrogenic as well as somnogenic (see Chapter 5). Since both MPs and IL1 are phylogenetically ancient substances, it seems likely that these substances may have been incorporated gradually as autonomic modulators of body temperature and SWS during the evolution of homeotherms. The coevolution of SWS with warm-bloodedness may have contributed to the minimizing of energy expenditure in the latter. Thus it may be reasonable to conclude that similar and/or identical substances become involved in the regulation of both sleep and temperature.

C. The Sleep Regulatory System — Still Underdeveloped?

It might be speculated that sleep control mechanisms are still developing. A number of different substances are utilized to regulate sleep. Even bacterial products, MPs, are adopted as a sleep vitamin, as mentioned above and in Chapter 5. Therefore, no single sleep-specific humoral mechanism has been established. All putative sleep substances seem to be inadequate for regulating all aspects of sleep alone. They act by modifying the induction, maintenance, and/or development of sleep. Some of them affect both SWS and PS, while other modify either SWS or PS. A further evolutionary step seems to be required to achieve a true sleep-regulating hormone that is no longer influenced by circadian mechanisms, body temperature regulation, and the circulating nutrient level.

II. THE BEHAVIOR-MODULATORY PROPERTY OF SLEEP SUBSTANCES IN LOWER ANIMALS

A. Sleep Substances and Neural Activity in Invertebrates

Putative sleep substances have been investigated in mammals. Although they do not seem to exhibit a notable species specificity, at least among the animals tested, little information

Table 1
DISTRIBUTION OF BENZODIAZEPINE RECEPTORS IN THE NERVE AND BRAIN TISSUES IN VARIOUS ANIMALS[6]

Phylum	Species investigated	^{3}H-diazepam receptors
Invertebrates		
Annelida	Earth worm	Absent
Mollusca	Squid (*Loligo*)	Absent
Arthropoda—Crustacea	Woodlouse, lobster (*Nepros norvegicus*)	Absent
Arthropoda—Insecta	Locust (*Locusta migratoria*)	Absent
Vertebrates		
Agnatha	Hagfish (*Myxine glutinosa*)	Absent
Pisces	Eel, plaice, codfish	Present
Amphibia	Toad, frog (*Rana esculenta*)	Present
Reptilia	Turtle, lizard	Present
Aves	Pigeon, hen, gull	Present
Mammalia	Mouse, hamster, cat, dog, pig, cow	Present

is available as to their effects on lower vertebrates and invertebrates. Since there is much debate about the existence of sleep in nonhomeothermic animals,[4,8-12] it is of considerable interest from a comparative physiological viewpoint to know whether or not the sleep substances detected in mammals can exert any activity-modifying effect in lower animals.

Dolezalova et al.[13] reported that piperidine (see Chapter 8) accumulated in the central ganglia of dormant snails (*Otala lactea*). On the basis of this observation, it is suggested that this substance may regulate the hibernation of invertebrates. On the other hand, in the early stages of the purification and screening procedures for sleep-promoting substance (SPS, see Chapters 2 and 4), it was proven that some SPS fractions suppressed the neuronal activity of the crayfish abdominal ganglion and stretch receptor.[14,15] The SPS fraction containing uridine was effective in suppressing the spontaneous discharge of the stretch receptor in crayfish.[16] Ozaki et al.[17] has demonstrated that low doses (5 to 50 n*M*) of uridine, one of the active components of SPS, caused either a rapid, short-term suppression or a low, long-term suppression of neural firing activity in the crayfish stretch receptor *in vitro,* whereas large doses (higher than 1 μM) slowly enhanced firing activity. In their screening bioassays, Sargsyan et al.[18] found that some synthetic delta-sleep-inducing peptide (DSIP) derivatives suppressed the spontaneous discharge activity of cerebral ganglia in the snail (*Helix lucorm:* see Chapter 3).

These facts suggest that sleep substances are involved in the modulation of behavioral and neural activities in invertebrates. Furthermore, there may exist a close relationship between the reduced behavioral and neural activities in invertebrates and sleep in mammals. In this connection, it should be mentioned here that sleep substances have a wide spectrum of extra-sleep properties, including the modulation of behavior in mammalian species.

B. Uridine and Locomotor Activity in Insects

Uridine is commonly distributed in living organisms as an RNA constituent. Although its suppressive activity on discharges of crayfish neurons was established, no literature deals with its behavioral effect in invertebrates. Accordingly, an attempt was made to analyze the effects of uridine on the behavior of insects.[2,19,20]

The rhinoceros beetle, *Allomyrina dichotoma* (see Figure 2), is known as a night-active insect. Uridine at doses of 1 to 10,000 pmol was intra-abdominally injected through a small window made on the medican edge of the right elytra 3 h before the onset of the dark period. The locomotor activity was electronically detected as the vibrations of insect-containing cage during a 12-h nocturnal period. As summarized in Figure 3, uridine dose-dependently exerted

FIGURE 2. Male rhinoceros beetles.

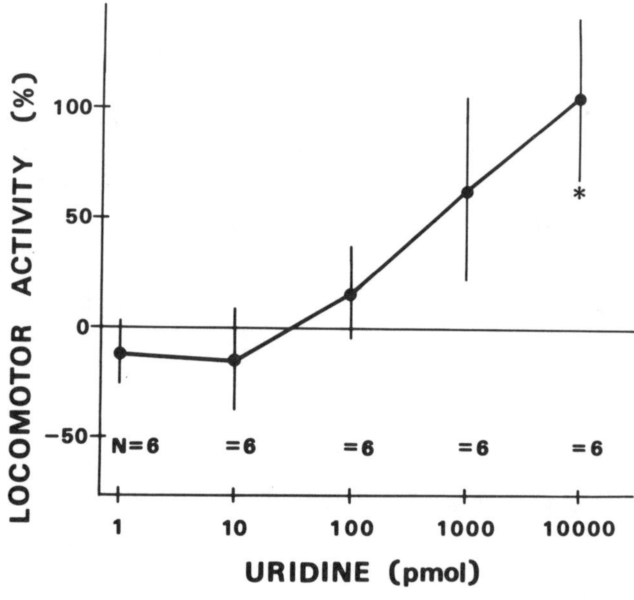

FIGURE 3. Dose-response relationship of the effects of uridine on the nocturnal locomotor activity in rhinoceros beetles. The locomotor activity after uridine administration is compared with that after saline injection (mean ± SEM). $*p < 0.05$. (From Inoué, S., Honda, K., Okano, Y., and Komoda, Y., *Zool. Sci.*, 3, 727, 1986. With permission.)

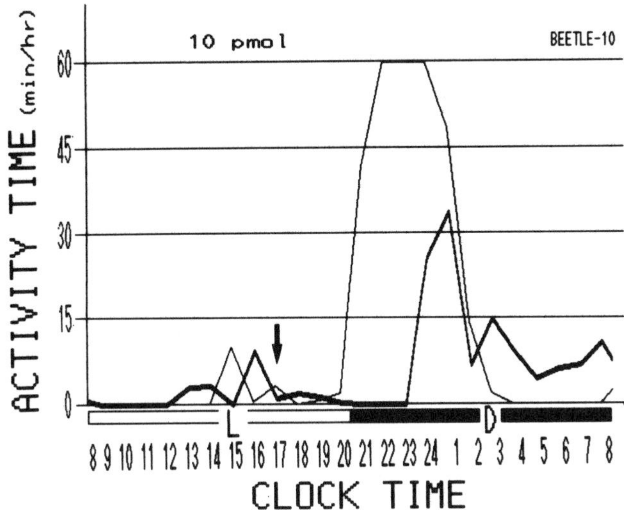

FIGURE 4. Hourly locomotor activity in a single rhinoceros beetle on the day of saline injection (thin curve), and on the next day of 10 pmol uridine injection (thick curve). The time of injection is indicated by an arrow. L and D represent the light and dark periods, respectively. (From Inoué, S., Honda, K., Okano, Y., and Komoda, Y., *Zool. Sci.*, 3, 727, 1986. With permission.)

a biphasic effect on the nocturnal locomotor acticity in rhinoceros beetles. The lower dosage (1 to 10 pmol) caused a slight reduction of the activity duration, although the change was not statistically significant. The larger dosage (0.1 to 10 nmol) brought about a prolongation of the active time. Figures 4 and 5 show typical examples.

C. Sleep Substances and Swimming Activity in Fish

Olson et al.[21] developed a bioassay for detecting the passage through the blood-brain barrier of various neuropeptides in goldfish (*Carassius auratus*). They found that an intracranial or intraperitoneal (i.p.) injection of 80 μg/kg of substance P caused an activation of motor activity, whereas that of D-Ala^4-DSIP, α-endorphin, γ-endorphin, α-melanophore-stimulating hormone (α-MSH), neurotensin, and somatostatin (SRIF) induced hypoactivity in goldfish.

We examined the effects of several sleep substances on the swimming activity of two species of cyrinids, the crucian carp, *C. carassius buergeri,* and the topmouth gudgeon, *Pseudorasbora parva.*[2,19,22] Their swimming activity was monitored continuously by optical switches placed around a small glass container with tap water at 25 ± 1°C (see Figure 6) in a 12-h light-dark cycle. Hourly activity counts revealed that they were day-active fish. The fish were exposed for 24 h to DSIP, a phosphorylated analog P-DSIP, uracil, or uridine by replacing the tap water in the container with solutions of these substances. All these substances were effective in either lowering or activating swimming activity (see Table 2 and Figure 7). A dose-response relationship was evident for uridine: the maximum reduction was induced by 100 μ*M*, while lower or higher doses were less effective or even caused an increase in swimming activity. The activity-reducing action was comparable to the quiescence induced after a 24-h rest-deprivation period through slowly rotating water in the container by means of a magnetic stirrer (see Figure 6 and Table 3). The results suggest that the rest-activity regulating mechanism in fish might be basically similar to the mammalian sleep-waking regulating mechanism.

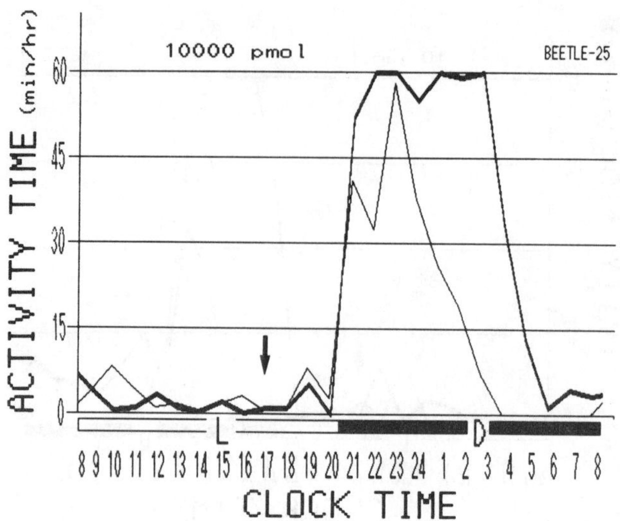

FIGURE 5. Hourly locomotor activity in a single rhinoceros beetle on the day of saline injection (thin curve) and on the next day of 10,000 pmol uridine injection (thick curve). For other explanations, see Figure 4. (From Inoué, S., Honda, K., Okano, Y., and Komoda, Y., *Zool. Sci.*, 3, 727, 1986. With permission.)

III. LINKAGE BETWEEN SLEEP SUBSTANCES AND TIME-KEEPING MECHANISMS

A. Circadian Phase-Shifting Activity of Sleep Substances

If we suppose that the state of rest in lower animals is the precursor of sleep in higher animals, then the sleep-promoting property of uridine and other substances in mammals may well be related to the behavior-modulating property in lower animals. It might be hypothesized that an endogenous substance that is related to the regulation of rest-activity rhythms may become involved in the regulation of sleep-waking cycles in the course of the evolution of sleep, and finally but still partially differentiated into a sleep substance. Some sleep-wakefulness-related substance such as neuropeptide Y, SRIF, VIP, and vasopressin are known to exist in the suprachiasmatic nucleus (SCN), i.e., a biological time keeper in the mammalian brain.[23] Thus, these substances may be involved in the regulation of circadian rhythm-dependent sleep, exhibiting a closer relationship to PS than to SWS.

Graf et al.[24-26] demonstrated a phase reversal by repeated daily i.p. injections of DSIP and P-DSIP in the evening in rats under a 24-h, light-dark cycle or constant light (see Chapter 3). Redman et al.[27] found the phase-entraining effect of melatonin (1 mg/kg) when subcutaneously (s.c.) administered regularly at a fixed time of day in rats with a free-running circadian rhythm. Turek and Losee-Olson[28] reported that a short-acting BZ, triazolam, i.p. injected at a certain free-running circadian time, induced a definite phase advance or delay in the locomotor activity rhythm of golden hamsters kept under either constant light or constant darkness. The phase-shifting effect of traizolam seems to be mediated by the γ-amino butyric acid (GABA)-containing neuron system in the SCN, since BZ receptors form a complex with GABA receptors (see Chapter 8). Thus, exogenously supplied sleep substances may shift the phase of circadian rhythms. However, with 24-h environmental time cues, a single administration of sleep substances usually induces a transient alternation of rest-activity rhythms without causing a phase shift of circadian rest-activity cycles.[29]

On the other hand, a specific protein seems to be involved in the regulation of a circadian

FIGURE 6. Experimental setup for monitoring the swimming activity of cyprinid fish by means of optic switches attached around the glass wall of the container. This was also used for depriving the fish of rest by slowly rotating a magnet bar (1 cycle/s) inside the water container by the magnetic stirrer.

rhythm in the eye of *Aplysia californica.*[30] Although no circadian rhythm was detected in the protein synthesis in rat SCN,[31] protein synthesis inhibitors can induce phase-dependent phase shifts of the circadian oscillator commonly found in widely different species, including microorganisms, invertebrates, and vertebrates.[32,33] Thus, the biochemical mechanisms that generate circadian oscillations in mammals may share common features throughout distantly and closely related phylogenetic groups. Sleep may also share such features as a common background.

B. Sleep Substances and Hibernation-Inducing Substances

Prechel et al.[34] suggested an involvement of AVT, which seems to be more or less related to the regulation of sleep (see Chapters 1 and 7), in the time-keeping mechanism of annual cycles, since the pineal AVT immunoactivity showed a marked yearly variation in golden hamsters and rats. Another pineal sleep-related hormone, melatonin (see Chapter 8), may participate in the regulation of hibernation. Ralph et al.[35] reported that a pinealectomy performed in the golden-mantled ground squirrel (*Citellus* [= *Spemophilius*] *lateralis*) and Richardson's ground squirrel (*C. richardsonii*), before the animals entered hibernation, resulted in the inability of the animal to sustain the normal depth and duration of hibernation in the second over-wintering period. However, an s.c. implant of melatonin (30 mg in beeswax pellets) could no more affect the time-course events in pinealectomized Richardson's ground squirrels, presumably because of insufficient circulatory levels of melatonin. Inter-

Table 2
EFFECTS OF EXPOSURE TO SOLUTIONS OF SLEEP SUBSTANCES ON THE SWIMMING ACTIVITY OF THE CYPRINID FISH

Treatment	Fish	n	Reduced	Unchanged	Increased
DSIP:	Topmouth gudgeon	2	2	—	—
90 n*M*	Crucian carp	3	1	1	1
P-DSIP:	Topmouth gudgeon	2	2	—	—
1.8 μ*M*	Crucian carp	3	—	1	2
Uracil:	Topmouth gudgeon	2	—	1	1
50 p*M*	Crucian carp	3	—	2	1
Uridine:					
1 n*M*	Topmouth gudgeon	2	—	2	—
2 n*M*	Crucian carp	6	—	2	4
20 n*M*	Crucian carp	3	—	1	2
1 μ*M*	Crucian carp	5	2	3	—
10 μ*M*	Crucian carp	5	2	1	2
100 μ*M*	Crucian carp	5	4	1	—
1000 μ*M*	Crucian carp	5	1	3	1

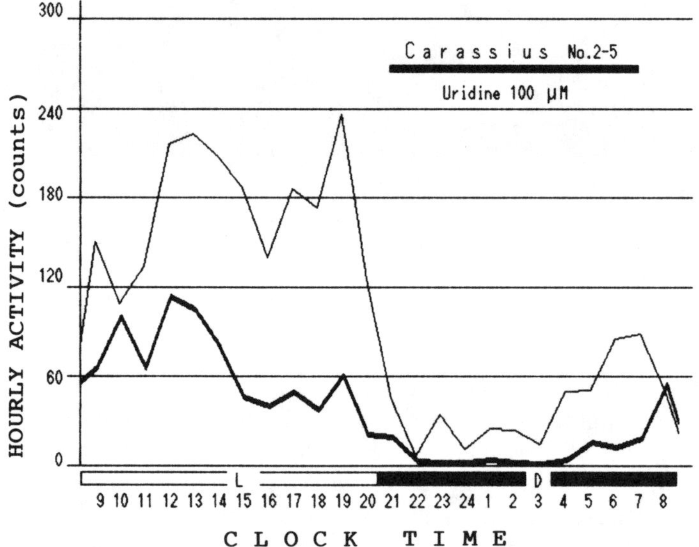

FIGURE 7. Hourly swimming activity in a single fish kept in tap water (thin curve) one day and in 100 μ*M* uridine solution (thick curve) the next. For other explanations, see Figure 4.

estingly, Stanton et al.[36] successfully prolonged the duration of hibernation in golden-mantled ground squirrels by administering an i.c.v. infusion of melatonin (200 or 400 ng/h for several days during hibernation) by an osmotic minipump. Finally, it should be recalled here that the brain content of DSIP in the golden-mantled ground squirrel, *C. suslicus*, was elevated significantly in a Januray through February period as compared with the other seasonal period, whereas that of ACTH was increased significantly during short-term arousal in winter as compared with torpor.[37] Further, the pancreatic levels of insulin and SRIF were reduced significantly during hibernation in golden-mantled ground squirrels.[38]

Table 3
EFFECTS OF 24-H REST DEPRIVATION ON THE SWIMMING ACTIVITY OF THE CYPRINID FISH

Fish	n	Reduced	Unchanged	Increased
Topmouth gudgeon	6	6	—	—
Crucian carp	9	5	1	3
Total	15	11	1	3

IV. A BRIEF REVIEW

Integrating the past and present knowledge of putative sleep substances may lead to the following conclusions and comments.

1. There are a considerably large number of endogenous sleep substances, if the definition and criteria for them are not strictly determined.
2. Each sleep substance has a specific sleep-modulatory activity. However, the action of an individual sleep substance may be modified by various factors, such as the time of day, the nutritional conditions, the coexistence or preexistence of other sleep substances, the preceding state of vigilance (i.e., sleep deprivation, stress, insomnia, etc.), age, the body and environmental temperatures, and so forth.
3. Multiple sleep substances can exert either synergistical or interfering interactions and harmonize the induction, maintenance, and termination of sleep. However, the molecular mechanisms underlying the interactions are almost unknown, especially at the level of receptor sites. Meanwhile, the role of sleep substances is frequently discussed with reference to a nonspecific activity of a certain neurotransmitter(s), which cannot always explain the regulatory process of sleep convincingly.
4. Sleep substances can modulate the neural activity of the central nervous system that is related to the regulation of sleep and wakefulness. Thus, a neuroendocrine complex system is responsible for the regulation of sleep. This may include the basal forebrain-brainstem reticular formation system, the gut-brain system, the hypothalamo-hypophyseal-adrenal/gonadal system, the retina-suprachiasmatic-pineal system, and others. The neural circuits and sleep substances may supplement each other to orchestrate the process of sleep-waking.
5. Most sleep substances are more or less related to the so-called extra-sleep physiological functions, i.e., the regulation of hormone secretion, temperature, metabolism, immunoreaction, circadian rhythm, learning and memory consolidation, hibernation or daily torpor in certain cases, etc. Thus, sleep substances may contribute to normalize the bodily function through the homeostatic regulation of the state of vigilance.
6. Some sleep substances enhance both SWS and PS, whereas others modulate either SWS or PS or even wakefulness. Some sleep substances are concerned mainly with triggering sleep, while others serve to maintain or potentiate sleep. However, each sleep substance has only a partial sleep-modulatory role in the total regulatory process. No absolutely specific and powerful sleep regulator has yet differentiated into existence, since natural physiological sleep does not require a narcosis- or coma-like unconscious state and the sleep-regulatory mechanism is too young from the viewpoint of the evolution of sleep.

REFERENCES

1. **Inoué, S.,** Sleep substances: their roles and evolution, *Seikagaku,* 55, 445, 1983.
2. **Inoué. S.,** Sleep substances: their roles and evoluation, in *Endogenous Sleep Substances and Sleep Regulation,* Inoué, S. and Borbély, A. A., Eds., Japan Scientific Societies Press, Tokyo/VNU Science Press BV, Utrecht, 1985, 3.
3. **Inoué, S. and Sekiguchi, T.,** Neuroendocrine control systems and systems analyses of the brain-hypophyseal-gonadal mechanism, in *Neuroendocrine Control,* Yagi, K. and Yoshida, S., Eds., University of Tokyo Press, Tokyo, 1973, 361.
4. **Karmanova, I. G.,** *Evolution of Sleep,* S. Karger, Basel, 1982.
5. **Horne, J. A.,** *Why We Sleep: Perspectives on the Function of Sleep in Humans and Other Animals,* Oxford University Press, Oxford, 1988.
6. **Nielsen, M., Braestrup, C., and Squires, R. F.,** Evidence for a late evolutionary appearance of brain-specific benzodiazepine receptors: an investigation of 18 vertebrate and 5 invertebrate species, *Brain Res.,* 141, 342, 1978.
7. **Krueger, J. M.,** Muramyl peptides and interleukin-1 as promoters of slow wave sleep, in *Endogenous Sleep Substances and Sleep Regulation,* Inoué, S. and Borbély, A. A., Eds., Japan Scientific Societies Press, Tokyo/VNU Science Press BV, Utrecht, 1985, 181.
8. **Allison, T. and Van Twyver, H.,** The evolution of sleep, *Nat. Hist. N. Y.,* 79, 56, 1970.
9. **Durie, J. B.,** Sleep in animals, in *Psychobiology of Sleep,* Wheatley, D., Ed., Raven Press, New York, 1981, 1.
10. **Meddis, R.,** The evolution of sleep, in *Sleep Mechanisms and Functions,* Mayes, A., Ed., Van Nostrand Reinhold (UK), Wokingham, 1983, 57.
11. **Tobler, I.,** Introduction of phylogenetic approaches to the function of sleep, in *Sleep 1982,* Koella, W. P., Ed., S. Karger, Basel, 1983, 126.
12. **Kaiser, W.,** Comparative neurobiology of sleep — the honey bee model, in *Sleep '84,* Koella, W. P., Rüther, E., and Schulz, H., Eds., Gustav Fischer Verlag, Stuttgart, 1985, 225.
13. **Dolezalova, H., Stepita-Klauco, M., and Fairweather, R.,** The accumulation of piperidine in the central ganglia of dormant snails, *Brain Res.,* 72, 115, 1974.
14. **Nagasaki, H., Iriki, M., and Uchizono, K.,** Inhibitory effect of the brain extract from sleep-deprived rats (BE-SDR) on the spontaneous discharges of crayfish abdominal ganglion, *Brain Res.,* 109, 202, 1976.
15. **Uchizono, K., Higashi, A., Iriki, M., Nagasaki, H., Ishikawa, M., Komoda, Y., Inoué, S., and Honda, K.,** Sleep-promoting fractions obtained from the brain-stem of sleep-deprived rats, in *Integrative Control Functions of the Brain,* Vol. 1, Ito, M., Tsukahara, N., Kubota, K., and Yagi, K., Eds., Kodansha, Tokyo, 1978, 392.
16. **Komoda, Y., Ishikawa, M., Nagasaki, H., Iriki, M., Honda, K., Inoué, S., Higashi, A., and Uchizono, K.,** Uridine, a sleep-promoting substance from brainstems of sleep-deprived rats, *Biomed. Res.,* 4(Suppl.), 223, 1983.
17. **Ozaki, T., Nishie, H., and Uchizono, K.,** Crayfish single unit firing activity and sleep substance, *Annu. Rep. Natl. Inst. Physiol. Sci., Okazaki,* 8, 449, 1987.
18. **Sargsyan, A. S., Sumskaya, L. V., Aleksandova, I. Y., Besrukov, M. V., Mikhaleva, I. I., Ivanov, V. T., and Balaban, P. M.,** Synthesis and some biological properties of δ-sleep inducing peptide and its analogs, *Bioorg. Khim.,* 7, 1125, 1981.
19. **Inoué, S., Honda, K., Okano, Y., and Komoda, Y.,** Behavior-modulating effect of sleep substances in fish and insects, *Sleep Res.,* 14, 84, 1985.
20. **Inoué, S., Honda, K., Okano, Y., and Komoda, Y.,** Behavior-modulating effects of uridine in the rhinoceros beetle, *Zool. Sci.,* 3, 727, 1986.
21. **Olson, R. D., Kastin, A. J., Montalbano-Smith, D., Olson, G. A., Coy, D. H., and Michell, G. F.,** Neuropeptides and the blood-brain barrier in goldfish, *Pharmacol. Biochem. Behav.,* 9, 521, 1978.
22. **Inoué, S., Honda, K., and Komoda, Y.,** Effect of sleep substances on circadian swimming activity in fish, *Int. J. Biometeorol.,* 29 (Suppl. 1), 118, 1985.
23. **Moore, R. Y.,** Suprachiasmatic nucleus, in *Encyclopedia of Neuroscience,* Vol. 2, Adelman, G., Ed., Birkhäuser, Boston, 1987, 1156.
24. **Graf, M. V., Christen, H., Tobler, H. J., Baumann, J. B., and Schoenenberger, G. A.,** DSIP a circadian 'programming' substance?, *Experientia,* 37, 624, 1981.
25. **Graf, M. V., Christen, H., Tobler, H. J., Maier, P. F., and Schoenenberger, G. A.,** Effects of repeated DSIP and DSIP-P administration on the circadian locomotor activity of rats, *Pharmacol. Biochem. Behav.,* 15, 717, 1982.
26. **Graf, M. V., Christen, H., and Schoenenberger, G. A.,** DSIP/DSIP-P and circadian motor activity of rats under continuous light, *Peptides,* 3, 623, 1982.
27. **Redman, J., Armstrong, S., and Ng, K. T.,** Free-running activity rhythms in the rat: entrainment by melatonin, *Science,* 219, 1089, 1983.

28. **Turek, F. W. and Losee-Olson, S.,** A benzodiazepine used in the treatment of insomnia phase-shifts the mammalian circadian clock, *Nature,* 321, 167, 1986.
29. **Honda, K. and Inoué, S.,** Sleep substances and sleep-related rhythms, in *Comparative Aspects of Circadian Clocks,* Hiroshige, T. and Honma, K., Eds., Hokkaido University Press, Sapporo, 1988, 157.
30. **Yeung, S. J. and Eskin, A.,** Involvement of a specific protein in the regulation of a circadian rhythm in *Aplysia* eye, *Proc. Natl. Acad. Sci. U.S.A.,* 84, 270, 1987.
31. **Glotzbach, S. F., Randall, T. L., Radeke, C. M., and Heller, H. C.,** Absence of a circadian rhythm of protein synthesis in the rat suprachiasmatic nucleus, *Neurosci. Lett.,* 76, 113, 1987.
32. **Takahashi, J. S. and Turek, F. W.,** Anisomycin, an inhibitor of protein synthesis, perturbs the phase of a mammalian circadian pacemaker, *Brain Res.,* 405, 199, 1987.
33. **Turek, F. W.,** Pharmacological probes of the mammalian circadian clock: use of the phase response curve approach, *Trends Pharmacol. Sci.,* 8, 212, 1987.
34. **Prechel, M. M., Audhya, T. K., and Simmons, W. H.,** Pineal arginine vasotocin activity increases 200-fold during August in adult rats and hamsters, *J. Pineal Res.,* 1, 175, 1984.
35. **Ralph, C. L., Harlow, H. J., and Phillips, J. A.,** Delayed effect of pinealectomy on hibernation of the golden-mantled ground squirrel, *Int. J. Biometeorol.,* 26, 311, 1982.
36. **Stanton, T. L., Daley, J. C., III, and Salzman, S. K.,** Prolongation of hibernation bout duration by continuous intracerebroventricular infusion of melatonin in hibernating ground squirrels, *Brain Res.,* 413, 350, 1987.
37. **Kramarova, L. I., Kolaeva, S. H., Yukhananov, R. Y., and Rozhanets, V. V.,** Content of DSIP, enkephalins and ACTH in some tissues of active and hibernating ground squirrels *(Citellus suslicus), Comp. Biochem. Physiol.,* 74C, 31, 1983.
38. **Bauman, W. A., Meryn, S., and Frolant, G. L.,** Pancreatic hormones in the nohibernating and hibernating golden mantled ground squirrel, *Comp. Biochem. Physiol.,* 86A, 241, 1987.

INDEX

A

B

C

D

E

F

G

H

I

J

K

L

M

N

O

P

Q

R

T

U

V

W

Y